Ethnobotany

A Phytochemical Perspective

Edited by B. M. Schmidt and D. M. Klaser Cheng

WILEY Blackwell

Registered Offices
John Wiley & Sons, Inc., 111 River Street, Hoboken, NJ 07030, USA
John Wiley & Sons Ltd, The Atrium, Southern Gate, Chichester, West Sussex, PO19 8SQ, UK

Editorial Office
111 River Street, Hoboken, NJ 07030, USA

For details of our global editorial offices, customer services, and more information about Wiley products visit us at www.wiley.com.

Wiley also publishes its books in a variety of electronic formats and by print-on-demand. Some content that appears in standard print versions of this book may not be available in other formats.

Library of Congress Cataloging-in-Publication data applied for

ISBN - 9781118961902

Cover Image: Courtesy of Professor Ilya Raskin
Cover Design: Wiley

Set in 10/12pt Warnock by SPi Global, Pondicherry, India

10 9 8 7 6 5 4 3 2 1

Contents

List of Contributors *ix*
Foreword *xi*
Preface *xiii*
Acknowledgments *xvii*

Part I   Introduction to Ethnobotany and Phytochemistry *1*

1 Ethnobotany *3*
B. M. Schmidt
Key Terms and Concepts   3
Ethnobotany throughout History   4
References and Additional Reading   18
Current Topics in Ethnobotany   18
References   22
Taxonomy: Plant Families   23
References and Additional Reading   90
Plant Anatomy and Architecture: The Highlights   93
References and Additional Reading   105
Keys for Plant Identification   106
Herbaria, Plant Collection, and Voucher Specimens   107
References and Additional Reading   109

2 Phytochemistry *111*
D. M. Klaser Cheng
Primary and Secondary Metabolites   111
Databases   126
References and Additional Reading   127
Extraction and Chromatographic Techniques   127
References   133
Evaluation of Biological Activities   134
References   139

Part II   Case Studies   141

3   Introduction   143

4   Africa   145
    M. H. Grace, P. J. Smith, I. Raskin, M. A. Lila, B. M. Schmidt, and P. Chu
    Introduction   145
    Northern Africa   149
    Southern Africa   151
    The African Diaspora   156
    References and Additional Reading   159
    Achillea millefolium: Re-Exploring Herbal Remedies for Combating Malaria   160
    References   164
    Vanilla: Madagascar's Orchid Economy   166
    References   170
    Traditional Treatments for HIV in South Africa   171
    References   176
    Duckweed as a More Sustainable First-Generation Bioethanol Feedstock   178
    References   181

5   The Americas   183
    J. Kellogg, B. M. Schmidt, B. L. Graf, C. Jofré-Jimenez, H. Aguayo-Cumplido, C. Calfío, L. E. Rojo,
    K. Cubillos-Roble, A. Troncoso-Fonseca, and J. Delatorre-Herrera
    Introduction   184
    North America   185
    Central America and the Caribbean   189
    South America   194
    References and Additional Reading   198
    Phlorotannins in Seaweed   199
    References   202
    Agave: More Than Just Tequila   203
    References   208
    Quinoa: A Source of Human Sustenance and Endurance in the High Andes   209
    References   213
    Maqui (*Aristotelia chilensis*): An Ancient Mapuche Medicine
    with Antidiabetic Potential   215
    References   220
    Betalains from *Chenopodium quinoa*: Andean Natural Dyes with
    Industrial Uses beyond Food and Medicine   222
    References   226

6   Asia   227
    P. Li, W. Gu, C. Long, B. M. Schmidt, S. S. Ningthoujam, D. S. Ningombam,
    A. D. Talukdar, M. D. Choudhury, K. S. Potsangbam, H. Singh, S. Khatoon, and M. Isman
    Introduction   228
    Central Asia   229
    Western Asia   231

South Asia *233*

Southeast Asia *237*

East Asia *240*

References and Additional Reading *243*

Ethnobotany of Dai People's Festival Cake in Southwest China *244*

References *247*

The Ethnobotany of Teeth Blackening in Southeast Asia *247*

References *251*

Artemisia Species and Human Health *252*

References *257*

Traditional Treatment of Jaundice in Manipur, Northeast India *258*

References *265*

Ethnobotany and Phytochemistry of Sacred Plant Species *Betula utilis* (bhojpatra) and *Quercus oblongata* (banj) from Uttarakhand Himalaya, India *267*

References *270*

Neem-Based Insecticides *272*

References *275*

7   Europe   *277*

T. B. Tumer, B. M. Schmidt, and M. Isman

Introduction *278*

References and Additional Reading *285*

Differential Use of *Lavandula stoechas* L. among Anatolian People against Metabolic Disorders *285*

References *289*

Mad Honey *291*

References *294*

Indigo: The Devil's Dye and the American Revolution *295*

References *302*

Insecticides Based on Plant Essential Oils *303*

References *306*

8   Oceania   *309*

B. M. Schmidt

Introduction *310*

References and Additional Reading *317*

Banana (*Musa* spp.) as a Traditional Treatment for Diarrhea *317*

References *322*

Pharmacological Effects of Kavalactones from Kava (*Piper methysticum*) Root *323*

References *328*

Botanical Index *329*

Subject Index *339*

List of Contributors

C. Calfío
Facultad de Química y Biología
Universidad de Santiago de Chile
Santiago, Chile

D. M. Klaser Cheng
Visiting Scientist
Rutgers University
New Brunswick, New Jersey, USA

M. D. Choudhury
Department of Life Science and
Bioinformatics
Assam University
Silchar, Assam, India

P. Chu
Department of Plant Biology and
Pathology, SEBS
Rutgers University
New Brunswick, New Jersey, USA

K. Cubillos-Roble
Facultad de Química y Biología
Universidad de Santiago de Chile
Santiago, Chile

H. Aguayo-Cumplido
Facultad de Recursos Naturales
Renovables
Universidad Arturo Prat
Iquique, Chile

J. Delatorre-Herrera
Facultad de Recursos Naturales
Renovables
Universidad Arturo Prat
Iquique, Chile

M. H. Grace
Plants for Human Health Institute
North Carolina State University
Kannapolis, North Carolina, USA

B. L. Graf
Department of Plant Biology and
Pathology, SEBS
Rutgers University
New Brunswick, New Jersey, USA

W. Gu
Key Laboratory of Economic Plants and
Biotechnology
Kunming Institute of Botany
Chinese Academy of Sciences
Kunming, China.

M. Isman
Faculty of Land and Food Systems
University of British Columbia
Vancouver, British Columbia, Canada

C. Jofré-Jimenez
Facultad de Química y Biología
Universidad de Santiago de Chile
Santiago, Chile

J. Kellogg
Plants for Human Health Institute
North Carolina State University
Kannapolis, North Carolina, USA

S. Khatoon
Pharmacognosy and Ethnopharmacology
Division
CSIR-National Botanical Research Institute
Lucknow, Uttar Pradesh, India

P. Li
Key Laboratory of Economic Plants and
Biotechnology
Kunming Institute of Botany
Chinese Academy of Sciences
Kunming, Yunnan, China

M. A. Lila
Plants for Human Health Institute
North Carolina State University
Kannapolis, North Carolina, USA

C. Long
Kunming Institute of Botany
Chinese Academy of Sciences
Kunming, Yunnun, China

B. M. Schmidt
L'Oreal USA
Clark, New Jersey, USA

D. S. Ningombam
Department of Life Sciences
Manipur University
Imphal, Manipur, India

S. S. Ningthoujam
Department of Botany
Ghanapriya Women's College
Imphal, Manipur, India

K. S. Potsangbam
Department of Life Sciences
Manipur University
Imphal, Manipur, India

I. Raskin
Department of Plant Biology and
Pathology, SEBS
Rutgers University
New Brunswick, New Jersey, USA

L. E. Rojo
Facultad de Química y Biología
Universidad de Santiago de Chile
Santiago, Chile

H. Singh
Plant Diversity, Systematics and
Herbarium Division
CSIR—National Botanical Research
Institute
Lucknow, Uttar Pradesh, India

P. J. Smith
Division of Pharmacology
University of Cape Town
Medical School
Cape Town, South Africa

A. D. Talukdar
Department of Life Science and
Bioinformatics
Assam University
Silchar, Assam, India

A. Troncoso-Fonseca
Facultad de Química y Biología
Universidad de Santiago de Chile
Santiago, Chile

T. B. Tumer
Faculty of Arts and Sciences
Department of Molecular Biology and
Genetics
Çanakkale Onsekiz Mart University,
Terzioglu Campus,
Çanakkale, Turkey

Foreword

Science, especially with the rise of the -*omics* technologies, has made amazing strides to decipher and elucidate the capacity of plants to synthesize and accumulate natural products that are uniquely capable of interfacing with human therapeutic targets to prevent or treat chronic human diseases. However, these revelations made in the modern scientific community would hardly come as a surprise to many indigenous communities, who have relied on well-known, highly potent medicinal plants for centuries. Modern science has only recently allowed us to characterize the phytochemical structures that provide anti-inflammatory or chemopreventive relief to humans, or to discern their mechanisms of action, but the traditional ecological knowledge of native communities has both channeled the search for phytoactive compounds and guided tests for safety and efficacy based on long history of human use. This book illustrates how traditional ethnobotany enriches the scientific discovery process, and in turn, how science validates and reinforces the wisdom of the elders.

Mary Ann Lila
Director, Plants for Human Health Institute
David H. Murdock Distinguished Professor
Food Bioprocessing & Nutrition Sciences
North Carolina State University
North Carolina, USA

Preface

The Purpose of This Book

Both lead authors/editors and all contributing authors work in the fields of ethnobotany and/or phytochemistry. We wrote this book to help bridge the gap between the two fields of study and to bring about new collaborations and discoveries. A range of applications from around the world is presented, illustrating many areas of relevance and importance to phytochemical research. In industry, there is an ongoing initiative to discover new molecules from nature with a specific usefulness, be it nutritional, cosmetic, pharmaceutical, dyes, and so on. But there are not many reference books that describe the traditional use of a plant and also explain the chemistry at work "behind the scenes." Existing ethnobotany textbooks and reference books tend to focus on the anthropological side of ethnobotany, often ignoring phytochemistry altogether. They offer information on how a plant was used, but fall short of describing why a plant produces the observed effects. There are many edited reference books available that contain only case studies, with each contributing author following their own format. Occasionally the case studies stray far from the topic at hand or provide little background information. There are a few quality ethnobotany reference books on the market with a focus on phytochemistry, but the case studies need to be updated. This book features relevant, modern case studies in a uniform format that will be easy for the reader to follow. It also contains basic information on ethnobotany and phytochemistry, as tools to understand the information presented in the case studies and published ethnobotany research articles in general.

Target Audience

This book is intended as a reference book for an upper-level undergraduate- or graduate-level ethnobotany or phytochemistry course. Basic organic chemistry knowledge will be required to fully understand the phytochemistry. However, the book is still useful for readers without a chemistry background who are interested in ethnobotany. Industry scientists may use this book to gain a basic understanding of ethnobotany and phytochemistry and to become familiar with recent research for their position in the pharmaceutical, personal care, food, nutrition, or raw materials industry.

Inclusions and Exclusions

One of the most difficult tasks when writing a book is deciding where to draw the boundaries. What information will be included and what will be excluded? This book focuses on examples of ethnobotany where phytochemistry plays a role in the observed effects or activities. Therefore, aspects of ethnobotany that cannot easily be explained by science (such as supernatural or religious uses) and aspects that have only limited connections to phytochemistry (such as art and building materials) are not covered in depth. Unfortunately, fungi, microalgae, and other "lower plants" are not covered, although there are case studies on macroalgae and duckweed. There are thousands of cultural groups that all deserve mention for their ethnobotanical ingenuity, but there are not enough pages in this book to mention everyone. Readers may notice that this book does not include many long lists of bioactivities described for a specific plant. It seems that, even when studying the use of one plant by one cultural group, the list of uses is extensive, in some cases ranging from snakebite treatment to liver tonic with 20 other remedies in between. The task of deciding which activities to include usually came down to how often the plant and activity in question were mentioned in respected publications (journal articles, pharmacopoeias, historical texts, etc.). Please use the provided references for additional in-depth information, as this book provides just the tip of the iceberg on many topics.

The first section of the book contains a mini-course on botany and phytochemistry written by the lead authors. It should provide the reader with tools to understand the case studies and other related literature. The largest section in the ethnobotany chapter is "Taxonomy" (page 21). Families are arranged using the Angiosperm Phylogeny Group (APG) III phylogenetic classification system. The chapter describes hundreds of ethnobotanical plants, why they are important, their primary uses and/or activities, and active phytochemicals. Understanding taxonomy is essential if you want to make new phytochemical discoveries. You may notice that plants in the same family and closely related families tend to produce similar phytochemicals. So, for example, if you were searching for new isoquinolone alkaloids, you would be wise to investigate plants from families where they are known to occur such as Berberidaceae, Euphorbiaceae, or Ranunculaceae. We made every attempt to include the class of molecule with chemical names, to aid in further research.

Some may consider the case studies in Part II the most interesting part of the book. Most of the case studies were written by contributing authors in their field of expertise. The lead authors wrote the regional introductions and a few of the case studies as well. Many of the cases are cutting-edge research at the time of publication. Others tell in-depth stories of plants that had a huge impact on society.

Words of Caution

Scientific literature quickly becomes out of date and this textbook is no exception. What may have been state of the art or common knowledge at the time of publication may give way to new ways of thinking. Taxonomy is one important example. Plant names and classifications change quickly. Please take care to check the currently accepted names using a respected database, such as The Plant List by Royal Botanic Gardens Kew and Missouri Botanical Garden. Another problem area is plant centers of diversity/origin. Many scientists still use Russian botanist Nikolai Vavilov's 1940 work, The theory of origins of cultivated plants after Darwin as a reference. But new discoveries in archaeology, paleobotany, and molecular genetics have proved a number of Vavilov's assumptions incorrect. Along the same lines, the authors have carefully attempted to provide accurate information regarding historical use of plants. Quite frequently, date agreement is a problem. Occasionally, a reference claimed that an ancient culture used a plant species plant that was not native to the region and was not introduced until hundreds to thousands of years later. This problem typically occurs when an author is either not familiar with botany or when there has been disagreement as to a plant's origins.

The final lesson for anyone reading this book comes from a quote attributed to Albert Einstein: "The more I learn, the more I realize how much I don't know." Both ethnobotany and phytochemistry are immense fields of study with so much more to be discovered. We hope readers will go away somewhat dissatisfied, with a desire to learn more about a particular plant or culture.

Acknowledgments

We would like to give special thanks to our academic advisors who helped guide us as young scientists, especially David Seigler at the University of Illinois, Mary Ann Lila at North Carolina State University, and Ilya Raskin at Rutgers University. Alain Touwaide from the Institute for the Preservation of Medical Traditions at the Smithsonian Institution provided valuable information and editing for the first chapter of the book. Furthermore, this book would not have been possible without the generous contributions from all the contributing authors, photographers, and technical experts. Please accept our sincere gratitude for the work you put into your chapters and your willingness to share your botanical photos with the world. Finally, we would like to thank our families for their patience and support as we spent many hours working on this book.

Part I

Introduction to Ethnobotany and Phytochemistry

1

Ethnobotany
B. M. Schmidt

Key Terms and Concepts

Ethnobotany is the scientific study of the relationship between plants and people. It includes traditional and modern knowledge of plants used for medicine, food, fibers, building materials, art, cosmetics, dyes, agrochemicals, fuel, religion, rituals, and magic. A broader definition also includes how people classify, identify, and relate to plants along with reciprocal interactions of plants and people. In many ways, ethnobotany is a product of European curiosity with the New World and other native peoples encountered during their exploration voyages starting in the fifteenth century. Before the American botanist John Harshberger coined the term "ethnobotany" in 1896, "aboriginal botany" was used to describe European interest in the way aboriginal people used plants for medicine, food, textiles, and so on. Many European exploration missions were undertaken with the sole purpose of exploring natural and cultural wonders (see Captain Cook's breadfruit voyages, Chapter 1, the section titled "Moraceae: Mulberry Family"), often followed by colonial or imperialist expeditions. Early explorers, missionaries, clergy, physicians, traditional healers, historians, botanists, anthropologists, and phytochemists have all contributed to the field of ethnobotany. **Botany** is the study of plants (Kingdom Plantae) including physiology, morphology, genetics, ecology, distribution, taxonomy, and economic importance. Sometimes fungi (Kingdom Fungi) are included in botany, but for the purposes of this book, they will not be covered.

Ethnobiology is a multidisciplinary field that studies the relationships of people and their environment, which includes plants and animals. Ethnobotany could be considered a specialized branch of ethnobiology. There are several specialized branches of ethnobotany that focus on one particular aspect of the field. **Ethnomedicine** focuses on traditional medicine including diagnostic and healing practices along with herbal medicines. **Ethnopharmacology** is the study of the uses, modes of action, and biological effects of plant-based medicines, stimulants, or psychoactive herbs. **Economic botany** is closely related to ethnobotany. The main distinction is that economic botany focuses on applied economic, agricultural, or commercial aspects of human uses of plants, but does not deeply explore traditional cultures, the "ethno" side of ethnobotany. Economic botany studies often have the goal of developing new plant-derived products, which

Ethnobotany: A Phytochemical Perspective, First Edition. Edited by B. M. Schmidt and D. M. Klaser Cheng.
© 2017 John Wiley & Sons Ltd. Published 2017 by John Wiley & Sons Ltd.

may or may nor be based on traditional uses, while ethnobotany studies may simply document facts about plant use when there is no prospect of commercial gain.

Ethnobotanists use a variety of tools for their scientific investigations including historical texts, surveys, interviews, and field observations of human–plant interaction. They typically collaborate with indigenous people or local scientists to make an inventory of local natural resources, identifying which plants are useful and in what way. **Biocultural diversity** is the total variety exhibited by the world's natural and cultural systems. It includes both the **biodiversity index** (the diversity of plants, animals, habitats, and ecosystems), and the **cultural diversity index** (diversity of human cultures and languages). Biodiversity is measured by dividing the number of distinct species in an area by the total number of individuals in the area. Cultural diversity can be calculated by dividing the number of distinct languages, religions, and ethnic groups in an area by the number of total individuals in the area. Hot spots of biocultural diversity include Central Africa, Malesia, and the Amazon Basin.

Phytochemistry is the study of plant natural products. Natural products include both primary metabolites (e.g., amino acids, carbohydrates, and fats) and secondary metabolites (e.g., alkaloids, carotenoids, and polyphenols). Phytochemistry also encompasses plant biosynthetic pathways and metabolism, plant genetics, plant physiology, chemical ecology, and plant ecology. It can be considered either a branch of chemistry or botany, depending on whether the scientist and/or research program focuses more on the plant or the chemicals.

Historically, there has been a significant gap between the fields of ethnobotany and phytochemistry. Ethnobotanists are often great anthropologists, with rich knowledge of traditional cultures, texts, and historical context. They provide valuable plant inventories in vulnerable areas and are particularly interested in the cultural role of plants. But when it comes to phytochemistry, the fundamental nature of how a plant works as a biologic (drug, stimulant, etc.), what properties make it a good building material or fiber, or why one natural dye requires a mordant and another does not, they frequently provide superficial answers. Therefore, their publications or presentations will often stop short of describing the chemistry behind the traditional use. Phytochemists, on the other hand, have a thorough grasp of the chemical nature of plants, from the biosynthetic pathways to the effects of the environment on the production of secondary metabolites, to the metabolism of phytochemicals in the human body. Often, they have a background in botany, with a picture of how plants are related and function on the basis of their common chemistries. But they lack training in anthropology or linguistics, with limited awareness of the historical and cultural context of plants. Collaboration between these two groups of scientists is essential to present the whole story of how valuable plants are to our society. More and more, university programs are preparing ethnobotanists and phytochemists with tools from both disciplines. Together, with the common goal of preserving biocultural diversity and promoting social well-being, we can make the best use of our natural resources.

Ethnobotany throughout History

Humans have been using plants since before recorded history. Our earliest relatives gathered plants to use as food, medicine, fibers, and building materials, passing on their knowledge through oral traditions. Agriculture, the practice of producing crops and raising livestock, came about independently in different regions of the world 10,000–15,000 years ago. Botanical knowledge was a great advantage in ancient civilizations, as it conferred a greater chance of survival. Many ancient scholars took a keen interest in botany, publishing herbals that contained botanical information, as well as plants' usefulness. With this information, a person could identify a plant in the wild or in a garden and also know how to use it.

Ethnobotany as a science did not come about until more modern times. While people historically had a close connection to plants and many scholars studied botany, few studied the botanical knowledge of a social group until the twentieth century. The following are a few of the influential botanical scholars and texts that helped disperse ethnobotanical knowledge throughout the ages.

Egypt, Greece, and Rome

The mortuary temple of **Queen Hatshepsut** of Egypt (c. 1508–1458 BCE) at Deir el-Bahri depicts a trade expedition to the region of Punt. This is one of the earliest documentations of botanical trade. Her ships returned with myrrh trees, among other treasures. Scholars believe she died of bone cancer, which may have been caused by inadvertently using a carcinogenic skin salve composed of palm oil, nutmeg apple oil, and creosote. Creosote contains benzo(a)pyrene, which is highly carcinogenic. Her story provides a glimpse into the extent of botanical knowledge over 3500 years ago.

Around the same time (c. 1500 BCE), the Egyptian **Papyrus Ebers** (Figure 1.1) was written. It is considered the oldest book of botanical knowledge, a collection of old folk medicine that was likely copied from books that were hundreds of years older. The papyrus was found in a tomb along with another medical text, the Edwin Smith Papyrus.

Figure 1.1 Papyrus Ebers, column 38. © 2009, Einsamer Schütze.

Numerous herbal remedies are listed, such as *Acanthus*, aloe, balsam, barley, beans, caraway, cedar, castor oil, coriander, dates, figs, garlic, grapes, indigo, juniper, linseed, myrrh, onions, palm, pomegranate, poppy, saffron, turpentine, watermelon, wheat, willow, and zizyphus lotus.

Hippocrates (c. 460–350 BCE) was a Greek physician, often called the Father of Western Medicine. The *Hippocratic Collection*, a compilation of treatises containing medical remedies and most notably the Hippocratic Oath, was attributed to Hippocrates in ancient times. However, scholars now believe the *Hippocratic Collection* was actually written by several authors, perhaps medical scholars from different schools.

Theophrastus (c. 372–287 BCE) was the student of Aristotle and is known as the Father of Botany. After Aristotle's death, he inherited Aristotle's library and garden. Two of his most notable works are *Peri phytôn historia* also known by the Latin title, *De historia plantarum*, "Inquiry into Plants" (ten books) and *Peri phytôn aitiôn*, "The Causes of Plants" (eight books). Theophrastus described roughly 500 plant species, classifying them into four groups: herbs, undershrubs, shrubs, and trees. He noted many anatomical differences and separated flowering from non-flowering plants. Many of the names he gave to plants are still used today.

Pliny the Elder's publication of *Historia Naturalis* (c. 77–79 CE) built upon Theophrastus's work, but *De Materia Medica* (c. 70 CE) by the Roman physician **Pedanios Dioscorides** (c. 40–90 CE) (Figure 1.2) became the authoritative text on botany and medicinal plants for the next 1500 years. Dioscorides traveled as a surgeon with the Roman army, which allowed him to study the features of many medicinal plants. He advocated observing plants in their natural environment, during all stages of growth. *Materia Medica* describes 600 species of medicinal plants, including 100 not described by Theophrastus. These plants include opium and *Mandragora* (mandrake root) as surgical anesthetics, willow for pain relief, and henna for shampoo. Numerous drugs, spices, oils, cosmetics, and beverages still in use today were mentioned in *Materia Medica*. Dioscorides' work became the foundation for modern botany. He formed the connection between plants and medicine, which eventually gave rise to ethnobotany. *De Materia Medica* was widely copied and translated into Arabic and Latin through the Middle Ages.

Claudius Galen (c. 129–199), a Greek physician who brought about "Galenic medicine," is another notable figure of this time period. He wrote *De Alimentorum Facultatibus*, "On the Properties of Foodstuffs," a physiological treatise rather than a *materia medica*. He described a wide range of plants for food and medicine, for all classes of citizens. Some of his works were translated into Arabic and influenced Islamic medicine.

China

Chinese botanical legends date back farther than Queen Hatshepsut, but lack the same documentation. Chinese **Emperor Shen Nong** (Figure 1.3) was a legendary ruler of China. Although he may or may not have been a true historical figure, he is considered the founder of Chinese Herbal Medicine. Legend says that he wrote *Pen Ts'ao Ching*, "Great Herbal," in 2700 BCE. Modern researchers believe it was actually a compilation of oral traditions, written around 200–300 CE. The original text is no longer in existence, but it was said to be a catalog of 365 medicines from plants, minerals, and animals that formed the foundation of Chinese medicine. Emperor Shen Nong is credited with

Figure 1.2 Portrait of Dioscorides receiving a mandrake root in an early sixth-century copy of *De Materia Medica*.

the discovery of tea when a tea leaf accidentally landed in his pot of boiling water. He taught his people to plow the land and cultivate grains. Legend says that his love for plants lead to his demise, as he was poisoned by a toxic plant.

The oldest known traditional Chinese medicine text, *Huangdi Neijing*, "Yellow Emperor's Cannon" (c. 300–100 BCE), predates Dioscorides' *De Materia Medica*. Two influential herbalists emerged during the Han Dynasty (c. 202 BCE–220 CE), Zhang Zhongjing and Hua Tuo. **Zhang Zhongjing** (c. 150 CE–219 CE) was a physician, the Hippocrates of China and the Father of Medical Prescriptions. His text *Shang Han Lun*, "Treatise on Cold Damage Disorders," contains remedies still used in Chinese medicine today. It is one of four books students of Chinese medicine are required to study. Zhang Zhongjing advocated treatment according to symptoms, using a combination of acupuncture and herbs. **Hua Tuo** (c. 140–208 CE) was a physician best known for introducing the use of wine and hemp (Ma Fei San) for surgical anesthesia. There is speculation among scholars that Ma Fei San may have actually contained more potent anesthetics such as opium or other powerful alkaloids from *Datura, Aconitum*, or *Mandragora* species.

Figure 1.3 Shennong, one of the mythical emperors of China, Indian ink on silk by Xu Jetian.

India

Ayurveda, Indian naturalistic medicine, came about sometime during the sixth century BCE. The Saṃhitâs or "collections" are the main source of recorded knowledge for Ayurveda. The chronology remains unclear, but three primary Sanskrit texts were written c. 100 BCE–600 CE: the Charaka Samhita (c. 100 BCE–100 CE), Suśrutha Saṃhitâ (c. 300–400 CE), and Bheda Samhita (c. 600 CE). Excerpts from the Bheda Samhita are found in the medical portions of the Bower Manuscript (c. 400–600 CE), a birch bark document discovered by British intelligence officer Hamilton Bower in 1890. It also remains unclear if **Charaka**, **Suśrutha**, and **Bheda** were historical figures or divine beings, but they were likely not the authors of the manuscripts that bear their names.

The Middle Ages (500–1500 CE)

Europe and the Arabo-Islamic World

Little progress was made in European botany during the Middle Ages, as manuscripts were destroyed during wars and the fall of the Roman Empire. But it was during this time period that Islamic botany began to thrive. Islam was widespread, and there was extensive travel throughout northern Africa, India, and the Middle East. Abû Ḥanîfa Dînawarî or "**Al-Dinawari**" (c. 828–896) is considered the founder of Arabic botany for his publication *Kitab al-nabat*, "Book of Plants." The first section of the book contains an alphabetical list of plants, mostly from the Arabian Peninsula. The second section contains monographs of plants and their uses. Al-Dinawari's "Book of Plants" became the authority on Arabic plant names. **Ibn Juljul** (c. 944–1009) built upon the work of Dioscorides, publishing a supplement titled *Maqalah* containing 60 plants not mentioned by Dioscorides. Ibn Sina "**Avicenna**" (c. 981–1037) was considered the father of early modern medicine. His *Qanun (Canon) of Medicine* (1025 CE) (Figure 1.4) is an encyclopedia of medicine based in part on Galen's work from the first century. Book two is a *materia medica* that describes, among other things, plant-based drug treatments for disease. Book five is a formulary of compounded drugs.

Al-Ghafiqi was born near Córdoba, Spain (c. 1100–1165) during the Muslim rule of the Iberian Peninsula. He was an influential physician and medical author, publishing *Kitab al-jami' fi 'l-adwiya al-mufrada*, "Book of Simples," and a *Materia Medica* manual. "Book of Simples" included plants not mentioned in any Greek text or Middle Eastern publications. His *Materia Medica* is considered one of the best from the Middle Ages. **Al-Idrisi** (c. 1100–1166) was born in Morocco, lived and studied in Spain, and traveled extensively throughout the region. He knew many languages and botanical names. In his *Jami' on Materia Medica*, he names botanical drugs in Spanish, Arabic, Berber, Hebrew, Latin, Greek, and Sanskrit. **Ibn al-Suri** (c. 1177–1242) was another physician botanist that traveled extensively. Accompanied by an artist, he documented plants throughout the region, especially in the Lebanon range. His *Materia Medica* contained paintings of plants at different stages in their life cycles and as they looked dried on a pharmacist's shelf. It was the first Arabic book illustrated in color. **Ibn al-Baytar** (c. 1197–1248) was an outstanding Islamic herbalist of the Middle Ages. Born in Malaga Spain, he studied in Seville and published several notable books including *Al-Mughni fi al-Adwiyah*, "The Sufficient"; *Al-Kitab 'l-jami' fi 'l-aghdiya wa-'l-adwiyah al-mufradah*, "The Comprehensive Book of Foods and Simple Remedies"; and a *Materia Medica*. He concentrated on "simples," one-ingredient drugs and remedies. "The Comprehensive Book of Foods and Simple Remedies" contained 3,000 simples listed in alphabetic order. As its title suggests, it was the most comprehensive encyclopedia of simples in the Middle Ages.

There are few (non-Islamic) European botanists worth mentioning from the Middle Ages. **Hildegard of Bingen** (1098–1179) was a German nun who wrote medical texts including *Physica*, based on her experiences in the monastery herb garden. **Matthaeus Platearius** (?–1161), an Italian physician from the medical school in Salerno, wrote *Liber de Simplici Medicina* or "Book of Simple Medicines." This book of simples was an influential guide to herbal medicine throughout the Middle Ages and was used as a prototype for modern pharmacopeias. German scientist and theologian **Albertus Magnus** (c. 1205–1280) was known as Albert the Great and Doctor Universalis. His work *De Vegetabilis et Plantis* contains a detailed descriptions of plant morphology and physiology, distinguishing dicots from monocots. English scholar **Bartholomaeus Anglicus** (c. 1203–1272) wrote the encyclopedia *De Proprietatibus Rerum*, "On the Properties of Things." The encyclopedia contained a section on natural sciences with descriptions of plants and their uses.

China

The Chinese empire also made advances throughout the Middle Ages. **Sun Simiao** (c. 581–682), the King of Medicine, wrote two famous books, *Bei Ji Qian Jin Yao Fang*, "Essential Recipes Worth a Thousand Gold for Any Emergency," and *Qian jin Yi fang*,

Figure 1.4 *Kitâb al-Qânûn fî al-ţibb* (The Canon on Medicine) by ibn Sînâ.

"A Supplement to Recipes Worth a Thousand Gold." The first book focused on women and children's health and listed over 4500 medical formulas. It was later regarded as the first encyclopedia of clinical practice. The supplement added a further 800 ingredients. He is also well known for his interest in identification and preparation of medicinal herbs. He stressed the importance of collecting genuine plants during the appropriate season:

> Without knowing where the medicines are from, and whether or not they are genuine and fresh, they cannot cure five or six patients out of ten ... If you do not know the proper seasons when they should be placed in the shade or in the sun to dry, the result will be that you know their names but do not obtain their intended effects. If you gather them at an improper time, they will be good for nothing just like rotten wood, and you will have made a futile effort.

After the invention of the printing press in 1440, there was a surge in herbal publications and sharing of botanical information in general. All of the botanical scholars during this time period cannot be named in this book owing to space constraints. The following are some of the most influential.

China

Li Shizhen (c. 1518–1593) was a physician during the Ming Dynasty and is still a household name in China. Li published 17 books, including his most well-known *Bencao Gangmu*, "Compendium of Materia Medica." The illustrated book contained 374 new drugs. He classified drugs under 16 headings: water, fire, earth, metal, stone, plant, cereal, vegetable, fruit, tree, products derived from "garments and tools," insect creatures with scales, creatures with shells, birds, quardrupeds, and products of human origin. He organized plants by habitat and special features, which differed from previous more arbitrary classifications.

Europe

Otto Brunfels (1489–1534), **Hieronymus Bock** (1498–1554), and **Leonard Fuchs** (1501–1566) were described as the German Fathers of Botany. They were the first European botanists since Dioscorides to illustrate or describe plants using live specimens in gardens, rather than copying from other books. Brunfels wrote *Herbarium vivae eicones*, "Living Picture of Plants" (1530), and *Contrafayt Kräuterbuch*. The books make up for weak scientific merit through the detailed, realistic illustrations. Bock's original herbal, *Kreuterbuch* (1539), was not illustrated. Therefore, he was forced to give very detailed plant descriptions. A later version (1552) contained hand-colored woodcuts. Fuchs wrote *De Historia Stirpium* (1542) and *Plantarum Effigies* (1549). He copied many of the plant descriptions from other books, but again the illustrations are what made the books outstanding. *Plantarum Effigies* was published as a pocket-sized field manual, with primarily just the woodcut illustrations.

Nicolás Monardes (1493–1588), a Spanish physician, holds the distinction of popularizing tobacco. In his 1571 publication on useful plants from the West Indes, he describes tobacco as a remarkable plant that curbs hunger, reduces fatigue, and treats insomnia (Figure 1.5). It is capable of curing 20 ailments, including headaches, parasitic worms, joint pain, inflammation, skin wounds, toothache, and cancer. From the 1577 English translation by John Frampton, *Joyfull Newes our of the Newe Founde Worlde*:

> In the venomous Carbuncles [cancer], the Tobacco being put in the manner as is said does extinguish the malice of the venom, and it does that which all the works of Surgery can do, until it be whole. – Nicolás Monardes

Pierandrea Mattioli (1501–1577), personal physician to Roman Emperors Ferdinand I and Maximillian II, published a popular commentary on Dioscorides' De Materia Medica, *Commentarii in sex libros Pedacii Dioscoridis* (1544). It was filled with beautiful colored woodcuts, a first of its kind, as previous publications were illustrated in black and white. A dispute developed between Mattioli and Swiss naturalist **Conrad Gessner** (1516–1565), first over the poisonous nature of the apothecary drug *doronicum*, and later over Mattioloi's woodcut of *Aconitum primum* featured in *Commentarii*.

Figure 1.5 Nicolás Monardes describes the use of tobacco plant. © 2016 Wellcome Library, London.

Gessner claimed he described *Aconitum primum* in his publication *Historia animalium* (1551–1558) and was the first to publish a true picture of *Aconitum primum* in his 1555 book on rare plants, *lunaria*. He claimed Mattioli's picture was fictitious, designed to fit Dioscorides' description. The matter was never resolved, but both men retained their reputations.

England produced several notable renaissance botanists as well, including **John Gerard** (1545–1612), **William Turner** (c. 1510–1568), and **Nicholas Culpeper** (1616–1654). Gerard was the head gardener for many estates around London, including the estate of Elizabeth I's Secretary of State, William Cecil. He dedicated his exemplary work, *The Herball* (1597) (Figure 1.6), to Cecil. Turner is considered the Father of English Botany. His works include *Libellus De Re Herbaria Novus* (1538), *The Names of Herbes* (1548), and *A New Herball* (1551). *A New Herball* was the first herbal written in vernacular English, allowing people to easily identify English plants. Nicholas Culpeper was an English physician and apothecary who dedicated his life to serving the poor and

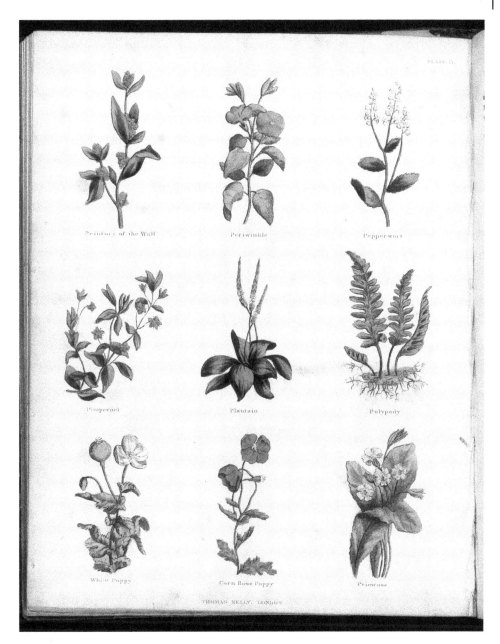

Figure 1.6 A page from Nicholas Culpeper's *The Complete Herbal*. Plant names (left to right from top left): Sweet Flag, Foolston, Fumitory, Frogbit, Fleabane, Yellow Flag, Feverfew, Fluellin, Alkanet, Purging Flax, and Pellitory of Spain.

speaking out against political oppression. He translated Latin medical books to English to make them more accessible to the masses. *The English Physician* (1653), later known as "Culpeper's Herbal," was his most successful publication. It contained thorough descriptions of nearly all the medicinal herbs known at the time, along with their healing properties and directions on how to extract and compound them into medicine.

Joseph Pitton de Tournefort (1656–1708) was Chief Botanist to Louis XIV, Professor of Botany in charge of the Jardin des Plantes in 1683, botanical explorer, and physician. He went on several botanical expeditions around France and the rest of Europe, bringing back not only botanical inventories, but information about people, cities, and cultures as well. He was the first person to divide plants to the level of genus, developing an easy-to-use system of classification.

Several members of the **de Jussieu family** from France were influential botanists during the eighteenth century. Brothers Antoine (1686–1758), Bernard (1699–1777), and Joseph (1704–1779) were sons of a French apothecary, each graduating with a medical degree. Antoine and Bernard had prestigious careers at Paris' botanical gardens. Joseph was the most adventurous of the three brothers, making significant ethnobotanical discoveries for France. In 1735, he traveled on an expedition headed by Charles de la Condamine to Peru. Joseph's notes mention a Jesuit being cured of fever by a Peruvian bark (Linnaeus named this bark *Cinchona* in 1732; its active ingredient quinine is still used to treat malaria). Joseph also discovered that coca leaves had powerful analgesic properties, after watching the natives. La Condamine traveled back to France in 1745 with precious samples of rubber and quinine. But Joseph remained in Peru for 36 years, working as a physician and collecting botanical samples. He sent back dried coca leaves and seeds of the first *Heliotropium* ever introduced to Europe.

New World

Around the same time that Turner and Gerard were writing their herbals in England, two Aztecs at the College of Santa Cruz in Tlaltilulco, **Martinus de la Cruz** and **Juannes Badianus**, wrote the Badianus Manuscript (1552) (Figure 1.7). It is the oldest known

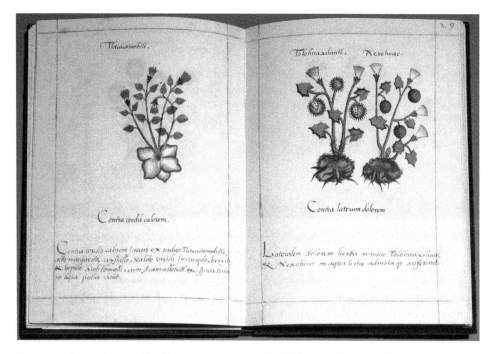

Figure 1.7 A page from the Bandianus manuscript. © 2016 Wellcome Library, London.

American herbal, containing a wealth of native Mexican medicinal plants and remedies. De la Cruz wrote the manuscript in the Aztec language Nahuatl and was likely the illustrator; Badianus translated it into Latin. The original is housed in the Vatican Library. It contains 184 beautiful color illustrations of native medicinal plants. Badianus used mostly the Aztec plant names, as he was unaware of their Latin botanical names. Under each illustration is a name of a disease or condition the plant was used to treat and how it should be prepared.

As the Spanish conquered Central and South America, their botanists were busy taking inventory of the local flora. **Francisco Hernández de Toledo** (c. 1514–1580) was a Spanish naturalist and physician who led one of the first scientific expeditions to the New World. He traveled to Mexico in 1570 with the intention of studying Mexican flora. He returned to Spain in 1577 with a number of live specimens and began work on *Rerum Medicarum "Quatro libros de la naturaleza y virtudes de las plantas y animals."* Years after Hernández's death, Dominican priest Francisco Ximénez published the book in Mexico in 1615.

The Eighteenth Century through Modern Day

As plants from the New World were introduced to Europe, botanists published herbals describing new medicinal plants. One notable example is Scottish botanist and illustrator **Elizabeth Blackwell** (c. 1707–1758). She wrote the richly illustrated *A Curious Herbal*, depicting "curious" medicinal plants from the New World. With her husband in debtor's prison, she needed to find a way to support herself and her child and also secure her husband's release. Using her training in drawing and painting and her imprisoned husband's skills in medicine, she sought to fill a gap in the marketplace with a desk reference work for apothecaries. She drew specimens from the Chelsea Physic Garden, a medicinal plant garden established for apothecaries to identify medicinal plants, including exotic specimens from the New World. Each week from 1737 to 1739, she took drawings of the plants to her imprisoned husband for identification and translation into different languages. Then she would engrave copper plates for printing and hand-color the images. She published four plates per week for a total of 500 images.

Some botanists were not content to remain in Europe publishing herbals or studying plant specimens brought from the New World. Instead, they became plant explorers. Spaniards **Hipólito Ruiz López** (1754–1816) and **José Antonio Pavón Jiménez** (1754-1840) set out on an expedition to explore the flora of Peru and Chile in 1777. They collected 3000 specimens over the course of 10 years and brought back knowledge of medicinal plants such as *Cinchona*. Their publication *Flora Peruviana et Chilensis* included 10 volumes with rich illustrations.

Sir Joseph Banks (1743–1820) was a wealthy British explorer and botanist who sailed with the likes of Captain James Cook. His first voyage was aboard the H.M.S. *Niger* in 1766 to Newfoundland and Labrador. In 1768, he sailed aboard the *Endeavour* with Captain Cook, stopping in Rio de Janeiro, Tierra del Fuego, Tahiti, New Zealand, and eastern Australia, most notably, Botany Bay and Endeavour River. Banks not only took notes on the plants of Australia, but also on Aboriginal customs. After he was elected president of the Royal Society in 1778, he directed and financed many collection expeditions to the Cape of Good Hope, West Africa, East Indes, South America, India, Australia, and around-the-world–voyages, including William Bligh's voyage to the South Seas. Despite all of his enthusiasm, adventurous voyages, and thousands of specimens collected, he published very little except for a few short articles. But he is considered the Father of Australia because of his exploration and devoted interest in the colony.

Sir Joseph Dalton Hooker (1817–1911) was an English botanist, son of Sir William Jackson Hooker and Maria (nee Turner), daughter of English botanist Dawson Turner. Joseph earned a medical degree from Glasgow University, where his father was a botany professor. As a boy, he attended his father's botany lectures and dreamed of the day that he could travel to far-off lands and encounter penguins, as depicted in Captain Cook's *Voyages*. He became assistant surgeon on the HMS *Erebus*, launching his career as an explorer. They visited Antarctica, New Zealand, Tasmania, and the small islands around Antarctica. Hooker collected plants to bring back to England, while attending to his surgical duties. He considered **Charles Darwin** a role model and had the opportunity to review proofs of *Voyage of the Beagle* before it was published. His father had become the first director of the Royal Botanic Gardens at Kew and helped secure a grant to publish Joseph's first book, *Botany of the Antarctic Voyage*. The book contained six volumes, two each for *Flora Antarctica*, *Flora Novae-Zelandiae*, and *Flora Tasmaniae*. The publications were not a financial success, and Hooker was having no luck securing permanent employment in England. Charles Darwin sent Hooker a letter congratulating him on the books and asking if he would help classify plants from the Galapagos Islands. This was the beginning of a long relationship between Darwin and Hooker. Hooker continued his travels, to India and Nepal, where he collected 7000 species and landed in jail for ignoring warnings not to enter Tibet. He published *Rhododendrons of the Sikkim-Himalaya* and brought back 25 new species of rhododendron, much to the delight of English gardeners. After his father's death in 1865, Joseph became director of Kew Gardens. His father had substantially increased the size and reputation of Kew. Under his management, Kew played an important role in transferring *Cinchona* trees from South America to India for quinine production. Portraits of William Hooker and Sir Joseph Dalton Hooker are found throughout Kew Gardens today. Both men are buried in St. Anne's Churchyard, Kew Green.

Chemical investigations into the healing properties of herbal remedies began in Europe during the late eighteenth and early nineteenth centuries. Reverend **Edward Stone** (1702–1768) had no formal training in chemistry but was curious about the medicinal properties of willow. In 1763, he conducted a successful clinical trial by administering a powder made from willow twigs to patients with malaria fever. The powder had astringent properties and cured "agues," fever associated with malaria. The active compound, salicin, would not be isolated for another 65 years by the German pharmacologist **Johann Büchner** (1783–1852). In 1804, pharmacist **Friedrich Sertürner** (1783–1841) was busy isolating morphine from poppies. He conducted animal studies to confirm the hypnotic and analgesic properties of the "sleeping agent" and described its alkaline properties. He is considered the first alkaloid chemist. Following in Sertürner's footsteps, the German chemist **Friedlieb Ferdinand Runge** (1795–1867) identified the anticholinergic properties of belladonna by accident when a drop of the extract splashed in his eye. He repeated this dilation experiment on his cat for Goethe in 1819. Impressed, Goethe offered Runge coffee beans, a rare gift at the time, and asked him to figure out what made them so intoxicating. Runge isolated caffeine a few months later. He eventually isolated atropine from belladonna and quinine from cinchona as well. However, French alkaloid chemists **Pierre Joseph Pelletier** (1788–1842) **and Joseph Bienaimé Caventou** (1795–1877) are sometimes credited with the first isolation of quinine and caffeine. Along with Polish chemist **Filip Neriusz Walter** (1810-1847) they isolated a number of alkaloids from plants including emetine from *Carapichea ipecacuanha* and strychnine from *Nux vomica*.

In 1831, carotin was isolated from carrots by pharmacist **Heinrich Wilhelm Ferdinand Wackenroder** (1798–1854). The small red crystals were oil soluble and imparted a beautiful yellow color to butter. Early essential oils work by **Jean Baptiste André Dumas** (1800–1884) and **J.J. Houton de la Billardiere** (1755–1834) led to the characterization of several terpenes including camphene by **Marcellin Berthelot** in 1885, camphor by **Julius Bredt** in 1883, α-pinene by **Egor Egorevich Vagner** (AKA Georg Wagner), and citral by **Johann Carl Wilhelm Ferdinand Tiemann** and **Friedrich Wilhelm Semmler** in 1885. Nobel Prize winner **Otto Wallach** (1847–1931), called the Messiah of Terpenes, introduced the isoprene rule: all terpene compounds are constructed of isoprene units in a head to tail configuration. Simplified line drawings of chemical structures are attributed to these pioneering terpene chemists, who first used the shorthand drawings in the 1800s.

There are many other modern chemists who made significant contributions to the field of phytochemistry. **Richard Martin Willstätter** (1872–1942) won the Nobel Prize in Chemistry in 1915 for his work on chlorophyll. But he also did extensive work on anthocyanins and carotenoids, identifying lycopene in tomatoes and describing xanthophylls, including lutein. The Hungarian physiologist **Albert Szent-Györgyi** (1893–1986) won the Nobel Prize in Physiology or Medicine in 1937 for his work on vitamin C and scurvy, but he is also known for his work on "vitamin P." He postulated that vitamin P was helpful in repairing capillaries and reducing inflammation in skin disorders and frostbite. Vitamin P would later be known as a whole class of compounds, the flavonoids. During the course of his flavonoid anticancer research, Szent-Györgyi discovered rutin. The Chinese scientist **Tu Youyou** (Figure 1.8) won the Lasker Medical

Figure 1.8 Tu Youyou and her teacher Lou Zhicen working in the laboratory in 1951.

Research award in 2011 for her discovery of the antimalarial drug artemisinin from *Artemisia annua* in 1971. After reading ancient Chinese ethnobotanical texts, she realized that a cold extraction method was required to preserve the active constituents. Scientists such as **Tu Youyou**, **Arthur Schwarting**, **Jack Beal**, and **Norman Farnsworth** brought together the fields of phytochemistry and botany, forming modern day pharmacognosy. Farnsworth noted that 25% of all prescriptions dispensed from community pharmacies in the United States from 1959 to 1980 were plant-based drugs. The percentage is even higher in other countries, especially Asia, where a majority of people rely on plant-based medicine. **Richard Evans Schultes**, considered the father of modern ethnobotany, studied indigenous cultures in South America while collaborating closely with chemists. He discovered the source of curare and identified several hallucinogenic compounds. Countless others continue to make contributions to the closely related fields of phytochemistry and ethnobotany, preserving our knowledge of plants and people.

References and Additional Reading

Arber, A.R. (1986) *Herbals: Their Origin and Evolution: A Chapter in the History of Botany*, 3rd edition. Cambridge University Press, Cambridge.

Barlow, H.M. (1913) Old English Herbals, 1525–1640, *Proceedings of the Royal Society of Medicine*, **6**, 108–49.

Collins, M. (2000) *Medieval Herbals: The Illustrative Traditions (British Library Studies in Medieval Culture)*, University of Toronto Press, Scholarly Publishing Division, Toronto.

Dharmananda, S. (2001) *Sun Simiao: Author of the Earliest Chinese Encyclopedia for Clinical Practice*, Institute for Traditional Medicine, Portland.

The Foundation for Science, Technology and Civilisation, Botany, Herbals and Healing in Islamic Science and Medicine, http://muslimheritage.com/node/621 (accessed 21 April, 2015).

Needham, J. (2000) *Science and Civilization in China*, vol. 6, Cambridge University Press, Cambridge.

Palin, M. (2015) *About Joseph Hooker*, Royal Botanic Gardens, Kew, London.

Prioreschi, P. (2002) *A History of Medicine*, Horatius Press, Omaha.

Wexler, T.A. (2006) Tobacco: From Miracle Cure to Toxin, *YaleGlobal*, 12 June.

Current Topics in Ethnobotany

Since the beginning of the twentieth century, the focus of ethnobotany switched from plant exploration and simple plant inventories to the role of plants in society. Several new topics came into view as we learned more about plant biodiversity and indigenous cultures. These topics include conservation of botanical resources along with their habitats, preservation of indigenous cultures and their specialized knowledge of plants, ethics of commercializing plant natural resources, and the efficacy of ethnomedicines.

Modern ethnobotany reflects the rapid change in our society and the environment. There is an urgency to protect both botanical knowledge and the plants themselves from extinction. In many areas of the world, habitat is threatened not only by development (housing, commercial centers, hydroelectric dams, clearing for agriculture, etc.), but also natural disasters such as drought, fire, desertification, and floods. Habitat destruction played a role in the extinction of several species including the Saint Helena olive tree [*Nesiota elliptica* (Roxb.) Hook.f.] and the Jamaican guava (*Psidium dumetorum* Proctor). Another form of natural disaster is widespread destruction by pest, disease, or invasive species. For example, chestnut blight (*Cryphonectria parasitica*) killed an estimated three billion American chestnut [*Castanea dentata* (Marshall) Borkh.] trees, critically reducing the number and genetic diversity of chestnut trees in the wild. The same virus is currently threatening the European chestnut (*C. sativa* Mill.) in Azerbaijan (Wall and Aghayeva 2014). Overharvesting of ethnobotanical plants is a common problem. It presumably brought about the extinction of the arunachal hopea tree [*Hopea shingkeng* (Dunn) Borr], a species harvested to make fence posts in India. Ongoing war in Northern Africa and Western Asia has destroyed plant habitats and displaced millions of people. Along with the people, their culture, botanical knowledge, and local plant varieties are disappearing. The Syrian war has prompted the first withdrawal from the Svalbard Global Seed Vault in Norway (Wagner 2015). This vault was built in 2008 to protect the world's most precious crop plants from total destruction. The International Center for Agricultural Research in Dry Areas (ICARDA) asked to withdraw a portion of the 116,000 samples of wheat, barley, and grasses adapted to arid regions that they had banked as war broke out in 2012. Their Aleppo seed bank was damaged in the war and relocated to Beirut, Lebanon. Other ways ethnobotanists are preserving the world's plant diversity are by publishing plant inventories in threatened areas. These inventories can be used to document the existence of threatened plants and persuade local governments to protect the habitat.

A second area of conservation efforts is preservation of botanical knowledge and mindfulness of the rights of indigenous people. Habitat destruction through war or development can displace or eradicate ethnic groups as mentioned above. Another reason for the destruction of indigenous cultures is that they have valuable natural resources that are desired by a nation-state or corporation, such as land, minerals, or oil. As the remaining members of ethnic groups die or assimilate into other cultures, their ethnobotanical knowledge is lost. Some ethnic groups face extinction not through habitat loss, but through the influence of Western culture. Young members may not be interested in carrying on the traditional ways of their elders, and so ethnobotanical knowledge is not passed on to the next generation. Economic development programs and educational opportunities with good intentions may accelerate the loss of ancestral culture.

Ethics and intellectual property rights for traditional knowledge are two highly debated topics in ethnobotany. If scientists are bioprospecting for a new plant-based drug either in the ethnobotanical literature or by direct contact with indigenous people, who owns the rights to the research and (potential) commercialization? The Convention on Biological Diversity (CBD) is a multilateral treaty opened for signature at the Earth Summit in Rio de Janeiro in 1992 to help answer this question. The treaty sets forth strategies for conserving biodiversity, sustainable use of biodiversity, and equitable benefit sharing of genetic resources. It encourages member countries to "... respect, preserve and maintain knowledge, innovations and practices of indigenous and local communities embodying traditional lifestyles relevant for the conservation and sustainable use of biological diversity ..." The Nagoya Protocol on Access to Genetic Resources and the Fair and Equitable Sharing of Benefits Arising from their Utilization to the Convention on Biological Diversity (the "Nagoya Protocol") is a supplement to the Convention on Biological Diversity, outlining a transparent legal framework for benefit sharing of genetic resources. Under the Nagoya Protocol, a company wishing to commercialize traditional knowledge and related genetic resources, such as selling drugs or cosmetics that were originally from a plant used by an indigenous group or local community, must obtain prior written informed consent from the group. In addition, the company would be required to work out a revenue sharing plan where the group receives a portion of the profits. The same obligations are applied to research and (noncommercial) development of genetic resources, covering universities, museums, botanical gardens, herbaria, and so on. However, international patent laws, particularly the Agreement on Trade Related Aspects of Intellectual Property Rights (TRIPS) of the World Trade Organization, seem to be in direct conflict with the Convention on Biodiversity. TRIPS recognizes patents covering traditional knowledge and genetic resources without requiring benefit sharing or informed consent (Wan Izatul *et al.* 2012). Some argue that TRIPS favors industrialized countries that wish to capitalize on genetic resources, whereas CBD favors countries rich in biodiversity and traditional knowledge (Wan Izatul 2013).

To some extent, the aforementioned treaties assume that traditional knowledge is exclusive to one group. On the surface, this would seem to be the case when reading ethnobotanical inventories of plants and their uses by one group of people, as there is rarely footnote saying, "this information is already known by dozens of other communities." Not clarifying whether or not the knowledge is known elsewhere generates the false notion of exclusivity. Furthermore, the extent or importance of the knowledge within the group is often not determined. For example, a typical ethnobotanical study will include a long list of plants that have 5–10 medical uses each, but it does not state how frequently each plant is prescribed for each of the uses. It also may not clarify whether these uses are known by everyone or only by traditional healers. In the same light, it is often not known which plants are the first line of defense for a particular condition and which plants are used secondarily. Ethnobotanical dereplication should be employed to determine the range or exclusivity of plant uses, clearly identifying uses that are unique to the community in question (Elvin-Lewis 2011). Dereplication can streamline the process of drug discovery by validating pharmacopeias in many ways, differentiating novel information from general knowledge. Numerous ethnobotanical databases have been developed to share plant use information such as:

- Survey of Economic Plants for Arid and Semi-Arid Lands (SEPASAL) maintained by Kew Royal Botanic Gardens
- Plants for a Future
- HerbMed (subscription required)
- Dr. Duke's Phytochemical and Ethnobotanical database

Another topic of current interest in ethnobotany is the complexity of plant extracts used for medicine and the variability of biological responses. Most indigenous people have developed an array of traditional medicines derived from plants. These medicines are often crude extracts made from wild harvested plants prepared by proprietary methods yielding nonstandardized mixtures [see kava case study in the section titled "Pharmacological Effects of Kavalactones from Kava (*Piper methysticum*) Root" in Chapter 8 as an example]. Adding to the opacity of traditional medicine preparation methods is the reluctance of some healers to share critical aspects of the methods with ethnobotanists. Phytochemists know that even if the correct plant is used with precise extraction protocols in a lab setting, plant extracts always have a large degree of variability. This variability comes from the plant material itself. Since plants frequently produce secondary metabolites in response to stress or other environmental factors, the phytochemical content can vary greatly from one season or location to the next. Time of harvest can have a large impact on the phytochemical profile as well. For example, sage (*Salvia officinalis* L.) produces more essential oil while it is in bloom compared to its vegetative state (Amr and Đorđević 2000). Biological responses to plant remedies are highly variable as well. Even with single-component pharmaceuticals, there is a range of biological responses depending on an individual's metabolism, genetics, and so on. When dozens if not hundreds of phytochemicals are included in a single remedy, the biological response becomes even more unpredictable. So, we tend to see conflicting clinical trial results even when a standardized plant extract is administered under highly controlled conditions. Putting aside weak clinical trials with small sample size, poor statistical design, and so on, what conclusions can be drawn regarding the efficacy of traditional medicines that have undergone the rigors of modern science and yet still fail to provide conclusive results? Many of our modern pharmaceuticals are based on phytochemicals, and some revolutionary drugs did not perform well in initial tests (see Artemisia case study in the section titled "Artemisia Species and Human Health" in Chapter 6). So certainly one has to be careful not to "throw the baby out with the bathwater" when evaluating the potential of traditional medicines. There is definitely more work to be done to bridge the gap between anecdotal stories of plant medicine efficacy and rigorous clinical trials that often come up short.

Although most ethnobotanists have a moral obligation to preserve traditional knowledge and natural resources, in some ways, ethnobotany is unintentionally at odds with conservation and the rights of indigenous people. Publishing descriptions of valuable plant species and traditional knowledge as a free exchange of scientific information could contribute to economic exploitation of the area and rob the people of potential economic benefits. The question of whether or not traditional knowledge should be made public is a topic of much debate. Information in the public domain is free for development by commercial interests unless a patent application is filed before publication. So academics must err on the side of caution when publishing potentially sensitive information. Some feel economic exploitation that greatly benefits society as a whole (e.g., new drugs, more sustainable building materials, biofuels) is justified, especially if the local community is in favor of the development and receives economic benefits. Others feel the risk of poaching and disruption of traditional communities is too high, petitioning the scientific community to leave critical information out of the public domain to guard proprietary traditional knowledge. Critical information could include GPS coordinates of plant species or villages, details of plant preparation methods, and specific identity of plant species. The struggle to find a balance between preserving biocultural diversity (ethnobotany) and the responsibility to provide benefit for society as a whole (economic botany) will continue to be a hot topic for years to come.

Finally, what can we expect from ethnobotany in the future? Without plants, humans would not exist, making ethnobotany a science of survival. Indigenous cultures use unconventional, innovative agricultural methods and raise or collect a broad diversity of food crops compared to the highly mechanized cultivation of monocultures epitomized in North America and Europe. It is estimated that humans historically used over 3000 species for food, but only 150 were ever commercially exploited. Three main crops: rice, wheat, and corn, were pushed into arid regions, particularly Central Africa, where they were difficult to grow. Over time, they displaced traditional crops like sorghum and millet that were well suited for arid regions. It is likely that some local cultivars are lost forever. Knowledge of regional crops and varieties, famine foods, wild crops, and landraces adapted to extreme conditions may be the key to expanding the world's limited food supply and adapting to global climate change. Crop yields and hardiness could be improved by breeding with wild relatives (or through genetic engineering), diversifying the gene pool and increasing resistance to pests, disease, and drought. Potentially, developing some of the 2850 "lost" food crops would have a large impact on the food supply, especially in arid areas where many of the world's food crops are difficult to grow. Learning more sustainable farming and wild crafting methods like pest control, mulching, cultivation, and irrigation could result in further agricultural improvements worldwide. Another lesson to be learned from indigenous people is how to live sustainably, reducing our fossil fuel dependence for energy and petrochemicals. Traditional fibers may replace industrial petroleum-based fibers, as they are often cheaper with similar performance (see Agave case study in the section titled "Agave: More Than Just Tequila" in Chapter 5). In addition, biofuels like plant oils and ethanol are being used for cooking and heating in tropical regions, reducing reliance on fossil fuels and wood. Natural rubbers, polymers, emulsifiers, and waxes could be used to replace synthetic compounds. The hunt for new plant-based medicines may seem to be exhausted; however, there is still much to learn from traditional medicine. Ultimately, as we prepare for space exploration with long missions (e.g., trip to Mars), scientists are developing innovative methods of growing plants for food. They are testing varieties that will not only grow well in outer space, but also provide necessary nutrients and phytochemicals to keep the astronauts healthy. Scientists are also investigating a bioregenerative life support system, where plants scrub excess CO_2 and provide oxygen. Many of these initiatives fall under the umbrella of general plant science. But just like plant explorers of the eighteenth and nineteenth centuries, the choice of which plants humans will take with them on a long voyage into space will be a fascinating chronicle in ethnobotany.

References

Amr, S. and Đorđević, S. (2000) The investigation of the quality of sage (*Salvia officinalis* L.) originating from Jordan. *Working and Living Environmental Protection*, **1** (5), 103–108. 8 (5 Suppl), 13–26.

Elvin-Lewis, M. (2011) Dereplications can amplify the extent and worth of traditional pharmacopeias. *African Journal of Traditional, Complementary, and Alternative Medicines*, **8** (5 Suppl), 13–26.

Wagner, L. (2015) Syrian civil war prompts first withdrawal from doomsday seed vault in the Arctic. *National Public Radio*, 23 September, http://www.npr.org/sections/thetwo-way/2015/09/23/442858657/syrian-civil-war-prompts-first-withdrawal-from-doomsday-seed-vault-in-the-arctic

Wall, J. and Aghayeva, D.N. (2014) The practice and importance of chestnut cultivation in Azerbaijan in the face of blight, *Cryphonectria parasitica* (Murrill) Barr. *Ethnobotany Research and Applications*, **12**, 165–174.

Wan Izatul, A.W.T. (2013) Protection of the associated traditional knowledge on genetic resources: beyond the Nagoya Protocol. *Procedia – Social and Behavioral Sciences*, **91**, 673–678.

Wan Izatul, A.W.T., Norhayati M.T., and Mohd, L.H. (2012) Traditional knowledge on genetic resources: safeguarding the cultural sustenance of indigenous communities. *Asian Social Science*, **8** (7), 184–191.

The term *gymnosperm* is derived from the Greek words *gymnos*, meaning "naked," and *sperma*, meaning "seed." The seeds develop on the surface of scales (such as conifer cone scales), leaves, or stalks as in *Ginkgo*. Conifer cones are a type of strobilus. The female cones bear the seeds and are typically woody, while the male cones that produce pollen are often herbaceous and inconspicuous. Gymnosperm wood lacks xylem vessels and companion cells in the phloem, except for the gnetophytes.

There are four divisions (phyla) of gymnosperms: Pinophyta (Coniferae), Ginkgophyta, Cycadophyta, and Gnetophyta. Pinophyta contains many well-known conifer families including cedars, cypress, firs, hemlocks, pines, redwoods, and spruces. Ginkgophyta contains only one living member, *Ginkgo biloba* L. (Figure 1.9). Cycadophyta contains palm-like or fern-like plants with mainly ornamental value. Gnetophyta has a few living members (*Welwitschia*, *Gnetum*, and *Ephedra*) with unique wood that contains xylem vessels.

All extant members of Pinophyta are woody coniferous trees with the exception of a few shrubs. There are six to eight families, all with significant ecological importance, and some with economic importance as well. Most produce resin to deter insect herbivory and pathogen attack. They typically have slender needle-like leaves in a spiral arrangement; however, there are exceptions, especially in the Cupressaceae family. Most members are evergreen except for a few deciduous trees such as *Larix*. All are wind pollinated, and most are monoecious.

Pinaceae: Pine Family

Trees in the Pinaceae family dominate the boreal forests of the Northern Hemisphere. *Pinus* (pines) is the largest genus in Pinaceae, with over 100 living members. Other notable genera include *Picea* (spruce), *Larix* (larches), *Pseudotsuga* (Douglas firs), *Aibes* (firs), *Cedrus* (cedar), and *Tsuga* (hemlock). Larches are the only deciduous members; all the rest are evergreen trees or rarely shrubs.

Pinaceae has significant economic importance for wood, paper, resins, and ornamentals. Softwood accounts for roughly 78% of the lumber industry. Pines, hemlocks, firs, spruces, and larches are all harvested for softwood lumber. The California redwood [*Sequoia sempervirens* (D.Don) Endl.] and bald cypress [*Taxodium distichum* (L.) Rich.] are also grown commercially for lumber; however, overlogging of native stands reduced their numbers. Structural timbers, poles, and piling are typically made from

Figure 1.9 *Ginkgo biloba* bark and leaves, autumn color.

softwood, since the trees tend to be straight, tall, with a large circumference. Softwoods like white pine, yellow pine, and Douglas fir account for most of the "sawmill" lumber such as boards, framing, doors, and so on. They are also used to make boxes, crates, and barrels, but have a lower value than hardwoods for furniture. Wood shakes and shingles are often made from red cedar, white cedar, redwood, cypress, and pine.

Pinaceae resins are a complex mixture of primarily terpenes, oils, and waxes. Also known as pinesap or pitch, they have been used as waterproof coatings and sealers since ancient times. In colonial North America, pine resins became known as "naval stores" since the Royal Navy relied heavily on these products to build and maintain their naval fleet. Native North Americans and European settlers alike used spruce resin to make spruce chewing gum. *The State of Maine Pure Spruce Gum* was produced and sold in 1848 by John B. Curtis. It was a mixture of spruce resin, paraffin wax, and flavoring. Turpentine is the volatile fraction of pine resin. It has been used in paints, lacquers, and solvents, and for medicinal purposes. After volatiles evaporate from pine resin, the remaining fraction is called rosin. Rosin (Figure 1.10) has several industrial uses, for inks, varnishes, adhesives, soldering flux, and sealants. It is also used to condition the bows of stringed instruments.

Cupressaceae: Cypress Family

The cypress family contains about 30 genera of trees and shrubs including cypress (*Cupressus*), juniper (*Juniperus*), giant sequoias (*Sequoiadendron*), and redwood (*Sequoia*). Cupressaceae contains the tallest [coast redwoods, *Sequoia sempervirens* (D.Don) Endl., Figure 1.11], largest [giant sequoia, *Sequoiadendron giganteum* (Lindl.) J.Buchholz], and stoutest (ahuehuete, *Taxodium huegelii* C.Lawson) trees in the world.

Several genera are economically important for wood, flavorings, and fragrances. *Thuja* species (arborvitaes) are not true cedars but are commonly called red cedar or white cedar. They are grown as ornamentals and harvested for their decay-resistant wood. Cedar storage chests are typically made from *Thuja* wood to repel moths and other insects. Thuja oil has been used as a topical medicine, and the leaves were valued as a tea by Native Americans. The high vitamin C content reduced the occurrence of scurvy in Native Americans and early European settlers. Hinoki wood comes from Japanese cypress (*Chamaecyparis obtusa* Endl). It is highly valued in Japan for building temples, shrines, castles, and palaces. The fresh, citrus-scented wood was used to construct Horyu-ji Temple and Osaka Castle. It contains the monoterpene thujaplicin hinokitiol

Figure 1.10 Block of rosin is used to condition stringed instrument bows.

Figure 1.11 Coastal redwoods (*Sequoia sempervirens*) in Muir Woods National Monument, CA, the United States, are considered the tallest living things. © 2015 Daniel Schmidt.

(β-thujaplicin), along with other related compounds commonly found in most members of Cupressaceae. *Calocedrus*, incense cedars, are closely related to *Thuja* species. Its essential oil is composed of mono- and diterpenes including α-pinene, totarol, and ferruginol, and it has antimicrobial properties. The wood has a spicy odor and is decay resistant, thanks in part to the presence of these compounds. Historically, it was valued for making coffins in China.

Taxaceae: Yew Family

The yew family is a small family of evergreen shrubs and small trees. The plants are dioecious; male and female cones are only a few millimeters long. The female cone holds only one seed. As it matures, the cone scale forms a fleshy aril around the seed. The aril is red, sweet, and edible; however, the seed is highly toxic due to the presence of cardiotoxic taxine alkaloids including taxine A & B, taxol A & B, 10-deacetyltaxol, baccatin III, 10-deacetylbaccatin III, 3,5-dimethoxyphenol, monoacetyltaxine, mono-hydroxydiacetyltaxine, triacetyltaxine, and monohydroxytriacetyltaxine. Taxine B is considered the most toxic of the group. All taxine alkaloids are cardiac myocyte calcium and sodium channel antagonists. Yews are considered some of the most toxic plants in the Northern Hemisphere. They have been used as arrow poison for hunting, for suicide, and have poisoned livestock, pets, and humans who unwittingly ingested the seeds, bark, or leaves of the plant. Yews are often planted as ornamental hedges and are relatively long lived. English yews (*Taxus baccata* L.) can live 2000 years.

Taxol

The pacific yew (*Taxus brevifolia* Nutt.) is the original source of the cancer drug paclitaxel. It was discovered as part of the US National Cancer Institute's (NCI) screening program. In 1962, Arthur Barclay and his graduate students were collecting plant samples for the USDA Agricultural Research Service in Washington State. They collected stems, fruit, and bark (Figure 1.12) from *Taxus brevifolia* B-1645 (Barclay's 1645th plant sample). It was a random sampling, not based on ethnobotany or phytochemistry, although the historical use of yew wood and alkaloidal poisons were known (Bolsinger and Jaramillo 1990). The samples were sent to Wisconsin Alumni Research Foundation for extraction and bioactivity testing. The extracts showed in vitro activity

Figure 1.12 Bark of the Pacific yew (*Taxus brevifolia*) contains taxol.

against human oral epidermoid carcinoma (KB) and in vivo leukemia L1210 in mice. So two years later, Barclay collected more bark.

Most labs were reluctant to pursue the project, since plant extracts with high in vitro cytotoxicity often prove too toxic for healthy cells as well. Barclay finally sent the bark to Monroe Wall at Research Triangle Institute (RTI). RTI performed bioassy-guided fractionation to discover the active diterpene, taxol. Mansukh Wani attempted to characterize the molecule, but it proved difficult. The yield of taxol was low, they did not have more raw material to extract, and they could not even characterize the structure or molecular weight. So the project was moved to low priority to focus on other more successful projects like the discovery and development of camptothecin from *Camptotheca acuminata* Decne.

Eventually, Wall and Wani used x-ray crystallography, mass spectrometry, and NMR spectroscopy to generate the molecular structure, publishing their results in the *Journal of the American Chemical Society* in 1971. Nearly a decade later, Susan Horwitz at the Albert Einstein College of Medicine at New York's Yeshiva University contacted NCI and Monroe Wall, asking for radiolabeled taxol so that she could perform mechanism of action research. The USDA was ordered to retrieve 7000 pounds of Pacific yew bark, much to the dismay of environmentalists concerned about the welfare of old-growth forests in the Pacific Northwest. Horowitz eventually received her taxol and discovered that its mechanism of action was unique, that is, different from other cancer drugs. Many cancer drugs arrest cell division by disrupting microtubule production through inhibition of tubulin polymerization. Taxol actually stimulated overproduction of microtubules to the point that it disrupted the coordination of cell division. Taxol overcame a few other obstacles before entering clinical trials in humans for ovarian cancer. It was a huge success. But to develop this drug and distribute to people with ovarian cancer, the NCI would need to produce 240 lbs of taxol, which would require the killing of 360,000 Pacific yews.

Scientists around the world began to work on taxol synthesis. Pierre Potier from France achieved semi-synthesis using 10-deacetylbaccatin III from English yew (*Taxus baccata* L.). Others followed, including Robert Holton at Florida State University, who developed a semi-synthetic procedure with twice the yield of Potier that could be scaled up for commercialization. In 1989, the NCI signed a cooperative research and development agreement (CRADA) with Bristol-Myers Squibb (only Bristol-Myers at the time), and Holton negotiated terms for the use of his semi-synthesis method. In 1990, Bristol-Myers Squibb gave taxol the generic drug name "paclitaxel." In 1992, the US Food and Drug Administration (FDA) approved the use of paclitaxel for refractory ovarian cancer, and later for breast cancer and AIDS-related Kaposi's sarcoma. Currently, it is approved for many different cancers around the world. At its peak in 2000, annual sales topped USD $1.6 billion. Since 1995, Bristol-Myers Squibb has been manufacturing paclitaxel in Germany using a proprietary *Taxus* plant cell culture method. This method eliminates reliance on any mature species of *Taxus*.

Gnetophytes: Ephedraceae, Gnetum, and Welwitschia

Gnetophytes are an ancient group of plants, with only three extant genera. There is much debate concerning their classification. A 2011 study using molecular data placed Gnetales as a sister group to the rest of the gymnosperms (Lee *et al*. 2011), contradicting the prevailing hypothesis that they are a sister group to the angiosperms. Gnetophytes do have some traits in common with angiosperms including xylem vessel elements, flower-like reproductive structures, and reduced gametophytes. But their scale-like leaves (*Ephedra* and *Welwitschia*) and reduced sporophylls resemble gymnosperms.

The family Ephedraceae contains only one genus: *Ephedra*. They are woody shrubs, 2–5 ft. tall and wide, found growing in arid regions. The stems contain ephedrine, pseudoephedrine, and other analogs with phenylethylamine alkaloidal structures. The herb má huáng comes from Asian species of *Ephedra*. It is grown on a large scale in China for export; however, a synthetic process is used to produce ephedrine and its analogs for pharmaceuticals. American species known as "Mormon tea" contain very low levels of alkaloids, although they have been used traditionally as a stimulant drink. Ephedrine stimulates adrenergic receptors and has a bronchodilating effect. It has traditionally been used to treat asthma and bronchitis. Pseudoephedrine is a component of many pharmaceuticals used to treat respiratory symptoms brought on by the common cold or seasonal allergies. Ephedrine is structurally similar to methamphetamine and can be used to manufacture this and other related street drugs.

Gnetum is a tropical genus with about 30 species. They have broad leaves with net venation, making them difficult to distinguish from angiosperms. They are not used extensively for any economic purpose, although the leaves are edible, and the plants have some medicinal properties, likely related to their stilbenoid content.

Darwin described *Welwitschia* as the platypus of the plant kingdom. Austrian naturalist Freidwich Welwitsch discovered *Welwitschia mirabilis* growing in the Namib Desert on the Angolan coast. Locally it was known as *otjitumbo*, which translates to "stump." It is extremely slow growing, with a long tap root, woody crown, and only two broad leaves. The leaves are not replaced throughout the plant's life, and so they become frayed and tangled. They are unpalatable to herbivores due to their high concentrations of lignin and calcium oxalate. Although quite bizarre and interesting, *Welwitschia* has no economic value and no reported ethnobotanical use.

Angiospermae

Angiosperms are the flowering, seed-producing plants. They are the most diverse group of land plants on earth. They differ from the gymnosperms in several important ways. First, they produce flowers with ovules produced inside carpels rather than on naked scales. They also have stamens which produce pollen that is carried by pollinators to the ovule for fertilization. The ovule becomes a seed and the ovary becomes fruit.

There are about 250,000–400,000 species of flowering plants classified into one of seven orders. The basal or ancestral angiosperms consist of Amborellales, Nymphaeales, and Austrobaileyales. The remaining species belong to Chloranthales, magnoliids, monocots, or eudicots. A review of each of the roughly 415 angiosperm families is beyond the scope of this book. Consequently, only families with significant economic or ethnobotanical value are included. The arrangement of families follows the APG III system of organization developed by the Angiosperm Phylogeny Group (APG 2009).

Annonaceae: Custard Apple Family

More than 130 genera belong to Annonaceae. It is a family of trees, shrubs, and a few lianas (woody vines) found mostly in the tropics. Some produce edible berry-like or aggregate fruit. *Annona reticulata* L. (custard apple), *A. cherimola* Mill. (cherimoya),

A. muricata L. (soursop, guanábana, Figure 1.13), *A. squamosa* L. (sweetsop, sugar apple), *A. diversifolia* (ilama), and *A. purpurea* Moc. & Sessé ex Dunal (soncoya) all produce similar annonaceous fruit that is consumed throughout the tropics. The custard-like flesh is eaten fresh as a dessert, or mixed into salads, ice cream, milk shakes, or sherbets. It is not very stable post-harvest and therefore is relatively unknown outside the tropics. Frequent consumption of *A. muricata* has been linked to a neuro-degenerative, Parkinsonian condition, caused by the annonacin. Annonacin is a type of polyketide belonging to the class of compounds known as ace-togenins. More than 400 acetogenins have been isolated from annonaceous species.

Asimina triloba (L.) Dunal (American pawpaw, Figure 1.14) is one of the rare members that grows in temperate regions of North America. It is a small understory tree with purple flowers in the spring and bearing fruit in the sum-mer. The fruit can be eaten fresh or added to ice creams and pies. The fruit also produces a yellow dye. *Cananga odorata* (Lam.) Hook.f. & Thomson (ylang-ylang) is another economically important species. The essential oils from the flowers are often used as a fra-grance for cosmetics. *Xylopia aethiopica* (Dunal) A.Rich. (Guinea pepper, negro pepper, and other local names) is a shrub

Figure 1.13 Unripe soursop (*Annona muricata*) fruit.

Figure 1.14 American pawpaw fruit (*Asimina triloba*).

from Western Africa that bears clusters of elongated fruit. The fruit is typically dried and smoked, ground, and used as a pepper substitute, although it imparts a bitter flavor. The tree contains *ent*-trachylobane, kolavane, and kaurane diterpenoids such as kaurenoic acid, as well as various terpenes. *Monodora myristica* (Gaertn.) Dunal (calabash nutmeg) seeds are a nutmeg substitute that originated in Africa. They were brought through the slave trade to the Caribbean, where it earned the name Jamaican nutmeg. Essential oil from the seed contains α-phellandrene and other terpenes.

Lauraceae: Laurel Family

There are nearly 3000 species of Lauraceae belonging to 68 genera. They are generally tropical aromatic, evergreen trees, with some exceptions. Well-known members of Lauraceae include *Sassafras*, cinnamon (*Cinnamomum* spp.), avocado (*Persea* spp.), and bay laurel (*Laurus* spp.). Sassafras species are deciduous and grow in temperate regions. Native Americans and early American colonists used sassafras for medicinal purposes to treat ailments such as fevers and sexually transmitted diseases (STDs). Essential oil from the root bark contains the phenylpropanoid safrole, and was once used as the flavoring for root beer. However, in 1960, the US FDA banned the use of sassafras in food or drugs after safrole was shown to have carcinogenic properties. Sassafras root is still used to make root beer or sassafras tea by hobbyists. Leaves from *Sassafras albidum* (Figure 1.15) are dried and crushed to make filé or gumbo powder, a spice used in creole cooking. Safrole has recently been used as a precursor for the illicit drug ecstasy.

Cinnamon

Cinnamon is an ancient spice produced from the inner bark of several *Cinnamomum* species. "True cinnamon" (*C. verum* J.Presl) is native to Sri Lanka. *C. cassia* (L.) J.Presl (Chinese cinnamon), *C. burmannii* (Nees & T.Nees) Blume (Indonesian cinnamon), and *C. loureiroi* Nees (Saigon cinnamon) are three other species marketed as cinnamon. Cinnamon has a rich ethnobotanical history. Essential oils from *C. verum* and *C. iners* Reinw. ex Blume (cassia) were used in Egypt as early as 2000 BCE to perfume bodies for burial and as anointing oils. Ancient Romans burnt cinnamon in funeral pyres. Arab

Figure 1.15 Leaves of *Sassafras albidum* growing in eastern North America.

traders brought cinnamon from India to the Mediterranean, where it became popular in the Middle Ages as a luxury spice. High demand and a short supply of cinnamon greatly influenced European exploration into Asia. Slow overland trade routes could only deliver small quantities to Europe, and traders controlling the trade routes set exorbitant prices. Once the Turks came to power in Asia Minor and Mamlûks controlled Egypt, trade in cinnamon dried up. Portuguese traders arrived in Ceylon (Sri Lanka) in the fifteenth century to find organized cinnamon production. They expanded cinnamon production and built their own fort. Now the Portuguese controlled cinnamon trade to Europe and reaped high profits. They ruthlessly controlled the native people, which allowed the Dutch to gain access to the Ceylon market, in exchange for their protection of the Ceylon king from the Portuguese. By 1640, the Dutch broke the Portuguese monopoly and took over their cinnamon factories. Until this point, only wild *C. verum* trees were used for cinnamon production. But to meet European demand, a Dutch East Indies Company employee named De Coke decided the trees must be cultivated for eight years before harvest. Anjarakkandy Cinnamon Estate was established as Asia's largest cinnamon estate. In 1796, the British arrived in Ceylon and displaced the Dutch. However, a tight monopoly became impossible as other species of cinnamon came to European markets from different growing regions.

Today, the top producers of cinnamon are Indonesia, China, Sri Lanka, and Vietnam. In modern practice, a cinnamon tree is coppiced after 2–3 years, producing tender shoots at its base. The stems are cut, outer bark stripped, and inner bark is unrolled. The inner bark curls upon drying and is sold as "quills" (Figure 1.16) or ground into powder. Essential oil made from cinnamon contains around 90% cinnamaldehyde, which is primarily responsible for the classic cinnamon flavor and odor. *C. verum* bark has trace amounts of coumarin, whereas the other species have much higher levels, prompting warnings from European health organizations about overconsumption of products with cinnamon. Coumarin is a benzopyrone with a sweet odor of freshly mown hay. It is hepatotoxic in animal models.

Figure 1.16 Thick-barked cinnamon sticks from *Cinnamomum cassia* are the most common type of cinnamon marketed in North America.

Figure 1.17 Fruit from a wild avocado (*Persea* sp.) tree in Belize.

Avocado (*Persea americana* Mill.) is a tree native to Central America (Figure 1.17). The fruit is a berry with one large seed. Although it may appear to be a drupe with a stony endocarp, the endocarp is a soft white layer over the hard seed coat. Avocados are high in monounsaturated fat and are popular throughout the Americas. The seed, leaves, bark, and fruit skin are all considered toxic to livestock and pets due to the presence of the acetogenin persin.

Bay laurel (primarily *Laurus nobilis* L.) is an evergreen tree native to the Mediterranean. The ancient Greeks and Romans used *L. nobilis* to make wreaths that were given as a symbol of high achievement or status. The terms *baccalaureate* and *poet laureate* are both derived from laurel. The aromatic leaves can be used as a spice for cooking or to produce soothing salves and oils in traditional medicine. Laurel oil is also a main ingredient for traditional Aleppo soap from Syria. The monoterpene eucalyptol makes up about 45% of the essential oil, with various terpenoids accounting for the remaining 55%. Camphor laurel [*Cinnamomum camphora* (L.) J.Presl] is the source of the terpene camphor. Camphor is the active ingredient in many commercial vapor-steam inhalants and has been used in traditional medicine as a decongestant. It is also burned in various Hindu religious ceremonies.

Piperaceae: Pepper Family

Piperaceae has only 13 genera; most of the species belong to *Piper* or *Peperomia*. Members of this family are trees, shrubs, woody vines, or herbaceous plants that grow in tropical or subtropical regions. Flowers are borne on spike inflorescences, and the fruits are drupelets. The woody vine *Piper nigrum* L. is the source of black pepper. It is native to South and Southeast Asia. Unripe drupelets are hand-harvested, sometimes boiled, and then sun-dried. As they dry, the fruit shrivels and browns. The result is black peppercorns. To produce white peppercorns, the fruit is fermented for several days to separate the pericarp from the white seeds or "peppercorns." Green peppercorns are unripe druplets preserved in a way that retains their green color, such as canning in vinegar or brine. The pyrrolidone alkaloid piperine and its isomers are responsible for the spicy, pungent flavor. *Piper longum* L. (long pepper) fruit has a similar flavor and chemical composition to *P. nigrum*, but the seeds are much smaller. Both peppers have various culinary and medicinal uses and were important components of the spice trade between Europe and Asia. As with cinnamon, a few parties monopolized the trade routes, charging exorbanant prices. Only after the discovery of *Capsicum* species (Solanaceae) and allspice [*Pimenta dioica* (L.) Merr., Myrtaceae] in the Americas did the demand for pepper subside.

Betel quids made with *Piper betle* L. have long been chewed for their stimulating properties. Ingredients in the quid stain the teeth a dark color and have traditionally been used for teeth blackening (see teeth blackening case study in the section titled "The Ethnobotany of Teeth Blackening in Southeast Asia" in Chapter 6). Kava (*Piper methysticum* G.Forst.) roots are used to produce a sedative tonic or tea. It is consumed throughout Polynesia, with the largest production in the Republic of Vanuatu [see kava case study in the section titled "Pharmacological Effects of Kavalactones from Kava (*Piper methysticum*) Root" in Chapter 8]. Yangonin, kavain, and methysticin are three of the psychoactive compounds. These kavalactones are arylethylene-α-pyrones, which are structurally similar to chalcones.

Araceae: Taro Family

This family of tuberous herbs produces heart-shaped leaves with calcium oxalate crystals called "raphides" and milky sap. It was previously known as the arum family, and so members are often called aroids. Some members are thermogenic to attract pollinators, such as skunk cabbage [*Symplocarpus foetidus* (L.) Salisb. ex W.P.C.Barton], and elephant foot yam [*Amorphophallus paeoniifolius* (Dennst.) Nicolson]. In Hawaii, taro corms [*Colocasia esculenta* (L.) Schott] (Figure 1.18) are steamed, crushed, and fermented to produce poi. They are rich in the soluble starch amylose. Giant or elephant

Figure 1.18 Flooded taro field on the island of Hawaii.

leaf taro [*Alocasia macrorrhizos* (L.) G. Don], and swamp taro [*Cyrtosperma merkusii* (Hassk.) Schott], were also consumed as a starchy crop in Southeast Asia and Oceania until they were more or less displaced by *Colocasia* spp. All types of taro root must be thoroughly cooked before consuming to remove the raphides. If the raphides are consumed, they produce burning and stinging of the mouth and throat along with edema and dysphagia. The symptoms can last for weeks.

Amaryllidaceae: Amaryllis Family

Several species from the type genus *Allium* have economic importance as vegetables, including onions and shallots (*A. cepa* L.), garlic (*A. sativum* L.), leek (*A. ampeloprasum* L.), chives (*A. schoenoprasum* L.), and Chinese chives (*A. tuberosum* Rottler ex Spreng.). There have been numerous studies investigating potential health benefits of *Allium* consumption. They contain relatively high levels of flavonoids, particularly quercetin, as well as pungent thiosulfinates. Garlic has been used for centuries in ethnomedicine. Ancient texts by Hippocrates, Dioscorides, and others regarded garlic as a panacea or "cure-all." Dioscorides suggested garlic for lung disorders, and Pliny the Elder suggested using it for consumption (tuberculosis). Upon tissue damage, garlic produces the thiosulfinate allicin (diallylthiosulfinate) from the non-protein amino acid alliin (S-allylcysteine sulfoxide). Allicin is responsible for the antibiotic activity of freshly crushed garlic. It is a broad-spectrum antibiotic, which may well explain its use as a "cure-all" throughout the ages.

Colchicaceae: Autumn Crocus Family

Several members of Colchicaceae have medicinal properties. Colchicine is a tyrosine-derived phenethylisoquinoline alkaloid produced in the seeds and corm of *Colchicum autumnale* L. (Figure 1.19) and other species from the genera *Colchicum*, *Gloriosa*, and *Sandersonia*. It is used to treat gout, a disorder of purine metabolism, and the inflammatory disorders Familial Mediterranean fever and Neuro-Behçet Syndrome. Saffron is derived from the stigmas of *Crocus sativus* L. It is used as a fabric dye, most notably to color the robes of Buddhist and Hindu monks (Figure 1.20). The yellow color primarily comes from crocin and other carotenoids. By weight, saffron is regarded as the most expensive spice, used to color and flavor foods from many different cultures. In biology, it is an ingredient in hematoxylin-phloxine-saffron (HPS) stain used to stain tissues such as collagen for higher visibility in bioimaging.

Figure 1.19 Wild autumn crocus (*Colchicum autumnale*) growing in Switzerland.

Liliaceae: Lily Family

Liliaceae was recently reorganized, with Alstroemeriaceae, Amaryllidaceae, Asparagaceae, Colchicaceae, and Xanthorrhoeaceae receiving family status. There is still disagreement; for example, some place Aloe in Xanthorrhoeaceae instead of Asparagaceae. With the current organization, tulips and lilies are arguably the best-known members of Liliaceae. *Fritillaria* spp. bulbs are used in TCM as an antitussive and expectorant. Depending on the species, they contain the steroidal alkaloids puqietinone, puqiedine, imperialine, peimisine, peimine, and peiminine.

Smilacaceae: Smilax Family

Smilax ornata Lem. is used to make Jamaican sarsaparilla, while other related *Smilax* spp. are used to make similar root beers and herbal drinks. *S. ornata* was used traditionally in America and Europe to treat syphilis. In TCM, *Smilax glabra* Roxb. is used to make the medicinal food Guîlínggâo. *Smilax* spp. produce a range of triterpenes and steroidal saponins, which could account for the purported bioactivity.

Orchidaceae: Orchid Family

Figure 1.20 Buddhist monk in Luang Prabang, Laos, wears robes dyed with saffron. The color fades from intense orange to pale yellow after long-term exposure to the sun. © 2013 Ilya Raskin.

Orchidaceae is the second largest flowering plant family, with 899 genera and over 27,500 species. Although they are highly valued ecologically, culturally, and ornamentally, relatively few have been used ethnobotanically. The one major exception is *Vanilla planifolia* Jacks. ex Andrews, the vanilla orchid (see vanilla case study in the section titled "Vanilla: Madagascar's Orchid Economy" in Chapter 4). This orchid is native to Mexico and produces seedpods that are extracted to produce vanilla flavor and fragrance. The extract contains several hundred compounds, but vanillin, a phenylpropanoid biosynthesized via the shikimic acid pathway, is the primary flavor component. Vanillin can be synthesized more cheaply from paper pulp lignin; however, a strong market still exists for vanilla extract from *Vanilla planifolia*, with its more complex flavor and aroma.

Several species of orchid are valued for their fragrance, and there are a few medicinal orchids worth noting, especially species that produce alkaloids. *Dendrobium* spp. produce terpenoid (e.g., dendrobine, nobiline, dendroxine, dendrowardine), indolizidine (e.g., crepidamine, dendroprimine), quinolizidine (e.g., crepidine), and a few other types of alkaloids. The plants are used in TCM and other systems of traditional medicine to reduce fevers and hypertension. *Phalaenopsis*, *Pleurothallis*, *Vanda*, *Vandopsis*, and *Liparis* spp. all produce mainly pyrrolizidine alkaloids (e.g., phalaenopsines from *Phalaenopsis*). They have a range of uses in traditional medicine.

Asparagaceae: Asparagus Family

Asparagaceae has several genera with ethnobotanical significance including *Agave* (formerly Agavaceae), *Yucca*, *Aloe*, and *Asparagus*. Agaves are fleshy rosettes that thrive in arid environments in North and South America. Traditionally, the Aztecs and other Native Americans used agave to make paper, cordage, thatch, thread and needles, and as food and beverage. Sweet sap (aguamiel) from *Agave americana* L. is used to make the alcoholic beverage pulque. Another alcoholic beverage, tequila, is mostly produced from the blue agave, *A. tequilana* F.A.C.Weber. It is native to Jalisco, Mexico, which is still the center of tequila production. Farmers called *jimadors* manually harvest the *piña* (heart of the agave). The *piña* is slow cooked at low temperature to avoid caramelization, then milled or crushed, and the sugars are extracted with water. The sugars are fermented and double distilled (see Agave case study in the section titled "Agave: More Than Just Tequila" in Chapter 5). Native Americans used *Yucca elata* (Engelm.) Engelm. fibers in a similar manner as agave. The roots contain saponins that were used as a shampoo and soap, hence the common name "soaptree."

Aloe vera (L.) Burm.f., a native of South Africa and Madagascar, has a long history of use as treatment for burns, wounds, skin disorders, and as a laxative. Glycoproteins and polysaccharides are the active components. Asparagus (*Asparagus officinalis* L.) is consumed as a vegetable and has been used as a diuretic. It contains asparagusic acid, an organosulfur compound derived from isobutyric acid that is metabolized into volatile odorous compounds (e.g., dimethyl sulfide, dimethyl disulfide, dimethyl sulfoxide, and dimethyl sulfone) found in human urine after asparagus consumption.

Figure 1.21 Cluster of nuts from the cohune palm, *Attalea cohune.*

Arecaceae: Palm Family

Most palms grow in tropical to subtropical regions and have compound evergreen leaves on top of an unbranched stem. Mesopotamians cultivated date palms over 5000 years ago. Since then, palms have been used in various ways, for food, oil, building material, and wax. The areca palm (*Areca catechu* L.) produces fruit called betel nut, which is actually a drupe. It is called betel because it is often sliced and wrapped in betel leaves (*Piper betle*, Piperaceae) for chewing (see teeth blackening case study in the section titled "The Ethnobotany of Teeth Blackening in Southeast Asia" in Chapter 6). Mayans used cohune palm (*Attalea cohune* Mart.) (Figure 1.21) kernel oil for cooking, lamp oil, and making soap. The heart of palm was considered a delicacy, and the leaves were used for thatching. Today, most commercial palm oil is derived from the fruit and kernels of *Elaeis guineensis*

Jacq., the African oil palm. As the name suggests, the palm originated in Guinea and was widely grown throughout tropical regions in western Africa for cooking oil and medicine. In 1848, the Dutch introduced it to Java. The British brought the palm to Malaysia sometime after that, establishing the first plantation in Selangor in 1917. Indonesia and Malaysia are currently the top global producers of palm oil. Of the 11 million hectares of palm oil plantations globally, Indonesia has about 6 million hectares, and Malaysia has about 4.5 (FAO 2013).

The coconut tree (*Cocos nucifera* L.) may have originated in the Americas or Indo-Pacific region, from fossil evidence. Coconuts are mainly cultivated for their large fruit (Figure 1.22), a one-seeded drupe weighing about 1.4 kg, comprising a husk (exocarp), a mesocarp with fibrous vascular bundles called coir, a stony endocarp, and the endosperm. The endosperm is made up of white flesh called copra and coconut "milk." The copra is dried and pressed to produce coconut oil, often used in personal care products. Coconut milk is a sweet beverage, but is also rich in cytokines used for plant tissue culture. Nata de coco is a smooth jelly-like dessert made by fermenting coconut milk with the aerobic bacterium *Acetobacter xylinum*. The bacteria converts glucose from the coconut water medium into cellulose. Nata de coco originated in the Philippines, but is produced throughout Southeast Asia. Coir fiber is used to make mats and rope and as a soil-less media for plant hydroponics.

The date palm (*Phoenix dactylifera* L.) produces edible fruits called dates. Dates have been a staple food in the Mediterranean since antiquity. They can be eaten fresh, stuffed, or processed into other foods. The date palm can also be tapped to produce palm syrup or palm sugar. In Asia and Africa, palm sap is collected from various types of sugar

Figure 1.22 A load of coconuts (*Cocos nucifera*) is shipped down the Mekong River.

palms (*Borassus flabellifer* L. and other spp.) and consumed as a fresh beverage or fermented. Palm wine is made by tapping the tip of the inflorescence of sugar palm trees for sap. The sweet sap is naturally fermented and distilled to increase the alcohol content. In India, the toddy or Palmyra palm (*B. flabellifer*) is tapped to produce sweet sap called neera or sweet toddy. For sweet toddy and sugar production, the inside of the sap collection vessel is rubbed with lime paste to prevent natural fermentation. The sap can be concentrated to produce palm sugar. Toddy or arrack is produced when the sweet sap is allowed to ferment.

The carnauba palm [*Copernicia prunifera* (Mill.) H.Moore] produces a waxy coating on the leaves that is harvested as carnauba wax. Carnauba wax is used in car waxes and shoe polish to produce a glossy finish. *Daemonorops* spp. produce rattan for furniture and basket making. The fruit is the source of "dragon's blood," a red resinous powder used as a traditional medicine and dye. *Dracaena cinnabari* Balf.f. (Asparagaceae) and a few other species produce similar substances also called "dragon's blood" by various cultures. Flavylium chromophores such as dracorhodin account for the brilliant red color in *Daemonorops* dragon's blood.

Bromeliaceae: Pineapple Family

Most bromeliads are epiphytic or xerophytice terrestrial herbs native to the New World tropics. They are Crassulacean acid metabolism (CAM) plants, well adapted to hot, dry climates. Olmecs, Incas, Aztecs, Mayas, and other Native Americans used bromeliads for food, fiber, medicine, and as ceremonial headdresses. The pineapple [*Ananas comosus* (L.) Merr.] (Figure 1.23) is a parthenocarpic aggregate fruit, the only bromeliad grown for food on a commercial scale. Ananas is a Guarani word meaning "fragrant excellent fruit." The fruit was named pineapple by European explorers, possibly Christopher Columbus, because it resembles a pinecone. Pineapples contain bromelain, a protease used commercially as a meat tenderizer.

Figure 1.23 Pineapple fruit growing wild near Lake Sandoval, Peru.

Poaceae: Grass Family

Poaceae is a large family, with over 11,000 species. It is considered the most economically important plant family on earth, highly valued for their primary metabolites (e.g., starches and oils). Poaceae provides food, fuel, building materials, and forage for animals. Domestication of grass species, especially wheat (*Triticum* spp.), corn (*Zea mays* L.), rice (*Oryza sativa* L.), rye (*Secale cereale* L.), barley (*Hordeum vulgare* L.), millet [*Pennisetum glaucum* (L.) R.Br., *Setaria italica* (L.) P.Beauv., *Eleusine coracana* (L.) Gaertn., and various other spp.], oats (*Avena sativa* L.) and sugarcane (*Saccharum officinarum* L.), provided the foundation for human civilization 10,000 years ago.

Each crop has a rich history of domestication and has made a substantial impact on society. For example, sugarcane (*S. officinarum*) (Figure 1.24) was domesticated in New Guinea from the wild relative *S. robustum* E.W.Brandes & Jeswiet ex Grassl. It spread throughout Southeast Asia into India and China. By 350 BCE, Indians discovered a way to crystallize sugar from the sweet stems. They traded sugar with people in Western Asia and China. China began operating sugarcane plantations by 650 BCE, while Arab chemists improved the manufacturing process of sugar substantially. By the fourteenth and fifteenth centuries, Europeans were producing sugar. They brought sugarcane to the New World, developing sugarcane plantations in the Caribbean and South America. Europeans sought cheap labor for their sugarcane plantations, leading to the slave trade between Western Africa and the Americas.

Several grain crops including rye, wheat, maize, and wild barley produce allelopathic benzoxazinoids as a defense against pathogens and herbivores. DIBOA [2,4-dihydroxy-2H-1,4-benzoxazin-3(4H)-one] and its C-7-methoxy derivative DIMBOA (2,4-dihydroxy-7-methoxy-1,4-benzoxazin-3-one) are the two most common benzoxazinoids. They are biosynthesized from indole and glucosylated for stability and increased solubility.

The fodder plants *Sorghum vulgare*, *S. bicolor*, *S. halepense*, and several other members of Poaceae produce a different allelopathic compound: dhurrin, a cyanogenic glycoside derived from tyrosine. Seedlings, frost-damaged, and drought-stressed plants often contain toxic levels of dhurrin, but healthy, mature plants tend to have lower levels.

Figure 1.24 Sugarcane field in Maui, Hawaii.

Grasses play host to several species of *Claviceps, Epichloë, Balansia,* and *Neotyphodium* fungi that produce ergot alkaloids derived from (+)-lysergic acid and indole-diterpenes. Indole triterpenes like paspaline, paspalicine, and paspalinine cause tremors in animals. In grazing livestock, paspalinine and paspalitrems cause "paspalum staggers," named after *Paspalum* spp. grasses infested with *Claviceps paspali.* Ergot alkaloids, principally from *C. purpurea,* cause ergotism or Saint Anthony's Fire in grazing animals and humans. They were responsible for the deaths of thousands of Europeans during the Middle Ages. Clinical features include gastronintestinal (GI) maladies, convulsions, psychosis, edema, gangrene, and eventually death. Ergot alkaloids have important medical uses. Ergometrine is used as an oxytocic to facilitate labor and to reduce bleeding after childbirth. Ergoline and ergotamine are used to treat migraines. Other (+)-lysergic acid derivatives have been used as treatments for cancer, diabetes, and Parkinson's disease. The semi-synthetic hallucinogen LSD is also derived from (+)-lysergic acid.

Musaceae: Banana Family

This small family with only two genera (*Ensete* and *Musa*) is best known for its edible fruit (bananas and plantains) and fibers. They are found throughout tropical forests, but the family originated in the Old World, most likely in the vicinity of Papua New Guinea. Both *Musa* and *Ensete* spp. have herbaceous stems rising from underground rhizomes resembling corms. They are nutritious and used to combat diarrhea in the tropics [see banana case study in the section titled "Banana (*Musa* spp.] as a Traditional Treatment for Diarrhea" in Chapter 8).

Zingiberaceae: Ginger Family

Plants in the ginger family are valued as ornamentals, aromatic spices (Figure 1.25), and medicine, especially as stimulants and carminatives (gas reducers). Rhizomes from ginger (*Zingiber officinale* Roscoe), turmeric (*Curcuma longa* L.), and galangal [*Alpinia galanga* (L.) Willd., *Kaempferia galanga* L., and other spp.], cardamom (*Amomum* spp. and *Elettaria* spp.) seed pods, and grains of paradise (*Aframomum melegueta* K.Schum.) seeds are some of the best-known spices from this family. They have been used medicinally for a range of disorders from GI issues to inflammatory disorders and infection. The chief active components are pungent gingerols and cucurminoids (e.g., curcumin). They are modified polyketides biosynthesized from phenylalanine via the phenylpropanoid pathway. In ginger, 6-, 8-, 10-, and 12-gingerol occur in the rhizome. Heating or drying ginger rhizomes converts the gingerols into zingerone and shogaols, respectively. Clinical trials support the use of ginger for dyspepsia and as an anti-emetic (Giacosa *et al.* 2015a, 2015b). It is included in pharmacopeias around the world. There have been some clinical trials investigating turmeric for digestive disorders like Helicobacter pylori stomach ulcers and inflammatory bowel disease, but the evidence is not as strong as it is for ginger. Likewise, cardamom has been used traditionally for GI disorders. While there have been animal studies supporting its gastroprotective effect, there have been no large clinical trials. Grains of paradise were traditionally used as a spice and medicine in West and North Africa and may have anti-inflammatory properties (Ilic *et al.* 2014).

Figure 1.25 Spices for sale including powdered ginger, turmeric, and galangal (foreground), cardamom pods (third row, right), and grains of paradise (third row, center). © 2014 M. Ameen.

Papaveraceae: Poppy Family

There are 41 genera and 920 recognized species in the poppy family. Most are lactiferous herbaceous plants with solitary, showy flowers. The type genus, *Papaver*, contains 55 species of poppies used as ornamentals, food, and perhaps most importantly, medicine.

Opium

Opium, the dried latex obtained from scoring unripe opium poppy seed capsules (*Papaver somniferum* L.) (Figure 1.26), contains about 25% alkaloids by weight. The poppy capsules contain about 0.5% alkaloids. These alkaloids are of great economic importance and ethnobotanical interest. As early as 3400 BCE, opium poppies were cultivated in Mesopotamia. The Sumerians, Assyrians, and the Egyptians all grew poppies. Eventually, trade routes were established, spreading the knowledge of opium to Europe (c. 1300 BCE), India (c. 330 BCE), and China (c. 300 CE). Throughout the Ottoman Empire (beginning in 1301), the Turks became addicted to black opium water and were the major supplier of opium to Europe. In the 1800s, there was a significant trade deficit between Britain and China. While the British could not get enough Chinese silk, tea, and spices, the Chinese Emperor had banned the import of most European goods, accepting only silver for payment. The British were running out of silver and needed a solution. Eventually, they found a product the Chinese did want: opium. Although the Emperor also banned opium, a black market emerged, followed by a great number of opium addicts. Around the same time, the value of Indian cotton had

(A)

(B)

Figure 1.26 (A) Poppies growing in a field in Helmand province, Afghanistan (US Marine Corps photo by Sgt. Pete Thibodeau); (B) Dried poppy seed capsules show scoring to extract opium (US Marine Corps photo by Lance Cpl. James Purschwitz).

dropped with the introduction of Egyptian cotton cloth. So the British turned India's cotton fields into poppy fields. British East India Company ships delivered opium to small Chinese islands for dissemination to the mainland. Addiction, two opium wars, and the relinquishing of Hong Kong to the British soon followed.

Modern opium production has not changed much since prehistoric times. Current centers of legal production include Turkey, India, and Australia. Britain has approved domestic production for pharmaceutical company use. There are numerous countries in the Eastern and Western hemispheres with significant production for illegal drug trade. As recently as 2010, Afghanistan produced around 90% of the world's illegal opium, but production has since collapsed due to war and political unrest. Opium is still harvested by hand, using a specialized tool to score the unripe capsule, cutting into latex tubes. A mixture of white milky latex is released that quickly browns (oxidizes) and thickens upon drying. The dried opium is scraped off and extracted using various techniques to produce morphine and other alkaloids.

To purify alkaloids from opium, crude opium is first dissolved in hot water and filtered to remove foreign objects and plant material. In some countries, the whole plant (poppy straw; Figure 1.27) is boiled, which is a lower-yielding but less labor-intensive process than collecting the opium. The opium extract is reheated to obtain a thick, sticky paste called "cooked" or "smoking" opium. The Hmong people like to smoke uncooked opium, but most others prefer smoking cooked opium. At this point, it may be sold as raw opium or further processed to obtain pure alkaloids. More than 40 alkaloids have been identified in opium, with six dominant structures: morphine (4–21%), codeine (0.8–2.5%), thebaine (0.5–2.0%), papaverine (0.5–2.5%), noscapine (4–8%), and narceine (0.1–2.0%) (US DEA 1992). They are all biosynthesized from the precursor (R)-reticuline. To obtain morphine, the cooked opium is dissolved in boiling water and precipitated with slaked lime (calcium hydroxide) to convert morphine into the soluble salt calcium morphenate. When the mixture is cooled, calcium morphenate remains in solution while the other alkaloids settle to the bottom in a brown sludge. Some residual codeine remains in solution, as it is somewhat water soluble. The calcium morphenate solution is filtered and reheated with ammonium chloride to raise the pH to 8–9. When the solution is cooled, morphine and residual codeine precipitate, and the resulting crystals are dried. To further purify the crystals, they are dissolved in dilute hydrochloric acid. Activated charcoal is added and the solution is reheated and filtered several times, yielding morphine hydrochloride. To obtain a crude alkaloid extract instead of morphine, sodium carbonate is added to the dissolved opium instead of

Figure 1.27 Afghani poppy farmers tend to their crops while US Marines hold position in Helmand province, Afghanistan (US Marine Corps photo by Gunnery Sgt. Bryce Piper).

slaked lime. Morphine can be separated from other alkaloids in the crude extract using benzene or other organic solvents. Benzene will solubilize the secondary alkaloids, leaving behind crystalline morphine.

The "sleeping agent" morphine was first isolated in 1804 by 21-year-old German pharmacist Friedrich Wilhelm Adam Sertürner (1783–1841). This was the first alkaloid ever isolated from plants and so Sertürner is considered the father of alkaloid chemistry. Morphine refers to Morpheus, the Greek god of sleep and dreams. It is one of the most powerful analgesics available, acting as a κ-, δ-, and μ-opioid receptor agonist. Strong binding to the μ-opioid receptor is primarily responsible for the analgesic and sedative effects. Morphine is the precursor to several other opiate drugs including codeine and heroin. Codeine has been used as an analgesic and antitussive, and as the precursor for synthesis of hydrocodone, oxycodone, and dihydrocodeine. Heroin is a highly addictive illicit drug made from morphine by acylation of both hydroxyl groups using acetyl chloride. It is less water soluble than morphine and therefore requires injection. Methadone is a synthetic opioid (opiate = derived from poppy; opioid = synthetic) used to treat heroin addiction.

Berberidaceae: Barberry Family

The barberry family contains 19 genera with about 750 species. The majority of species belong to *Berberis*, a genus of thorny deciduous or evergreen shrubs found in temperate and subtropical regions. *Berberis vulgaris* L. (European barberry) produces edible berries, found mainly in Iranian cuisine. The berries contain isoquinolone alkaloids such as berberine. Berberine is bright yellow in color, and so the berries and other parts of barberry plants have been used as a dye. In traditional medicine, berberine has been used as an antimicrobial. More recently, it has been studied for its beneficial affects against diabetes, high cholesterol, and a range of cancers. Mayapple (*Podophyllum peltatum* L.; Figure 1.28) contains the toxic lignan podophyllotoxin, traditionally used as an emetic. Today it is sold as a topical drug for warts.

Figure 1.28 Mayapple (*Podophyllum peltatum*) growing at the John J. Tyler Arboretum, Pennsylvania, the United States. © 2016 Derek Ramsey.

Ranunculaceae: Buttercup Family

Members of Ranunculaceae are typically herbaceous perennials, many with showy flowers. They are important garden ornamentals, as well as sources of isoquinoline and diterpene alkaloid neurotoxins. There are 65 recognized genera including *Aconitum* (monk's hood, wolfsbane), *Anenome*, *Aquilegia* (columbine), *Clematis*, *Delphinium* (larkspur), *Helleborus* (Lenten rose), *Hydrastis* (Goldenseal), and *Ranunculus* (buttercup). *Acontium* species are some of the most toxic plants ever encountered. *Aconitum ferox* Wall. ex Ser. roots are the source of the Nepalese bikh poison. The roots contain high levels of the diterpene alkaloids aconitine and pseudoaconitine that cause death in a mater of hours. *A. ferox* and other *Acontium* species have been used as arrow poisons for hunting and warfare throughout Asia and North America and to commit homicide or suicide. *Delphinium* also contains diterpene alkaloids, such as delphinine and methyllicaconitine, and is a common cause of livestock poisoning. Goldenseal (*Hydrastis canadensis* L.) is native to North America. Its yellow roots contain the isoquinoline alkaloids β-hydrastine, hydrastine, berberine, and canadine. Traditionally, goldenseal was used to treat inflammation, upper respiratory infections, GI disorders, and infection. Popularity as a dietary supplement has lead to overharvesting of wild species. It has been added to the Convention on International Trade in Endangered Species (CITES) Wild Fauna and Flora list.

Proteaceae is a family of the Southern Hemisphere, particularly Australia and South Africa. Australian aboriginals eat the fruit or seeds of several genera including *Persoonia*, *Gevuina*, and *Macadamia*. Gevuina nut oil is used in the cosmetics industry as a humectant and sunscreen. Macadamia nuts originated in Australia, but have become an important food crop in Hawaii since their introduction by William H. Purvis in 1882. Macadamia nut oil is also used as a cosmetic ingredient.

There are only two recognized genera in the goosefoot family, *Ribes* (currants) and *Grossularia* (gooseberries). Several species of currant were used by Native Americans as medicine. Roots, leaves, stems, bark, and berries were used for eye conditions, tuberculosis, common cold, gynecological aid, general tonic, and an antidote for poisoning. The fruit were also consumed as food by many tribes.

Vitaceae is a family of lianas with tendrils of temperate to tropical distribution. The genus *Vitis* supplies the world with grapes for wine, fresh consumption, and raisins. In 2012, over 67 million tons of grapes were produced worldwide; Italy, France, the United States, and Spain are the top grape producers (FAO 2013). Grapes (*Vitis vinifera* L.; Figure 1.29) were domesticated

Figure 1.29 A variety of white grapes nearly ready for harvest.

sometime during the Neolithic period (c. 8500–4000 BCE) in the mountains of Georgia, Armenia, and Azerbaijan. Shulaveri people from Georgia selected self-pollinating vines with a large, juicy fruit for wine production and learned how to propagate the best vines by grafting branches. They crushed the grapes to produce juice and, taking advantage of naturally occurring yeasts, fermented wine in large clay jars buried in the ground. From Georgia, viticulture spread to Europe and eventually the Western Hemisphere. In the late 1800s, William Thompson, a Scottish Immigrant to the United States, developed the *V. vinifera* variety "Thompson Seedless." This versatile variety is used for fresh consumption, juice, and raisin production in the United States. European *V. vinifera* cultivars such as Pino Noir, Merlot, Cabernet Sauvignon, and Chardonnay are used for wine production. Wine has cultural and religious significance around the world, but also may provide health benefits. Resveratrol, a stilbene, and flavonoids may be partly responsible for the beneficial effects.

Celastraceae: Bittersweet Family

Celastraceae is a family of mostly tropical trees, shrubs, and lianas used as ornamentals and medicines. Khat [*Catha edulis* (Vahl) Endl.] is native to the Arabian Peninsula and the Horn of Africa. It contains cathinone, a monoamine alkaloid stimulant. Leaves of the plant are chewed for the stimulating and euphoric effects, similar to amphetamines. The happy tree (*Camptotheca acuminata* Decne.) is the source of camptothecin, an important indole alkaloid anti-cancer agent. Camptothecin and camptothecin derivatives inhibit topoisomerase I and may also be useful against pathogenic protozoa such as *Leishmania donovani* and *Trypanosoma brucei*.

Erythroxylaceae: Coca Family

The South American shrub *Erythroxylum coca* Lam. is the source of coca leaves used to make the recreational drug cocaine. Chewing coca leaves or drinking coca tea (Figure 1.30) was and still is popular amongst native South Americans; however, oral absorption of cocaine is slow and inefficient compared to nasal inhalation or

Figure 1.30 Coca leaf tea is commonly served in the Andes.

insufflation of purified cocaine. Cocaine is hydrolyzed by stomach acids, catabolized by the liver, and is not readily absorbed by oral capillaries that constrict in the presence of cocaine. Therefore, it does not produce the same intense effects as inhalation or insufflation. Coca leaf extract was part of the formula for Coca-Cola up until the early twentieth century.

Formerly members of Clusiaceae, Hypericaceae contains 10 genera and 316 species. *Hypericum* contains several perennial flowering herbs with medicinal properties. *Hypericum perforatum* L. (Saint John's Wort, Klamath weed) was a source of livestock poisoning due to photosensitivity. Accordingly, it is listed as a noxious weed in western North America. But it is also one of the most popular herbal medicines worldwide. For thousands of years, it has been recommended to treat wounds and alleviate pain. But in 1525, the herbalist Paracelsus recommended it for depression. Active constituents include hypericin and hyperforin. Hypericin is a naphthodianthrone, an anthraquinone-derived highly photosensitive red pigment. Hyperforin is a prenylated phloroglucinol, also highly sensitive to light and oxygen. Extracts from the leaves and stems are prepared for pharmaceutical use in Europe. They are typically standardized for hypericin and/or hyperforin content. It was initially believed that hypericin was the primary active ingredient, acting as a monoamine oxidase inhibitor (MAOI). However, recent research has shown that activity is dependant on hyperforin concentration, with a novel mechanism of action.

Trees and shrubs in Clusiaceae have entire leaves with glands and a viscous exudate. *Garcinia × mangostana* L. (purple mangosteen) is a popular edible fruit in Southeast Asia. It can be eaten fresh or made into a juice. The rinds are sometimes steeped into a tea or made into an ointment for skin conditions. The purported anti-inflammatory, anti-diarrheal, antioxidant, and other medicinal properties are attributed to the xanthones like mangostin, garcinone E, 8-deoxygartanin, and gartanin.

There are 27 genera and 694 species in the family Passifloraceae. The best-known genus is *Passiflora* (Figure 1.31), with over 500 species of flowering vines and shrubs. *P. edulis* Sims fruit is locally known as passion, maracujá, maypop, or purple passionflower fruit. It was traditionally used to teat nervousness and insomnia in the Americas and Europe, and various other disorders around the world. Flavonoid glycosides including vitexin, schaftoside, and swertisin may be responsible for the observed activity; however, no rigorous clinical trials support its efficacy.

Euphorbiaceae is a large family with subfamilies Acalyphoideae, Crotonoideae, and Euphorbioideae. Many members of Crotonoideae and Euphorbioideae produce milky latex such as pointsettia (*Euphorbia pulcherrima* Willd. ex Klotzsch). Blind-your-eye mangrove tree (*Excoecaria agallocha* L.) has toxic latex that is used as a piscicide and

Figure 1.31 **Passionflower blooms.**

can cause blistering of the skin and temporary blindness if it comes in contact with the eye. The irritating agents are diterpene esters, some of which have been tested for activity against HIV. *Jatropha curcas* L. also has irritating latex, but it is nevertheless intensely cultivated for seed oil that is used as biofuel. The seed cake has potential as cattle feed, but contains toxic phobol esters (tetracyclic diterpenes) that must be removed. Castor oil is produced from castor bean (*Ricinus communis* L.) seeds. The oil is used as an industrial lubricant, in personal care products, and in medicine. It has long been used as a laxative and as a recipient for other drugs. The seeds contain a potent toxin, ricin. Ricin is a lectin that can be denatured by heat. The LD_{50} is 22 µg/kg if inhaled. It is a type 2 ribosome-inactivating protein, inhibiting protein synthesis and causing death by diarrhea and circulatory shock several days after exposure.

Vernicia fordii (Hemsl.) Airy Shaw seeds are the source of tung oil, traditionally used as a wood varnish or sealant. The tree is native to southern China and northern Vietnam, where the oil was used to waterproof wooden ships by the fourteenth century. Marco Polo is said to have brought tung oil to Europe from China, and the USDA brought seeds to the United States in 1905 from an agricultural exploration mission. Tung oil, primarily composed of eleostearic acid, dries quickly and polymerizes into a waterproof coating. The seeds are poisonous, causing vomiting, abdominal pain, and diarrhea, likely due to the presence of toxic saponins.

Over 11 million tons of natural rubber from *Hevea brasiliensis* (Willd. ex A.Juss.) Müll. Arg. latex was produced in 2012. *H. brasiliensis* is native to Brazil and was used by indigenous people of the Amazon and Mesoamerica to make rubber balls, shoes, and waterproof clothes. But today most natural rubber is produced in Asia: Thailand, Indonesia, and Malaysia. South American leaf blight hampers commercial production in South America. Rubber trees are tapped by hand, cutting the lacticifer vessels and collecting the latex in a bucket attached to the tree (Figure 1.32). Latex is primarily composed of polyisoprenes and proteins. The enzyme farnesyl diphosphate synthase polymerizes isoprene units into *cis*-polyisoprene. The protein fraction includes chitinases and lysozymes with activity against fungal cell walls and bacterial membranes. Thus, upon wounding, the resulting latex can seal wounds and fight pathogen invasion in wounded tissue.

Figure 1.32 Tapped rubber trees (*Hevea brasiliensis*) in Kerala, India. © 2006 M. Arunprasad.

Cassava (*Manihot esculenta* Crantz) is a starchy tuberous root cultivated as a staple crop in many tropical and subtropical areas, including its native South America. Dried cassava is known as tapioca, which is produced in powder and pearl form. It is made into alcoholic beverages, flour, and is prepared as a vegetable. It has also been used in some laundry starch products, as a biofuel, and as animal feed. One drawback of cassava is the presence of toxic cyanogenic glycosides lotaustralin and linamarin. But heat processing like boiling, drying, baking, steaming, and frying can reduce the cyanide concentration to safe levels, depending on the tuber cyanogenic glycoside concentration at harvest and processing method.

Salicaceae Willow Family

This small family of 2 genera and 25 species produces a pain relief compound used since antiquity. Hippocrates described the use of powdered willow bark and leaves (*Salix* spp.) to relieve pain, headaches, and fevers. Unbeknownst to Hippocrates, the active compound was the hydroxybenzoic acid salicylic acid. In 1897, a chemist at Bayer produced acetylsalicylic acid by acetylating salicylic acid with acetic acid. The compound was given the trade name Aspirin. It became a global success for treating pain and fevers.

Fabaceae Bean Family

Fabaceae (Leguminosae) is the third largest plant family with 917 genera and over 23,500 species. The Cronquist system lists Mimosaceae and Caesalpiniaceae as separate families from Fabaceae, but the APG system lists them as subfamilies. Members of Fabaceae have compound leaves and legume fruit, a carpel that splits along two seams. Some are indehiscent, some are winged, and some even form below ground (Figure 1.33).

Figure 1.33 A woman in Zambia holds an armful of freshly harvested peanuts (*Arachis hypogaea*). © 2012 Ilya Raskin.

The longest legumes are from the Central American liana *Entada gigas* (L.) Fawc. & Rendle, growing to 5 feet long. Table 1.1 lists several legumes with human use.

Many of the useful properties are related to the phytochemistry of the plant. Several natural gums are derived from members of Fabaceae (e.g., guar gum, gum acacia, gum tragacanth, and tara gum). These gums are a mixture of polysaccharides and glycoproteins with acidic groups. In the plant, gums often serve to seal wounds. People use gums for many purposes. The food, pharmaceutical, and cosmetic industries use gums for microencapsulation, to thicken and stabilize emulsions and liquids, form gels, prevent the formation of ice crystals (e.g., ice cream), as a binder for pressed tablets, and to maintain the quality of instant powder mixes. Gums have also been used as a paint binder, as a sizing for textiles, in paper production, as an oil drilling aid, a flocculant for mining, and binder for explosives.

Fabaceae is a source of isoflavonoids such as isoflavans, isoflavonols, rotenoids, coumestans, and pterocarpans. Isoflavonoids often have estrogenic activity; they can bind to estrogen receptor sites in mammals with positive or negative health effects. One hypothesis is that plants produce these compounds as a herbivory defense by reducing the fertility of herbivores. There have been numerous clinical trials on the health effects of soy isoflavones, especially genistein, daidzein, and glycitein. Rotenoids (e.g., rotenone) are found in many members of Fabaceae, especially the roots of *Derris* spp. They have broad-spectrum insecticidal and piscicidal activity. As the name implies, the Florida fish poison tree [*Piscidia piscipula* (L.) Sarg.] has been used as a fish sedative and poison. It is also used in herbal medicine as an analgesic. Likewise, crushed roots of devils shoestring [*Tephrosia virginiana* (L.) Pers.] were used by Native Americans to kill fish. Both plants contain various rotenoids and isoflavonoids that contribute to their activity.

Several types of indolizidine, quinolizidine, and pyrrolizidine alkaloids have been identified in members of Fabaceae, especially *Lupinus*, *Castanospermum*, and *Swainsona* spp. Quinolizidine alkaloids are present in the raw seeds and vegetative tissue of *Lupinus* spp. They are highly toxic to livestock. Castanospermine is an indolizidine alkaloid from the seeds of the Australian Morten Bay Chestnut tree, *Castanospermum australe*. It is an inhibitor of α-glucosidase I and II has antiviral properties against enveloped viruses like dengue fever. Swainsonine is another indolizidine alkaloid from the Australian plant darling pea [*Swainsona canescens* (Lindl.) F.Muell.] and the American spotted locoweed (*Astragalus lentiginosus* Hook. and other *Astragalus* spp.; Figure 1.34) and crazyweed (*Oxytropis* spp.). It is an α-mannosidase and mannosidase II inhibitor, causing lysosomal storage disease leading to wasting and death in livestock that chronically consume these species.

Table 1.1 Legumes with a history of human use.

Subfamily	Latin name	Common name(s)	Use
Papilionoideae	*Glycine max* (L.) Merr.	Soybean	Vegetable, oils, lecithin, tofu, soy sauce, miso, flour, soy milk, infant formula, biodiesel, cattle feed
	Phaseolus vulgaris L.	Common bean	Vegetables including kidney, pinto, white, pink, and black beans
	Vigna unguiculata (L.) Walp.	Cowpea	Vegetable
	Vicia faba L.	Broad bean	Vegetable
	Pisum sativum L.	Garden pea	Vegetable
	Lens culinaris Medik.	Lentil	Vegetable
	Cicer arietinum L.	Chick pea, garbanzo bean	Vegetable, flour, hummus, falafel
	Medicago sativa L.	Alfalfa	Livestock feed, traditional medicine
	Trifolium spp.	Clover	Livestock feed, traditional medicine
	Arachis hypogaea L.	Peanut	Vegetable, oil, flour, industrial uses
	Indigofera tinctoria L.	Indigo	Dye, rotenoids
	Haematoxylum campechianum L.	Logwood	Dye, haematoxylin, H&E stain
	Ceratonia siliqua L.	Carob	Chocolate substitute, beverages, sweetener, locust bean gum, traditional medicine
	Glycyrrhiza glabra L.	Licorice	Glycyrrhizin, sweetener, flavoring, traditional medicine
	Derris spp.	Tuba root	Rotenone insecticide and piscicide
	Astragalus spp.	Locoweed	Gum tragacanth; medicine in TCM and Persian medicine
	Cyamopsis tetragonoloba (L.) Taub.	Guar bean	Guar gum thickener
	Pueraria spp.	Kudzu	Traditional medicine
	Lupinus spp.	Lupine	Food and livestock feed
Mimosoideae	*Acacia senegal* (L.) Willd.	Gum acacia	Gum arabic thickener, binder
Caesalpinioideae	*Senna* spp.	Senna	Laxative, cassia gum thickener
	Caesalpinia spinosa (Molina) Kuntze	Tara tree	Tara gum, tannins for tanning leather, dye, traditional medicine
	Tamarindus indica L.	Tamarind	Flavoring, sauces, desserts, traditional medicine

Figure 1.34 Locoweed (*Astragalus lentiginosus*) growing in Red Rock Canyon, Spring Mountains, Nevada, the United States. © 2007 Stan Shebs.

Native Americans used mescal bean, *Dermatophyllum secundiflorum* (Ortega) Gandhi & Reveal, as a hallucinogen. The main constituent is the quinolizidine alkaloid cytisine. *Ormosia* seeds, called *Huayruro* in Peru, were used to make jewelry and are thought to bring good luck. They are highly toxic if consumed, due to the presence of quinolizidine alkaloids.

Rosaceae: Rose Family

Rosaceae is a large family, often divided into subfamilies, that contains many economically important temperate fruits, ornamentals, medicines, and timber. Based on FAO statistics, the farm gate value of Rosaceae crops tops $60 billion USD. Subfamily Amygdaloideae contains *Prunus* stone fruits: apricot (*P. armeniaca* L.), sweet cherry [*P. avium* (L.) L.], sour cherry (*P. cerasus* L.), plum (*P. domestica* L.), almond [*P. dulcis* (Mill.) D.A.Webb], peach and nectarine [*P. persica* (L.) Batsch], and black cherry (*P. serotina* Ehrh.). Members of Amygyloideae contain cyanogenic glycosides in their pits. Bitter almonds are especially high in the cyanogenic glycoside amygdalin. Amygdalin was historically used as a cancer treatment until clinical trials during the 1980s failed to demonstrate its effectiveness. Black cherry wood is one of the highest valued hardwoods. Native Americans and early settlers used its bark as a bitter stomach tonic.

Maloideae contains a variety of ornamentals and edible fruits: black chokeberry [*Aronia melanocarpa* (Michx.) Elliott], Japanese quince [*Chaenomeles japonica* (Thunb.) Lindl. ex Spach], European quince (*Cydonia oblonga* Mill.), Loquat [*Eriobotrya japonica* (Thunb.) Lindl.], hawthorn (*Crataegus* spp.), apple and crabapple (*Malus domestica* Borkh.) (Figure 1.35), and pear (*Pyrus* spp.). North American native chokeberries are high in antioxidants. They are used to color blended juice products, as a herbal tea component, and are made into wine in Lithuania. Quinces come from two closely related genera. Some are too astringent for fresh consumption, but production of non-astringent varieties is increasing in Central and Western Asia. Quince rootstock is valued for pear and loquat cultivar grafting. The loquat is native to China, where it is

Figure 1.35 Apples ready to harvest.

quite popular as a fresh fruit or mixed in desserts. The fruit has been used as a sedative and to stop vomiting, while the essential oils were used briefly as a perfume in France and Spain during the 1950s. Hawthorn (*Crataegus pinnatifida* Bunge and *C. cuneata* Siebold & Zucc.) is used as a stomach aid in traditional Chinese medicine and by Native North Americans. Apples are the third most important fruit crop worldwide after citrus and banana, in terms of production (FAO 2013). Their fruits are consumed fresh or processed into sauce, desserts, or cider. The wood is valued for specialty products such as pianos, croquet balls, and tool handles. Pears are the second most important Rosaceous fruit crop (FAO 2013). There are three species of primary global importance: *Pyrus communis* L. (European pear), *P. pyrifolia* (Burm.f.) Nakai (Japanese pear), and *P. ussuriensis* Maxim. ex Rupr. (Ussurian pear).

Subfamily Rosoideae contains strawberry (*Fragaria × ananassa*) and *Rubus* spp. (blackberry and raspberry) (Figure 1.36A). Wild strawberries [primarily *F. virginiana* Mill. and *F. chiloensis* (L.) Mill.] were a common source of fruit and medicine even before the domestication of *F. × ananassa* in eighteenth-century France. Native American tribes used strawberry fruit and the whole plant for diarrhea, stomachache, and for irregular menstruation. *F. virginiana* was brought from North America to Europe and became an important genetic component of the modern cultivated strawberry. *F. chiloensis* is the other species contributing to the modern strawberry. *F. chiloensis* is native to the west coast of North America from Alaska to central California and Chile in South America. The Picunche and Mapuche peoples of Chile cultivated *F. chiloensis* in gardens 1000 years ago. The Incas either stole or received specimens for their gardens. The fruit was eaten fresh and used as medicine. The Mapuche made a fermented drink, lahuen˜e mushca. Eventually, two distinct cultivars emerged, a large fruited white variety and a smaller fruited red variety. The modern *F. × ananassa* varieties genotypically align with the white form. The Spanish invasion spread strawberry production north into Cuzco, Peru, and Ecuador. It was not until 1712 that *F. chiloensis* strawberries were introduced to Europe. King Louis XIV of France commissioned Amédée-François Frézier to perform a reconnaissance mission to the Spanish empire in Peru and Ecuador. Although not a botanist, Frézier had a pension for collecting plants

(A)

(B)

Figure 1.36 (A) A woman in Sasquisili, Ecuador, selling strawberries and blackberries (notice how the blackberry receptacle remains in the fruit compared to raspberries, where it remains on the plant). © 2013 Ilya Raskin; (B) red raspberries.

and brought back several large fruited strawberry plants. Modern *F. × ananassa* varieties are a cross between *F. chiloensis* and *F. virginiana*.

Blackberries (*Rubus* spp.) were grown in Europe for over 2000 years for fruit, medicine, and as hedges. They are a genetically diverse group of brambles with complex origins. Chromosome numbers range from diploid to dodecaploid. They are organized into several subgenera and sections. Subgenera *Eubatus* contains sections *Moriferi* and *Ursini*. *Moriferi* species and *Caesii* subgenus are all native to Europe and Asia, whereas *Ursini* species and *Corylifolli* subgenus are all native to North America. "*Rubus fruticosus*" is sometimes used as an aggregate designation for all of the European cultivars. Up until the late 1800s, there was very little effort to improve or breed the varieties, since the wild fruit was abundant and pleasing, and the thorny branches were difficult to handle. Recent breeding programs have developed thornless varieties with an erect growth habit and firm, small-seeded fruit.

Most red raspberries (Figure 1.36B) are in the *Rubus* subgenus *Idaeobatus*. Canes have been found at ancient dig sites in Europe, Asia, and North America. Palladius documents their cultivation in the fourth century. In Greek mythology, Ida, the nursemaid of Zeus, pricked her finger while picking pure white raspberries, thus staining them red for eternity. The Latin name for the European red raspberry, *Rubus idaeus* L., means "bramble bush of Ida." Red raspberries have been used as food, dye, and medicine. Native American tribes made tea from the leaves to sooth menstrual cramps and digestive issues. The leaves have astringent properties and were used for dermatological issues as well. Black raspberries also belong to the subgenus *Idaeobatus*. *Rubus occidentalis* L. is native to

North America. They have been grown commercially since the 1800s. Chambord Liqueur Royale de France is made from red and black raspberry. Korean Bokbunja liquor is made from Korean black raspberries, *Rubus coreanus* Miq. However, there is considerable adulteration with *R. occidentalis*, as Korean production of *R. coreanus* cannot meet the demand for Bokbunja liquor and other Bokbunja products. Phenolic profile, flower color, and fruit pubescence can be used to correctly distinguish *R. coreanus* from *R. occidentalis*.

Rhamnaceae: Buckthorn Family

The buckthorn family includes *Ceanothus* ornamentals, Chinese jujube (*Ziziphus jujuba* Mill.), and *Rhamnus* spp. used for dyes. Jujube fruit are edible drupes; they have a long history of use as traditional medicine. The fruit are high in vitamin C, phenolics, flavonoids, triterpenic acids, and polysaccharides. Studies have looked at their anticancer, anti-inflammatory, antiobesity, immunostimulating, antioxidant, hepatoprotective, and gastrointestinal protective activities. *Rhamnus* dyes are derived from the berries and wood of several species. Ripe and unripe berries produce a yellow pigment that can be modified by the addition of mordants. Rhamnetin is the principal pigment, along with other flavonoids like kaempferol and quercetin.

Cannabaceae: Hemp Family

There are only around 100 species in the family Cannabaceae. *Cannabis* (hemp), *Humulus* (hops), and *Celtis* (hackberries) are the most well-known genera. *Cannabis sativa* L. (Figure 1.37A–D) is the source of hemp fiber, oil, seed, and a range of related products. It used medically as a treatment for nausea and pain and recreationally as a psychoactive drug. Tall-growing varieties are used for fiber, oil, and seed production, whereas low-growing varieties are used to produce medicinal products. Bast fibers associated with the phloem located directly under the bark are exceptionally long, growing in fiber bundles from 1–5 meters. To extract the fibers, stems are retted in the field by soaking them in water or by allowing microbes to rot away the outer layer of the stem and separate the inner core from the bast fibers. Decorticated crude fiber material is then subjected to steam and high pressure to further refine the fiber. The fibers are used to make specialty pulp for papers, plastic composites for autos, building materials like fiberboard and insulation, molded plastic products, horticulture products, and textiles. Shipbuilders used *Cannabis* bast fibers to make high-quality canvas sailcloth. In fact, the word *canvas* is derived from *Cannabis*. The achenes and their oils can be used as food, animal feed, cosmetics, and industrial oils. The oil is high in unsaturated fatty acids linoleic acid (50–60%), α-linolenic acid (20–25%), oleic acid (10–16%), and γ-linolenic acid (2–5%). Hemp oil has a short shelf life due to rancidity and must be protected from oxidative conditions. *C. sativa* was used medically and recreationally in China as early as 3000 BCE, but it was India where *C. sativa* was domesticated as a potent medicinal plant. Trichomes produce a resin containing psychoactive cannabinoids such as Δ9-tetrahydrocannabinol (THC), cannabigerol, cannabichromene, and cannabinol. Cannabinoids are terpenophenolics that bind to the Cannabinoid receptor type 1 (CB_1) and/or CB_2 receptors in the brain, producing effects such as euphoria, analgesia, sedation, memory and cognitive impairment, appetite stimulation, and anti-emesis.

Hops (*Humulus lupulus* L.) is a deciduous climbing vine (climbs by wrapping shoots around a support as compared to vines that use tendrils). It was cultivated for beer production as early as 1079 in Bavaria, Germany. Germany continues to be the largest producer of hops, followed by the United States. Female plants are grown for their strobili containing bitter α-acids like humulone and bitter β-acids like lupulone. The bitter acids impart flavor to the beer, while acting as an antimicrobial preservative.

(A)

(B)

Figure 1.37 (A) *Cannabis sativa*; (B) grow room at New Jersey, the United States medical *Cannabis* production facility; (C) harvested plants drying before extraction; (D) concentrated extract rich in cannabinoids. © 2016 Frank DiGiovanni.

(C)

(D)

Figure 1.37 (Continued)

Nearly all members of Moraceae are trees or shrubs that produce milky sap via laticifers. The family includes several genera of important fruit crops: *Ficus* (fig), *Morus* (mulberries), and *Artocarpus* (breadfruit and jackfruit). Breadfruit [*Artocarpus altilis* (Parkinson ex F.A.Zorn) Fosberg] (Figure 1.38) is native to Polynesia, where it spread to Southeast Asia. In 1787, breadfruit was Captain William Bligh's mission on the *HMS Bounty*. He was to transport breadfruit from Tahiti to the West Indies, where they were to become food for slaves. *The Bounty* was even outfitted with a makeshift greenhouse to house the potted plants. The 1787 mission ended in an epic mutiny with Bligh traveling thousands

Figure 1.38 Breadfruit (*Artocarpus atilis*).

of miles to safety on a small 23 foot boat. In 1791, Bligh set out on a more successful breadfruit mission, delivering 2126 breadfruit trees to the West Indies. Breadfruit trees produce prolific amounts of large, starchy, aggregate fruit that is typically cooked before eating. It is quite similar to breadnut (*Artocarpus camansi* Blanco) and is related to jackfruit (*Artocarpus heterophyllus* Lam.) (Figure 1.39A–B). Breadnut is starchy like breadfruit and is prepared much the same way. Jackfruit is considered the largest tree fruit, 30–100 cm long x 25–50 cm diameter. It has been cultivated in India since ancient times and can be eaten raw, pickled, or canned in sweet syrup. Flavonoid compounds from the heartwood such as artocarpin have been used to treat skin hyperpigmentation.

Ficus elastica Roxb. ex Hornem. is a large tree with buttress roots that was historically a source of natural rubber. It has since been replaced by *Hevea brasiliensis* (Euphorbiaceae). The common fig (*Ficus carica* L.) is native to Western Asia. It has been cultivated since prehistoric times, predating the domestication of wheat, barley, and legumes. It produces edible fruit that can be consumed fresh, dried, or cooked. The fruit contains a variety of phytochemicals and have been used in traditional medicine for a range of ailments.

Urticaceae: Nettle Family

This family of herbs includes stinging nettle (*Urtica dioica* L.) and *Boehmeria nivea* (L.) Gaudich., the source of ramie fiber. Stinging nettle has trichomes on the leaves that can pierce the skin upon contact and break off, while injecting histamine and acetylcholine. An immune response ensues, causing an itchy, burning rash that can last up to 12 hours. Nettles have also been used as food, fiber, and medicine. They have been cooked as a vegetable, their bast fibers were woven into fabrics during World War I, and they have been made into a tonic for rheumatism. Ramie fibers are made from white or green ramie bast fibers. Ramie fabrics date back to ancient times where they were used to make Egyptian mummy wraps. They never came into widespread commercial

(A)

(B)

Figure 1.39 (A) Woman in Nha Trang, Vietnam, peels jackfruit (*Artocarpus heterophyllus*); (B) Jackfruit tree.

production due to the labor-intensive retting process to extract the fibers. However, they have been used to make automobile plastic composites and are used to wrap traditional Vietnamese glutinous rice cakes.

Cucurbitaceae: Cucurbit Family

Cucurbits have a long history of human use as food, medicine, and poison. Many of the fruits can be dried and used as a storage vessel, kitchen utensil, musical instrument, jewelry, or decoration. The Vietnamese luffa [*Luffa cylindrica* (L.) M.Roem.] can be dried and used as a scrubbing sponge. The fruits can also serve as a source of water in arid regions, but many are toxic to animals and pose a hazard to grazing livestock. Several species have been used for arrow poisons in Africa [e.g., *Acanthosicyos naudiniana* Welw. ex Hook.f., *Citrullus lanatus* (Thunb.) Matsum. & Nakai, *Cucumis heptadactylus* Naudin, *Cucumis aculeatus* Cogn., *Oreosyce africana* Hook.f., *Zehneria scabra* Sond., *Momordica charantia* L., *M. balsamina* L., and *L. cylindrica*]. Cucurbitacins are bitter, highly oxygenated triterpenes mostly found in cucurbit roots and fruits. They have purgative properties and are highly cytotoxic.

Myrtaceae: Myrtle Family

There are currently 144 genera in Myrtaceae. Most are woody evergreens with essential oils, distributed in subtropical and tropical regions, especially Australia. The Australian *Eucalyptus regnans* F.Muell. trees are the world's tallest angiosperms. Several species have ethnobotanical and economic importance. Fruits from *Psidium guajava* L. (guava), *Acca sellowiana* (O.Berg) Burret (feijoa, guavasteen), *Eugenia* spp. (jambos, rose-apple, pitanga), and *Campomanesia* spp. (guabiroba) are all edible. Allspice [*Pimenta dioica* (L.) Merr., Figure 1.40] fruits are dried and used as a spice. Cloves are flower buds from *Syzygium aromaticum* (L.) Merrill & Perry. The tree is native to the Maluku Islands in Indonesia, known as the Spice Islands. Until the mid-1700s, clove production was tightly limited to the Spice Islands, which were controlled by the Dutch. The French East India Company commissioned horticulturist Poivre to smuggle seedlings of cloves

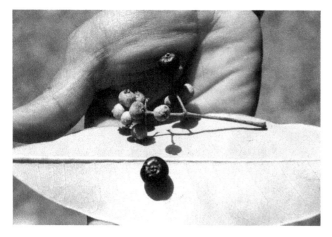

Figure 1.40 Leaf, unripe, and ripe fruits from the allspice tree, *Pimenta dioica*.

and nutmeg to the Seychelles and Mauritius. Although he was able to secure some plants, production never took off in Seychelles or Mauritius, and Indonesia remains the world leader in cloves production.

The leaves of *Melaleuca alternifolia* (Maiden & Betche) Cheel are the source of tea tree oil, used in personal care products. The essential oil is rich in monoterpenes like terpinen-4-ol and has been used topically in traditional Australian Aboriginal medicine for wounds, insect bites, itching, infections, and inflammation. *Leptospermum scoparium* J.R.Forst. & G.Forst. and *Kunzea ericoides* (A.Rich.) Joy Thomps. are grown in New Zealand for similar purposes, but have not been studied or exploited as extensively. Cajeput oil from *Melaleuca leucadendra* (L.) L. is high in cineole and is marketed as an antimicrobial treatment for aquarium fish and as a general antiseptic. Eucalyptus oil from *Eucalyptus* spp. is used in decongestants, cold remedies, and muscle pain relievers. In traditional Australian Aboriginal medicine, eucalyptus oil was used as an antiseptic to heal wounds and treat fungal infections. Tea made from eucalyptus leaves was used to reduce fevers. Joseph Bosisto established the Australian eucalyptus oil industry 150 years ago, with production peaking in 1940. Surgeons in the late nineteenth to early twentieth centuries used eucalyptus oil as an antiseptic to disinfect medical equipment. Currently, the medicinal type of oil has high cineole content and is primarily produced from two species: *E. polybractea* R Baker and *E. globulus* Labill. The perfumery type has high citrinellal content and is produced from *E. citriodora* Hook. and *E. staigeriana* F. Muell. Since eucalyptus is also grown for timber, eucalyptus oil may be a by-product.

Anacardiaceae: Cashew Family

Most members of the cashew family produce toxic phenols, mostly catechols and resorcinols, from resin canals, causing contact dermatitis. The compounds are a defense against herbivores and pathogens alike. Urushiol is a mixture of closely related 3-n-alk(ene)ylcatechols found in all members of *Toxicodendron* including poison ivy [*T. radicans* (L.) Kuntze], poison oak [*T. diversilobum* (Torr. & A.Gray) Greene], poison sumac [*T. vernix* (L.) Kuntze], wax tree [*T. succedaneum* (L.) Kuntze], and laquer tree [*T. vernicifluum* (Stokes) F.A. Barkley]. Cardol, a pentadecylcatechol, is found in cashew (*Anacardium occidentale* L.) fruit and nut shell along with anacardic acids and cardanol. Cashew nut shell liquid has industrial applications for coatings and adhesives. The nut does not contain toxic phenols and is a popular snack food and culinary ingredient. Likewise, mangos (*Mangifera* spp.) are an economically important tropical fruit (Figure 1.41). The fruit peels contain 5-alkyl and 5-alkenylresorcinols, which are mildly allergenic. The resinous sap, however, can cause dermatitis during harvesting. Pistachios (*Pistacia vera* L., Figure 1.42) are a popular nut crop that does not contain the same toxic phenols as cashews or *Toxicodendron* species. The hard endocarps of ripe pistachios split open at maturity, a desirable characteristic for human consumption. A related species, *Pistacia lentiscus* L., is the source of gum mastic resin, known as "tears of Chios" on the Greek Island of Chicos. It was traditionally used for stomach ulcers, to fill dental cavities, and as a chewing gum. It has also been used in varnishes, incense, and confectionaries.

Sapindaceae: Soapberry Family

There are many species of edible fruit in Sapindaceae including lychee (*Litchi chinensis* Sonn.), longan (*Dimocarpus longan* Lour.), and rambutan (*Nephelium lappaceum* L.); all are

Figure 1.41 Several varieties of mangoes for sale at a market in Tagatay, Philippines

Figure 1.42 Pistachios (*Pistacia vera*), cashews (*Anacardium occidentale*), and dried mango (*Mangifera indica*) are three members of Anacardiaceae being sold by this vendor in Dushanbe, Tajikistan. © 2014 Ilya Raskin.

native to Southeast Asia with a tropical to subtropical distribution. Guarana (*Paullinia cupana* Kunth) fruit contain more caffeine than coffee beans and have been made into a stimulant tea and used for weight loss. Like longan (also known as "dragon's eye"), guarana fruit resemble eyes, with a black seed and translucent white aril. In the Paraguayan Guarani language, guarana roughly translates as, "fruit like the eyes of the people." Maples (*Acer* spp.) used to be in their own family, Aceraceae, but have since been moved to Sapindaceae. Sugar maples (*A. saccharum* Marshall) are tapped in early spring for maple syrup. Approximately 40 L of sugar maple sap is boiled down to make 1 L of maple syrup. Maple trees are an important source of hardwood timber. Like maples, the Buckeye (also called horse chestnut, *Aesculus* spp.) was in the family Hippocastanaceae, but has also been moved to Sapindaceae. *Aesculus hippocastanum* L. seed extracts have historically been used for various ailments but recent research has focused on chronic venous insufficiency, hemorrhoids, and edema. One of the active components is aescin, a triterpene saponin with either α or β configuration.

Rutaceae: Citrus family

There are 148 genera in Rutaceae, but *Citrus* (Figure 1.43) is by far the most valuable. The origins of lemons [*C. limon* (L.) Osbeck], limes [*C. aurantiifolia* (Christm.) Swingle], sweet and blood oranges [*C. sinensis* (L.) Osbeck], sour orange (*C. aurantium* L.), pomelo [*C. maxima* (Burm.) Merr.], tangerine or Mandarin orange (*C. reticulata* Blanco), citron (*Citrus medica* L.), and kaffir lime (*C. hystrix* DC.) is still a mystery, but they were likely domesticated in parts of China, Southeast Asia, and India. Many species were not introduced to the Western Hemisphere until the nineteenth and twentieth centuries. Grapefruit (*C. paradisi* Macfad.) is likely a cross between pummelo and sweet orange, originating in the West Indies. It was described in Barbados in 1750. Kumquats were originally classified as *Citrus* spp., but then renamed as *Fortunella* spp. They can be eaten whole, in contrast to other hespiridiums that are typically peeled to remove the bitter, leathery rind. The peel is rich in flavonoids, particularly naringin and neohesperidin, as well as bitter limonoids like limonin and nomilin. The peel also contains terpenes like limonene in the essential oil. Limonene is commonly used as a citrus flavor or fragrance for personal care products, food, and household products.

Other family members worth mentioning are the curry leaf tree [*Murraya koenigii* (L.) Spreng.], white sapote (*Casimiroa edulis* La Llave), and jaborandi (*Pilocarpus* spp.). White sapote is native to Mexico, where the seeds, leaves, and bark were used as a sedative in folk medicine. Extracts have hypotensive properties in animal models. Several furoquinoline alkaloids, quinolones, polyphenols, and histamine-like compounds have been found in the seeds, roots, and bark. Leaves from the curry lead tree are used to impart flavor in curries, as the name implies. The leaves are aromatic and contain several carbazole alkaloids, such as bisgerayafolines, mahanimbins, girinimbine, koenimbine, and murray-amine-D. These alkaloids are being studied for various health benefits. The jaborandi tree (*Pilocarpus jaborandi* Holmes) is the source of the imidazole alkaloid and glaucoma drug pilocarpine. It is native to Northern Brazil where it was used as a diuretic tea to induce sweating. The Tupi people named it *jaborandi*, which means "slobber-mouth plant." They used it to induce salivation, and today it is available in tablet form to treat xerostoma, or dry mouth. Pilocarpine stimulates secretions of tears, saliva, and sweat by binding to muscarinic acetylcholine receptor M3. Eye application results in stimulation of the muscle that contracts the pupil, thus relieving intraocular pressure.

Figure 1.43 An orange tree at the Orangerie at Versailles, France.

Meliaceae: Mahogany Family

The mahogany family contains 52 genera of trees distributed throughout the tropics. Economically important products include timber such as mahogany (*Swietenia* and *Khaya* spp.), *Toona* spp., and *Guarea* spp. American mahogany production started in the West Indies, when Spanish, French, and British empires controlled the islands. Native people used mahogany for canoes. The Europeans used it for shipbuilding and eventually exported substantial quantities back to Europe for high-end furniture.

Neem oil is derived from *Azadirachta indica* A.Juss. fruits and seeds. *A. indica* is native to India but has a pantropical distribution. It grows so rapidly that it is considered a noxious weed in some areas of Africa and Australia. Neem has been used as medicine in India for over 2000 years to treat a range of ailments, particularly skin disorders. It is also an effective insecticide (see Neem case study in the section titled "Neem-Based Insecticides" in Chapter 6).

Superfamily Malvaceae now contains former members of Bombacaceae (e.g., *Durio*, *Ceiba*, *Adansonia*, and *Ochroma*) and Sterculiaceae (e.g., *Theobroma* and *Cola*). Members of Malvaceae are typically trees or shrubs with stellate hairs and mucilaginous sap. Some important genera in this cosmopolitan family are *Gossypium* (cotton), *Abelmoschus* (okra), *Theobroma* (cocoa), *Hibiscus*, *Durio* (durian), *Ceiba* (kapok fiber), *Ochroma* (balsa), and *Adansonia* (baobab).

Cocoa

Theobroma cacao L. is a small, evergreen, understory tree 5–8 m tall that grows in tropical rainforests. The small pink or white flowers are cauliflory (arising from the trunk) and are pollinated by midges. The cocoa fruit is a large egg-shaped berry often called a "cocoa pod," growing straight out of the trunk or large branches (Figure 1.44). Once the pods turn a rich, golden-orange, they are ready to harvest. The pods are hand-split, revealing 30–40 seeds, called "cocoa beans," surrounded by white mucilage. The beans are removed, and fermented in sweat boxes, baskets, or heaps to remove the mucilage. Then they are sun-dried or dried with hot air. The next step in the process involves winnowing, where the beans are cracked open to separate the shell from the nib. Nibs are roasted at 105–120°C, where they develop the characteristic chocolate aroma and flavor. Grinding nibs forms a chocolate liquid called "mass." It contains anywhere from 53% to 58% cocoa butter. Cocoa butter can be separated from cocoa solids using a press. The presscakes are pulverized into a fine powder used for beverages or baking.

The Olmecs of Mexico were the first cocoa farmers, starting plantations around 400 BCE. The Mayans and Aztecs made several concoctions out of cocoa for use in wedding and religious ceremonies such as cocoa powder in water with chilies (*Capsicum annuum* L.) or vanilla (*Vanilla planifolia* Jacks. ex Andrews) as added flavoring. Tejate is another ceremonial beverage made from cocoa and maize. The Latin name *Theobroma* means "food of the gods," and the species name *cacao* is derived from the Nahuatl word *xocolatl*, meaning "bitter water." Cocoa beans were traded as money, so only the royals and elite were wealthy enough to consume chocolate.

Cocoa beans contain around 0.5–2.0% of the xanthine alkaloid theobromine. Caffeine (0.2–0.9%) and theophylline (~0.4%) only develop once the beans are mature. This is not surprising, as theobromine can be converted into caffeine, and caffeine converted to theophylline during seed maturation.

Figure 1.44 Cauliflory flowers and fruit of a cocoa (*Theobroma cacao*) tree growing inside a Nestle research greenhouse.

Gossypium hirsutum L. (upland cotton) accounts for 90% of the world's cotton supply. It originated in Mexico thousands of years ago. Related cotton species *G. barbadense* L. (pima cotton) is native to South America around Peru and Ecuador, *G. herbaceum* L. (levant cotton) is native to Sub-Saharan Africa, and *G. arboreum* L. (tree cotton) is native to South Asia. The bolls of all species contain fluffy white fibers composed of nearly pure cellulose that is valued for fabric production. Cottonseed oil is pressed from cottonseeds and is used largely for the production of salad oils. The leftover presscake is sold as livestock feed, as it is high in protein. Gossypol is a yellow dimeric sesquiterpene found in pigment glands in the seeds, stems, and roots of *Gossypium* species. Gossypol is removed from cottonseed oil before it goes to market because of its toxic properties. It can become a livestock poison through chronic consumption of cottonseed meal. The precise mechanism of poisoning is unknown, although it is thought that gossypol binds to essential amino acids, rendering them unavailable, and can interfere with essential enzymatic reactions. Gossypol has also been tested as a male contraceptive.

Okra [*Abelmoschus esculentus* (L.) Moench, syn. *Hibiscus esculentus*] originated somewhere in Africa. The fruit is harvested immature and cooked before eating. The infamous mucilage is primarily composed of galacturonic acid, galactose, rhamnose, and glucose associated with proteins. Bast fibers from the stem are composed of α-cellulose, hemicellulose, and lignin. They can be used as reinforcement for plastic composites.

Hibiscus are well known for their ornamental value; however, they also are used as a medicinal tea. Hibiscus tea is prepared from the blossoms in a variety of ways, hot or cold, sometimes with other ingredients like sugars or citrus. The tea is bright red and has a naturally tart flavor. It contains vitamin C and was traditionally used as a diuretic. Preparations from roots, leaves, and flowers are ingredients in personal care products throughout Asia.

Durian (*Durio zibethinus* L.) is a tall tree, up to 40 meters, with bat-pollinated flowers. The large fruit (25 cm long x 20 cm diameter) is covered in spines (Figure 1.45). The seeds are covered with a creamy, edible aril. There are over 40 compounds that contribute to the sweet-rotten smell of durian that many find repulsive. Sulfides like ethyl methyl sulfide contribute to the unpleasant odor similar to rotten onions. Durian can be eaten fresh, but it is often used to flavor desserts.

Baobab trees (*Adansonia* spp.) are some of the largest and oldest trees on earth. Radiocarbon dating showed one species of *A. digitata* L. to be 1275 years old. Trees can reach 30 meters high and have a circumference of 7–11 meters. Aside from their cultural significance, baobab seeds and fruit powder have gained some economic significance as food additives and thickeners.

Balsa [*Ochroma pyramidale* (Cav. ex Lam.) Urb.] is well known for its lightweight wood (density = 100 kg/m^3), with high elastic moduli, strength, and energy absorption capacity. The trees are native to Central America, with Ecuador producing about 95% of the world's supply.

Brassicaceae: Mustard Family

The mustard family consists of annual, biennial, or perennial herbs. The flowers have four corollas arranged in a cruciform (hence the old family name, "Cruciferae"); the

Figure 1.45 A durian (*Durio* sp.) vendor's bicycle in Hanoi, Vietnam.

fruit is a silique or silicle. The plants typically contain glucosinolates, which are hydrolyzed to pungent isothiocyanates by myrosinase in damaged plant tissue. Glucosinolates are thioethers derived from an amino acid and glucose. The early part of the cyanogenic glycoside pathway forms an aldoxime from an amino acid. The aldoxime incorporates sulfur, is glucosylated, and undergoes side chain extension to form the glucosinolate. Side chain variation accounts for different levels of bioactivity and pungency of glucosinolates and resulting isothiocyanates.

Several species of Brassicaceae have considerable economic importance for their pungent flavor. *Brassica oleracea* L. (cabbage, Brussels sprouts, broccoli, cauliflower, kale, and kohlrabi); *Brassica rapa* L. (turnip), and *Raphanus raphanistrum* subsp. *sativus* (L.) Domin (radishes) are all cultivated vegetables. *Armoracia rusticana* P.Gaertn., B.Mey. & Scherb. (horseradish), *Wasabia japonica* Matsum. (wasabi), *Sinapis alba* L. (white or yellow mustard), *Brassica juncea* (L.) Czern. (Indian mustard), and *Brassica nigra* (L.) K.Koch (black mustard) are used as flavorings or spices. Table mustard originally contained a mix of black and white mustard. But since the 1950s, table mustard has contained white mustard and Indian mustard instead. The seeds are pressed for their oils, which are rich in isothiocyanates. Black and Indian mustard seeds contain the glucosinolate sinigrin that is converted to the pungent allyl isothiocyanate. Yellow mustard seeds contain the glucosinolate sinalbin that is converted to the slightly less pungent 4-hydroxybenzyl isothiocyanate. Like black mustard, horseradish root contains the glucosinolate sinigrin, but wasabi contains desulfosinigrin, which yields 6-(methylsulfinyl)hexyl isothiocyanate. True wasabi made by grating enlarged stems of *Wasabia japonica* is primarily available only in Japan owing to high domestic demand and difficulties in commercial cultivation. Wasabi paste served in North America and Europe is typically an imitation composed of horseradish, mustard, starch, and food colorings. Some retailers may add a small amount of wasabi leaves to add *W. japonica* to the ingredient list. Growers are cultivating *W. japonica* in North America, with a price tag of around $150USD/lb. fresh rhizome.

Isatis tinctoria L. is the source of woad, a blue dye produced from the leaves. It was cultivated since antiquity throughout Europe and Western Asia. Leaves were picked, crushed by a roller, and rolled into balls. The balls were left to dry while anaerobic bacteria at the center of the balls started to convert the indole alkaloids isatan A, isatan B, and indican to indoxyl. Indoxyl is eventually converted to the blue pigment indigotin, which is also called indigo. Eventually, woad was replaced by indigo extracts from Asian *Indigofera* spp. (Fabaceae), and finally both were replaced by synthetic indigo (see Indigo case study in the section titled "Indigo: The Devil's Dye and the American Revolution" in Chapter 7).

Cactaceae: Cactus Family

Cacti are well adapted to survive in dry, desert environments. They are succulent, with spines and a waxy coating to prevent water loss. Ripe *Opuntia* (prickly pear) fruit (Figure 1.46) are edible after the outer layer is removed. The stems or pads are eaten as a vegetable. Dragonfruit or pitahaya are sweet succulent fruit from the genus *Hylocereus*. The flowers bloom at night, and the plants are grown on supports (Figure 1.47A–D). Although native to the Americas, they are grown throughout the tropics. The peyote cactus [*Lophophora williamsii* (Lem. ex Salm-Dyck) J.M. Coult.] produces small stems called "bottons" that contain psychoactive alkaloids, most notably mescaline. Archaeological specimens indicate North Americans may have used peyote for its psychotropic properties 5700 years ago. Some tribes used it for speaking to the spiritual realm and considered the plant divine. Peyote buttons are either chewed dry or soaked to produce a liquid to drink. They were sometimes also smoked mixed with tobacco and/or cannabis. The phenylethylamine alkaloid mescaline is the predominant psychoactive found in peyote, occurring along with structurally related isoquinoline alkaloids. Mescaline is structurally related to ecstasy (3,4-methylenedioxy-methamphetamine MDMA). It interferes with dopamine and norepinephrine action in the brain, with psychotropic effects appearing within 1–2 hours of ingestion. Other cactus species including San Pedro cactus [*Echinopsis pachanoi* (Britton & Rose) Friedrich & G.D.Rowley], and several species of *Echinopsis* also produce mescaline.

Cornaceae: Dogwood Family

Figure 1.46 Ripe *Opuntia* fruit with aeroles and glochids in Arizona, the United States.

Members of Cornaceae are small trees and shrubs largely valued as ornamentals. The

(A)

(B)

(C)

(D)

Figure 1.47 (A) Dragon fruit (*Hylocereus* sp.) growing on supports; (B) immature flower; (C) red dragon fruit outside; (D) red dragon fruit cross section.

Chinese happy tree (*Camptotheca acuminata* Decne.) is a notable exception. Its bark is the source of the quinoline alkaloid camptothecin, an anti-cancer agent. American chemist Monroe Wall is credited with its discovery from a bioassay-guided plant extract screening program during the 1960s.

Sapotaceae: Zapote Family

The zapote family is made up of tropical evergreen trees and shrubs. They frequently contain milky latex in their bark and fruit and are valued for their edible fruit, seed oil, and/or latex. Notable members include the argan tree (*Argania spinosa* L.Skeels), shea tree (*Vitellaria paradoxa* C.F.Gaertn.), star apple (*Chrysophyllum cainito* L.), lucuma [*Pouteria lucuma* (Ruiz & Pav.) Kuntze, Figure 1.48], chicle [*Manilkara chicle* (Pittier) Gilly], and sapodilla [*Manilkara zapota* (L.) P.Royen]. Both argan oil and shea butter are native to Africa, valued the world as over as personal care ingredients. Chicle originated in Mesoamerica as a chewing gum to curb hunger and thirst. It was eventually developed into the modern chewing gum "chicklets." The star apple, lucuma, and sapodilla are all popular fruits in Central and South America. They have a variety of culinary uses.

Theaceae: Tea Family

Members of the tea family are typically evergreen trees or shrubs distributed throughout the tropics and subtropics, especially Southeast Asia. They have showy flowers

Figure 1.48 Lucuma tree growing in Peru.

and woody, capsular fruit. Leaves are thick, serrate, with small glands at the terminal tips. The most recognized genus is *Camellia*, both for its ornamental value and as the source of tea. Over 3 million tons of tea is grown each year. It has global cultural significance, from Japanese tea ceremonies, to the Boston tea party, and tea time in Britain.

Green, white, yellow, oolong, black, and puerh teas are all produced from *Camellia sinensis* (L.) Kuntze (Figure 1.49), but are processed differently. Green, white, and yellow teas are made from leaves that are harvested, steamed, and quickly dried before any oxidation can occur. This quick processing destroys polyphenol oxidases and peroxidases, reducing enzymatic browning of the leaves. White tea is made from young leaves and buds harvested from the growing tips, with the glabrous buds providing a white color. Yellow tea leaves are dried more slowly than green, producing a yellow color. Oxidation in tea production is often referred to as "fermentation." After harvest, tea leaves are withered with hot air. They may be rolled or cut, then placed in trays. Humid (98% RH), hot (29°C) air is circulated around trays of withered leaves for an amount of time determined by the type of tea desired. A tea master turns the leaves, monitoring the aroma and color, and adjusting environmental conditions appropriately. Oolong tea is only partially oxidized, leading to a light brown color. Black tea leaves are fully fermented, with a dark brown color. After oxidation is complete, the leaves are dried and packaged for shipment.

There are some teas that are fermented with microorganisms. Fuzhuan tea is fermented with a specific fungus, *Eurotium cristatum*. Pu'er is aged and fermented with bacteria that naturally occur on the plant. The harvested leaves are stored in large mounds that are stirred infrequently, like a compost pile. After fermentation, the leaves are transferred to a hot and humid room to be pressed into cakes and aged.

Figure 1.49 Women picking tea (*Camellia sinensis*) in Cisarua Bogor, West Java, Indonesia. © 2008 Danumurthi Mahendra.

Both oxidation and fermentation change the phytochemical profile of tea. Green, unoxidized tea contains several catechins, predominantly epicatechin (EC), epicatechin gallate (ECG), epigallocatechin gallate (EGCG), and epigallocatechin (EGC). Oxidation begins in the withering stage and is accelerated by mechanical rolling or cutting. This increases the surface exposure to oxygen, increasing the activity of polyphenol oxidases and peroxidases. Gallated catechins are converted to non-gallated catechins, increasing free gallic acid concentration and forming theaflavins. If oxidation continues, theaflavins are converted to thearubigins through interactions with other polyphenols. Theaflavins and thearubigins are responsible for the brown color of fermented teas. Thearubigins make up about 60% of the dry weight of a black tea infusion. Their red color gives black tea its reddish hue. Oxidation also converts flavonoid glycosides to their respective aglycones. In addition to polyphenols, tea contains the purine alkaloids caffeine and theobromine, which may also be reduced by the oxidation process.

Ericaceae: Heath Family

The heath family is a large family of woody shrubs distributed throughout temperate areas or high-altitude tropical regions. The genus *Vaccinium* includes *V. macrocarpon* Aiton (cranberry), *V. angustifolium* Aiton (lowbush wild blueberry, Figure 1.50), *V. corymbosum* L. (highbush blueberry), *V. myrtillus* L. (bilberry), and many other wild relatives. Cranberry has long been used to prevent or treat urinary tract infections. Proanthocyanidins reduce the ability of pathogenic bacteria to adhere to the bladder wall. *Gaylussacia baccata* (Wangenh.)K.Koch., the black huckleberry, grows throughout Eastern North America and closely resembles *Vaccinium* berries. Leaves from *Ledum* spp. are harvested for Labrador tea, a beverage consumed by native North Americans. Although the tea is a source of vitamin C, the presence of the sesquiterpene ledol and diterpene grayanotoxins in the essential oil makes toxicity a real concern.

Figure 1.50 Lowbush blueberries (*Vaccinium angustifolium*) growing in Maine, the United States. © 2002 Mary Ann Lila.

Rhododendrons are well known for their toxicity, yet have been used to treat pain and inflammation in many cultures. Bees that feed on *Rhododendron* flowers (especially *Rhododendron ponticum* L.) in the Black Sea region produce "mad honey" that contains grayanotoxins (see mad honey case study in the section titled "Mad Honey" in Chapter 7). Poisoning symptoms include hypotension, bradycardia, atrioventricular block, and autonomic nervous system symptoms occurring about 30 minutes after ingestion. In Turkey, mad honey is referred to as *deli bal*.

Rubiaceae: Coffee Family

Rubiacea contains several ethnobotanically important plants including coffee (*Coffea* spp.), *Cinchona* (source of quinine), noni (*Morinda citrifolia* L., Figure 1.51A–B), *Carapichea ipecacuanha* (Brot.) L.Andersson (source of ipecac), rose madder (*Rubia tinctorum* L.), kratom [*Mitragyna speciosa* (Korth.) Havil], yohimbe [*Pausinystalia johimbe* (K.Schum.) Pierre ex Beille], and woodruff [*Galium odoratum* (L.) Scop., source of coumarin]. Woodruff is a popular flavoring in Germany, where it is called "waldmeister." It is used to flavor sherberts, drinks, jams, and Berliner Weisse beer, among other things. Yohimbine is an indole alkaloid from *P. yohimbe* used in veterinary and traditional medicine. Kratom contains mitragynine and other indole alkalods with opioid receptor activity. Rose madder roots contain several anthraquinones including purpuroxanthin, quinizarin, purpurin, and alizarin. They produce an orange, red, or violet dye (depending on the pH of the extract) that has been used on fabrics and in paints since ancient times. Noni bark produces a purplish dye that has been used for making traditional batik in Java. The roots produce a yellowish dye used on fabric. The anthraquinone morindin is responsible for the vivid color. The plant is also used in traditional medicine. In Vietnam, for example, the roots are used for stiffness, tetanus, and

(B)

(A)

Figure 1.51 (A) Noni (*Morinda citrifolia*) fruit cross section; (B) noni fruit on tree.

arterial tension. The fruit gained popularity as a general tonic and is used traditionally for a range of ailments. However, the claims remain largely unsubstantiated.

Syrup of ipecac is an emetic derived from the roots and rhizomes of *C. ipecacuanha* and *C. acuminata*. Both are low-growing shrubs native to Central and South America, where ipecac has been traditionally used to treat dysentery. Ipecac was also mixed with powdered opium to produce Dover's powder, a diaphoretic taken for colds and flu. In the nineteenth and twentieth centuries, it was used to induce vomiting in cases of over-dose or accidental poisoning. The primary active compounds are the secologanin- and dopamine-derived monoterpenoid-isoquinoline alkaloids emetine and cephaeline. They have demonstrated anti-amoebic activity, and hence their success in treating dysentery. They also act upon the chemoreceptor trigger zone to induce vomiting.

Coffee

Coffea arabica L. is a small tree (2–8 m tall) native to dense tropical highland forests of Ethiopia. It grows optimally from 950 m to 1950 m above sea level. It was likely first domesticated and consumed as a stimulating chewed berry or beverage within the Horn of Africa. According to legend, an Ethiopian goat herder named Kaldi discovered coffee after his goats ate the berries and became so excited they could not sleep at night. The abbot at the local monastery concocted a drink from the berries and noticed he too was kept alert for many hours. Documentation of roasting coffee seeds to make a beverage is lacking until fifteenth-century Yemen. Linnaeus believed the species originated on the Arabian peninsula and hence named it *Coffea arabica*. From Yemen, the coffee plant was smuggled to India, followed by its spread to Europe, Southeast Asia, and eventually the Americas. *Qahveh khaneh*, or coffeehouses, soon emerged as places to socialize and discuss current events. European travelers brought back word of the "wine of Araby," an unusual black beverage. Clergy condemned the drink as "the bitter invention of Satan," but Pope Clement VIII enjoyed the taste and gave it his approval. By the mid-seventeenth century, coffeehouses became centers of social engagement in Europe. Although coffee was brought to New Amsterdam (New York City) in the mid-1600s, colonists continued to enjoy tea until the Boston Tea Party.

The Arabs were intent on maintaining their monopoly on coffee production, but the Dutch finally obtained some seedlings in the seventeenth century. They planted them in India, resulting in crop failure, but also in Java and Indonesia, where the plants thrived. The Dutch proudly presented King Louis XIV of France with seedlings for the Jardin royal des plantes in Paris. In 1723, Gabriel de Clieu, a young naval officer, obtained seedling(s) from this stock and took his prized cargo on an arduous voyage to the island of Martinique. He is credited with starting coffee cultivation in the Western Hemisphere, although other accounts suggest the Dutch had already taken coffee to their colony in Suriname, South America by 1718, and the French to their colony in Hispaniola by 1715. Francisco de Mello Palheta is credited with starting the billion dollar coffee industry in Brazil. He was sent to French Guiana to obtain coffee seedlings from the governor, who was unwilling to share. But, de Mello Palheta was able to charm the governor's wife. She gave him coffee seeds stuffed in a bouquet of farewell flowers.

Today, coffee is one of the most highly traded agricultural commodities. Although there are around 100 species of *Coffea*, the two most economically important species are *Coffea arabica* (arabica coffee) and *Coffea canephora* Pierre ex A.Froehner (syn. *C. robusta*, robusta coffee). Robusta coffee is also native to the highlands of Ethiopia, as well as West and Central Africa. French colonists introduced robusta to Vietnam in the nineteenth century. Today Vietnam is the largest producer of robusta coffee in the world, with Brazil leading the world in arabica coffee production. *Coffea liberica* Hiern (liberica coffee and excelsa coffee cultivars), native to Central and Western Africa, is grown on a smaller scale mostly in the Philippines and Indonesia. There are several differences between the species including optimum growing conditions, susceptibility to diseases, caffeine content, and flavor of the coffee.

Coffee is produced from the seeds of the drupe, often referred to as a coffee berry or cherry due to the red color and sweetness of the ripe fruit (Figure 1.52A–B). Each drupe bears two seeds. The raw seeds are a light brown color, about 10 mm long with a central groove, and are encased in a covering known as a pyrene (Figure 1.53). Arabica seeds are flat, while robusta seeds are more oval. Ripe fruit is usually picked by hand and processed using a dry or wet method. For the dry method, fresh fruit is spread in flat layers, and dried in the sun to 11% moisture (Figure 1.54). For the wet method, the fresh picked fruit are put through a pulping machine to remove the pulp, and floated through a water channel where the heavier ripe seeds sink and the unripe seeds float. Next, the ripe seeds are placed in drums to separate them by size and then placed in water-filled fermentation tanks for 12–48 hours to remove the mucilaginous coating. The seeds are rinsed and then dried in the sun to 11% moisture, just as in the dry method. Dried seeds are milled before export. This process involves hulling to remove any remaining endocarp, exocarp, or mesocarp; polishing (sometimes); and grading (Figure 1.55). Milled

(A) (B)

Figure 1.52 *Coffea arabica* unripe (A) and ripe (B) fruit.

Figure 1.53 Green *Coffea arabica* "beans."

Figure 1.54 Woman drying *Coffea arabica* outdoors in Laos. © 2013 Ilya Raskin.

seeds ready for export are referred to as "green coffee beans." Green coffee beans are roasted to an internal temperature of around 204°C. Pyrolysis causes sugars in the seeds to caramelize, producing desirable flavor compounds. Different roasting temperatures give rise to the many different flavors or "roasts" of coffee such as light roast, dark roast, French roast, American roast, Vienna roast, and so on.

The purine alkaloid caffeine is of fundamental importance to coffee's popularity. Along with theobromine and theophylline, caffeine is a derivative of xanthine with stimulant properties. They inhibit phosphodiesterase, resulting in an increase of cAMP

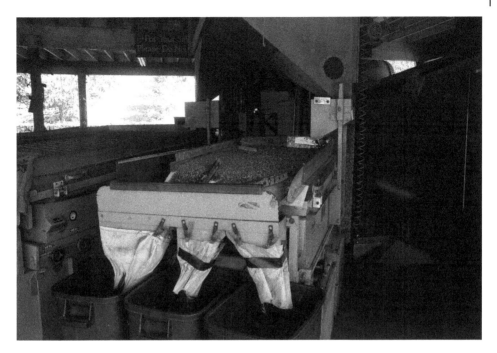

Figure 1.55 Coffee sorting machine in Hawaii.

and release of adrenaline. This leads to CNS stimulation, relaxation of smooth muscle, and diuresis. Coffee seeds contain from 1–2% caffeine with trace amounts of theobromine and theophylline. Low-caffeine varieties are being developed from breeding programs using low-caffeine land races. Caffeine occurs along with chlorogenic acid (5–7% dry weight of the seed). Roasting causes chlorogenic acid to break down into quinic acid and caffeic acid.

Quinine

Quinine, quinidine, cinchonidine, and cinchonine are all modified terpenoid indole alkaloids valued for their antimalarial properties. These quinoline alkaloids occur in the bark of the stems and roots of *Cinchona* trees (Figure 1.56). There are 26 species of *Cinchona*, but only two are valued for the production of quinine: *C. pubescens* Vahl and *C. officianalis* L. The trees are native to the tropical Andes, growing on steep mountain slopes. Native people used the bark to treat fevers, but it is the Spanish Jesuits who are credited with developing *Cinchona* bark as a treatment for malaria. South American *Cinchona* trees were overharvested, so the English developed plantations in India while the Dutch made plantations in Java. Just before WWII, around 95% of the world's supply of *Cinchona* bark was coming from Java.

Quinine is obtained by uprooting the trees when they are 8–12 years old and harvesting the bark. The bark yields anywhere from 4–14% alkaloids. During WWII, the Indonesian supply of quinine was cut off by the Japanese, so synthetic versions were developed. Quinine was largely replaced by chloroquine, primaquine, and mefloquine, all synthetics based on quinine's structure. They had fewer side effects than quinine and worked just as well. But recently quinine has been reintroduced. The offending

Figure 1.56 Hermann Adolf Köhler drawing of *Cinchona officinalis*.

organism, *Plasmodium falciparum*, developed resistance to the synthetics, especially chloroquine. *P. falciparum* is a protozoan transmitted by female *Anopheles* mosquitoes (see malaria case study in the section titled "*Achillea millefolium*: Re-Exploring Herbal Remedies for Combating Malaria" in Chapter 4). Five other species of *Plasmodium* also cause malaria, but most deaths are attributed to *P. falciparum*. The mechanism of action of quinoline alkaloids is not well understood, but it is hypothesized that they prevent polymerization of hemoglobin breakdown products, causing a buildup of cytotoxic heme in the protozoa.

Apocynaceae: Dogbane Family

The APG III system now includes Asclepiadaceae (milkweed family) as a subfamily of Apocynaceae. Members of Apocynaceae produce important drugs and notorious poisons. Dogbane (*Apocynum* spp.) produces cardiac glycosides and was used historically as a cardiotonic. *Cerbera odollam* Gaertn., known as the suicide tree or pong-pong, contains the cardiac glycoside cerberin. It has been used for suicides and homicides, especially young women living in Kerala, India, where the tree is native (Gaillard *et al.* 2004). *Nerium oleander* L. is a highly toxic garden shrub, producing the cardiac glycosides oleandrin and oleandrigenin. Oleander extracts have been used in traditional medicine in Mesopotamia, as far back as the fifteenth century BCE. In the eighth century CE, Arab physicians used oleander extracts to treat cancer. Research studies revealed some anticancer activity; however, there have been no successful clinical trials. Madagascar periwinkle [*Catharanthus roseus* (L.) G.Don] produces the antihypertension monoindole alkaloids ajmalicine and serpentine. It also produces two important anticancer drugs, the bisindole alkaloids vincristine and vinblastine. Indian snakeroot [*Rauvolfia serpentina* (L.) Benth. ex Kurz] produces the indole alkaloid reserpine, a drug once used as an antipsychotic and antihypertensive. It was used for centuries in India to treat mental illness and snakebite. It blocks the uptake of norepinephrine and dopamine. Milkweed (*Asclepias* spp.) has several ethnobotanical uses. The seed fibers have been used as mattress and pillow filling. Milkweed was elevated to the level of "wartime strategic material" in the United States during WWII. Seed fibers were used to fill Allied life jackets when the supply of kapok fibers [*Ceiba pentandra* (L.) Gaertn., Malvaceae; Figure 1.57] from Asia was cut off. The latex was studied as a wartime substitute for natural rubber, without success. African tribes have used the latex as an arrow poison. It has a high concentration of pregnane and cardiac glycosides including asclepin, frugoside, and gofruside. Ouabain is a cardiac glycoside from the seeds of the African plant *Strophanthus gratus* (Wall. & Hook.) Baill. The plant was traditionally used to treat heart conditions and as an arrow poison, depending on the concentration.

Convolvulaceae: Morning Glory Family

Members of Convolvulaceae are typically lactiferous climbing vines, often with showy tubular flowers. The genus *Ipomoea* contains some of the best-known members of Convolvulaceae: morning glory (various *Ipomoea* and *Rivea* spp.), sweet potato [*I. batatas* (L.) Lam.], water spinach (*I. aquatica* Forssk.), and moonvine (*I. alba* L.). The seeds of *Argyreia nervosa* (Burm. f.) Bojer, *Ipomoea violacea* L., and *Turbina corymbosa* (L.) Raf. contain psychedelic ergoline alkaloids derived from (+)-lysergic acid. Native Americans ground the seeds to make the hallucinogenic drug ololiuqui for religious ceremonies. Extracts from the seeds were ingested to communicate with the gods. The principle active component is ergine (d-lysergic acid amide or LSA). However, recent genetic evidence suggests that clavicipitalean fungi produce the alkaloids, not the plant

Figure 1.57 Spines on the bark of a *Ceiba* sp. tree.

itself. The fungi are seed transmitted and contain the biosynthetic equipment to produce ergoline alkaloids, but the plant does not. The alkaloids accumulate in the plant, not the fungal mycelium, suggesting a symbiotic relationship. When treated with fungicides, ergoline alkaloids are not detectable in the plant.

Solanaceae: Nightshade Family

Solanaceae is a family used for food, medicine, and poison. Most members are native to the New World, primarily Central and South America. The white or Irish potato (*Solanum tuberosum* L.; Figure 1.58), tobacco (*Nicotiana tabacum* L.; Figure 1.59), and tomatoes (*Solanum lycopersicum*) are native to the Andes Mountains in South America, and the chili pepper (*Capsicum annuum*) is native to Mexico. They were widely cultivated throughout the Americas by indigenous people. Eventually, European conquistadors brought the plants and knowledge of their use back to Europe.

Solanaceous plants produce a variety of simple and complex alkaloids including tropane, pyridine, pyrrolidine, piperidine, and steroidal glycoalkaloids. They also produce withanolides and physalins, both of steroidal origin. Many of these compounds are produced as a defense against herbivory or pathogen attack. Table 1.2 describes some of the most well-known members of Solanaceae, their origins, use, and relevant phytochemicals. Tropane alkaloids (e.g., hyoscyamine, scopolamine, and atropine) are anticholinergics that bind to acetylcholine receptors. Hyoscyamine is used to treat GI disorders by reducing motility. It can also be prescribed for Parkinson's disease, heart conditions, and excessive salivation. Atropine is used in ophthalmology to dilate the pupils and to increase the heart rate during extreme bradycardia.

Figure 1.58 Women selling different varieties of potatoes (*Solanum tuberosum*) in Peru.

Figure 1.59 Tobacco farm in Lancaster County, Pennsylvania, the United States.

Scopolamine (hyoscine) is used to treat motion sickness and postoperative nausea. Along with morphine, it induced "twilight sleep" anesthesia, a method that replaced chloroform in the early twentieth century. The "nicotinoids" nicotine, nornicotine, anabasine, and anatabine are simple molecules with complex origins. Nicotine and nornicotine are pyridylpyrrolidines, formed by the condensation of the pyridine ring of a nicotinic acid derivative (3,6-dihydronicotinic acid) and a pyrrolidine ring from ornithine (*N*-methylpyrrolinium cation). They are often simply called pyridine alkaloids or pyrrolidine alkaloids. Anabasine is produced from nicotinic acid in a similar manner, but lysine contributes a Δ^1-piperidinium cation, making it a pyridylpiperidine alkaloid. Synthesis of anatabine is comparable, but lysine does not appear to contribute the Δ^1-piperidinium cation. Solanaceous steroidal glycoalkaloids, such as solasodine, solanidine, α-solanine, α-chaconine and tomatine, are typically produced in stems and leaves but not roots. They can be used as precursors in the synthesis of therapeutic steroids. Potato tubers contain from 10–100 mg/kg glycoalkaloids. Acute toxicity occurs when 1–3 mg/kg body weight is consumed. The compounds alter membrane permeability; GI and neurological symptoms are the most common. Another notable alkaloid of mixed origin is capsaicin, the pungent compound in hot peppers. It is derived from the phenylpropanoid pathway and fatty acid synthesis. Phenylalanine contributes the aromatic portion of the molecule; the steps from ferulate to vanillylamine are unique to *Capsicum*. Valine contributes the branched chain fatty acyl-CoA tail via isobutyryl-CoA.

Table 1.2 Ethnobotanically important members of Solanaceae with their origin, use, and relevant phytochemicals.

Latin name	Common name	Origin	Use	Phytochemicals
Solanum tuberosum L.	White or Irish potato	Andes region of South America	Staple food crop	α-Solanine, α-chaconine
Solanum lycopersicum L.	Tomato	Andes region of South America	Food	Tomatine, carotenoids
Solanum melongena L.	Eggplant, aubergine	India	Food	Solanine, solanidine
Capsicum spp.	Pepper	Central and South America	Food; medicine for pain relief, digestive aid, antibiotic; sprays for personal defense	Capsaicin
Physalis spp.	Ground cherries, tomatillos	Africa, North and South America	Food	Phygrine, hygrine, tigloidine, physoperuvine, withanolides, physalins
Atropa belladonna L.	Belladonna, deadly nightshade	Europe, Africa, Asia, North America	Pupil dilation, pain relief, GI disorders	Scopolamine and hyoscyamine
Mandragora officinarum L.	Mandrake	Europe	Folklore, anesthetic for surgery, smoked hallucinogen, aphrodisiac, death-wine	Hyoscyamine, scopolamine, atropine, mandragorine, cuscohygrine, withanolides
Nicotiana tabacum L.	Tobacco	Central and South America	Insecticide, smoked as a sedative	Nicotine, nornicotine, anatabine, anabasine
Hyoscyamus niger L.	Henbane	Europe, Asia, and Africa	Pain relief, smoked for intoxication, fish poison	Hyoscyamine, scopolamine, tigloidine, tropine
Solanum americanum Mill.	Black nightshade	Europe	Cooked leaves and berries "famine food," sedative, various medicinal applications like dropsy, GI issues, herpes zoster	Solanine
Datura stramonium L.	Jimsonweed	North America and possibly Asia	Leaves smoked to treat asthma, anesthetic, seeds ingested as hallucinogen	Atropine, hyoscyamine, scopolamine

With high levels of tropane alkaloids, several Solanaceous species have been used as narcotics and hallucinogens. Jimsonweed (*D. stramonium*) is sometimes called Jamestown weed in reference to Bacon's Rebellion of 1676 where Virginia colonists took up arms and burned the capital, Jamestown. British soldiers were sent to crush the rebellion, but unwittingly became intoxicated for several days by ingesting *Datura* leaves that were boiled for their salads. *Datura* spp. are mentioned in the Badianus Manuscript as an Aztec narcotic drug for pain and various other ailments. The plant has been used for witchcraft and other spiritual rituals and still causes accidental poisoning, both livestock and human. The Delphic oracle in ancient Greece may have inhaled smoke from henbane (*H. niger*) to invoke prophecies. During the Middle Ages, witches' brews included deadly nightshade (*A. belladonna*), henbane, and jimsonweed. In the Americas, indigenous tribes used plants like borrachero (*Brunfelsia grandiflora* D.Don), culebra borrachero [*Brugmansia pittieri* (Saff.) Moldenke], angel's trumpet (*Brugmansia* spp.), chalice vine (*Solandra grandiflora* Sw.), kieli (*Solandra brevicalyx* Standl.), latua [*Latua publiflora* (Griseb.) Baill.], paguando [*Iochroma fuchsioides* (Bonpl.) Miers], and various *Datura* spp. for initiation rites, sacred hallucinations, divination, prophecy, and magic. Often their use was restricted to witch doctors and shamans, who would at times administer the plants as oral or topical medicines. The most typical preparation was tea made from the leaves, roots, and/or bark. Sometimes the tea would be taken by the witchdoctor to induce visions of how to treat an ailing patient.

Australian aborigines used pituri bush [*Duboisia hopwoodii* (F.Muell.) F.Muell.] as a stimulating chew. Often, they would roll a quid with ash to make the alkaloids nicotine and nornicotine more bioavailable. The ash was produced by particular species of *Acacia*, *Eucalyptus*, *Senna artemisioides* (all Fabaceae), *Grevillea*, *Hakea* (both Proteaceae), and *Ventilago viminalis* (Rhamnaceae). Tobacco was smoked and chewed by Native Americans long before European settlers arrived. It was typically smoked for special ceremonies, not as an everyday habit. The European settlers grew tobacco as a cash crop and used the profits to fund the revolutionary war. Tobacco cigarette smoking was not prevalent until after 1881, when James Bonsack invented an automated cigarette machine that could make 120,000 cigarettes per day. There are about 20 compounds in tobacco smoke that are definitive carcinogens, mostly polycyclic aromatic hydrocarbons (PAH) and 4-(methylnitrosamino)-1-(3-pyridyl)-1-butanone (NNK). Nicotine is not considered carcinogenic, but it is a highly addictive acetylcholine receptor agonist with stimulant and psychoactive effects. It is also a potent insecticide, but widespread use in agriculture has diminished. Imidacloprid is currently the most widely used insecticide on the market. It is a synthetic nicotine analog with a wide range of uses including crop protection, termite control, and flea treatment for household pets.

Oleaceae: Olive Family

Olea is the type genus for Oleaceae, and *O. europaea* L. is the source of olives (Figure 1.60A–B). The oleoside type secoiridoid glycoside oleuropein is one of several compounds that give green olives and extra virgin olive oil their bitter taste. It is very abundant in young, green fruit. As the fruit ripens and turns black, oleuropein content falls as oleuropein derivatives like elenolic acid glucoside and demethyloleuropein predominate. During commercial processing, olives are soaked in a lye solution to remove the bitterness. Enzymatic oxidation has been investigated as a method to reduce or

(A)

(B)

Figure 1.60 (A) Olive trees (*Olea europaea*) on the island of Mallorca. © 2015 Hans Braxmeier; (B) Ripe olives. © 2015 Jose Antonio Alba.

eliminate the need for the lye solution, which produces a considerable amount of wastewater. Olives contain 20–30% oil, which is pressed from whole olives. Olive oil has a range of uses in food, cosmetics, medicine, and religious ceremonies. It has been cultivated in the Mediterranean region since ancient times, as early as 2400 BCE. Recent research has investigated the health properties of olive oil, especially as a component of Mediterranean diets. It may help reduce cardiovascular disease, cancer, and inflammation.

Plantaginaceae: Plantago Family

The organization of Plantaginaceae has been up for debate, with molecular data suggesting it may be better described as nested inside Scrophulariaceae. Some have also proposed a return to the original family name, Veronicaceae. In any case, *Plantago* and *Digitalis* (formerly in Scrophulariaceae) are two genera with ethnobotanical significance. When the seed husks of *Plantago* become wet, they become mucilaginous, and are an effective laxative. Many over the counter "bulking agents" for constipation contain *P. psyllium* L. husk. It is commonly known as psyllium fiber or isabgol. *Plantago* leaves have been used as a poultice for a range of skin rashes and bites.

In 1775, William Withering evaluated a home remedy for dropsy (an abnormal buildup of fluid in the abdomen from congestive heart failure) made by Mrs. Hutton, an old woman from Shropshire, England. Of all the herbs contained in the home remedy, he discovered that extracts from *Digitalis* leaves could treat dropsy following scarlet fever or bad sore throats. Digitalis, known as foxglove, is a biennial herb native to Europe that produces potent cardiac glycosides. *D. purpurea* L. (Figure 1.61) and *D. lanata* Ehrh. are both currently cultivated for drug production. First year leaves are harvested and dried at 60°C to inactivate hydrolytic enzymes. In the past, a standardized preparation of mixed cardiac glycosides was produced from the leaves, but today individual cardiac glycosides are isolated. *D. lanata* differs from *D. purpurea* in that it produces lanatosides and their breakdown product, digoxin. Digoxin and digitoxin are the only compounds that have been routinely used as a drug; however, the leaves contain other cardiac glycosides such as gitoxin and lanatoside A-E. Digoxin is broken down to digitoxin in the body. Digitoxin increases the force and velocity of the heart's contraction by increasing the calcium concentration in cardiomyocytes. GlaxoSmithKline still produces digoxin from *D. lanata* leaves. According to the company, it takes 1000 kg of dried leaves to make 1 kg of digoxin.

Lamiaceae: Mint Family

Members of Lamiaceae are well known for their essential oils and culinary uses. The large family of herbs (245 genera with over 7800 species) has a cosmopolitan distribution. Notable genera include *Scutellaria* (skullcap), *Marrubium* (horehound), *Nepeta* (catnip), *Salvia* (sage), *Monarda* (beebalm), *Origanum* (marjoram, oregano), *Thymus* (thyme), *Mentha* (mint), *Rosmarinus* (rosemary), *Ocimum* (basil), *Pogostemon* (patchouli), and *Lavandula* (lavender).

Figure 1.61 *Digitalis purpurea* inflorescence.

Table 1.3 Major constituents of the essential oils from commonly used members of Lamiaceae (Kumari *et al.* 2014).

Latin name	Common name	Major oil constituents*
Marrubium vulgare L.	Horehound	Germacrene D, (-)-β-caryophyllene, thymol
Pogostemon cablin (Blanco) Benth.	Patchouli	Pogostone, patchouliol, a-guaiene
Lavandula angustifolia Mill.	Lavender	Linalyl acetate, linalool
Salvia officinalis L.	Sage	Cineole, borneol, thujone
Rosmarinus officinalis L.	Rosemary	Cineole, α-pinene, camphor
Ocimum basilicum L.	Sweet basil	Methyl eugenol, linalool, camphor, geranyl acetate
Mentha × piperita L.	Mint	Menthone, menthol
Origanum vulgare L.	Oregano	Thymol, *p*-cymene
Thymus vulgaris L.	Thyme	Carvacrol, thymol

* Highly variable depending on chemotype.

Glandular trichomes on the leaves synthesize and secrete essential oils, a complex mixture of terpenes and aromatic compounds. Major constituents of the essential oils from several commonly used species are listed in Table 1.3.

The personal care and food industries use essential oils for their pleasant aroma and preservative properties. These essential oils have antimicrobial properties and are a component of traditional medicines. For example, mint (*Mentha* spp.) has carminative properties and has been used by many cultures for digestive issues. Lemon balm (*Melissa officinalis* L.) was used to treat headaches and nervousness. Sage (*Salvia officinalis* L.) and lavender (*Lavandula* spp.) were used for respiratory illness. Skullcap (*Scutellaria baicalensis* Georgi) is used in TCM to treat cancer, hepatitis, and epilepsy. Recent research found that flavonoids baicalin, baicalein, and wogonin may be responsible for the activity. Catnip (*Nepeta cataria* L.) contains nepetalactones, known feline attractants. They may also be effective insect repellants.

Aquifoliaceae: Holly Family

The holly family has only three genera: *Ilex*, *Nemopanthus*, and *Prinos*; most species belong to *Ilex*. The English Holly (*Ilex aquifolium* L.) and American Holly (*Ilex opaca* Aiton) are popular ornamentals in temperate regions, with their unassuming flowers, spiny evergreen leaves, and red berries. *Ilex paraguariensis* A.St.-Hil. is a small tree native to higher elevations of South America. Its leathery leaves are made into a popular caffeinated beverage called yerba mate (Figure 1.62A–B). Historically, native people in Brazil and Paraguay consumed yerba mate and shared it Spanish explorer Juan de Solís in the sixteenth century. Its popularity grew, with Jesuit missionaries growing yerba mate plantations to keep up with demand. Leaves are toasted to deactivate enzymes and soften the leaves, creating a brown leaf tea with a smokey flavor. In addition to its stimulant properties, yerba mate was also used for weight loss, depression, and digestive aid. Active compounds include the xanthine alkaloids, caffeine,

Figure 1.62 (A) Dried *Ilex paraguariensis* leaves are used to make yerba mate; (B) Gourd for making yerba mate. During the mate gourd ceremony, yerba and hot water are added to the gourd. Then, the mate is sipped from the bombilla, with the gourd passed around a circle of friends © 2016 Eric Salem.

theobromine, and theophylline. The caffeine content is about 1% of leaf dry weight, which is comparable to coffee beans, but much lower than tea leaves. Some marketers claim their yerba mate does not contain caffeine, but instead contains the fictitious compound mateine. Leaves are also high in chlorogenic acid, which may account for some of the activity as well.

Asteraceae: Sunflower Family

Asteraceae (Compositae) is the largest flowering plant family, with over 1,900 genera and 36,700 species. Plants are distributed worldwide except for Antarctica. Most are herbaceous, but there are several notable examples of trees and shrubs. Their flowers occur as a capitulum with ray and disk florets. Fruits can be achenes (cypselae), as in sunflowers, but some species produce fleshy fruit resembling drupes. The family is probably best known for its ornamentals, such as edelweiss [*Leontopodium nivale* subsp. *alpinum* (Cass.) Greuter], *Senecio* spp., daisies (*Leucanthemum* and other genera), marigolds (*Tagetes*, *Calendula*, and other genera), chrysanthemum (*Chrysanthemum indicum* L.), and *Dahlia* spp. It also produces several noxious weeds such as thistle (*Cirsium* spp.), dandelion (*Taraxacum campylodes* G.E.Haglund and *T. erythrospermum* Andrz. ex Besser), and ragweed (primarily *Ambrosia* spp.).

Asteraceae also has several members with ethnobotanical importance. The genus *Artemisia* produces valuable essential oils. *A. dracunculus* L. (tarragon) is a culinary herb, and *A. absinthium* L. (Figure 1.63) is used to produce the potent alcoholic beverage absinthe. *A. arborescens* (Vaill.) L. is known in Arabic countries as *sheeba*. It was a popular medicinal tea in ancient Egypt, and is still consumed today, often mixed with mint leaves. *A. annua* L. has the real claim to fame in the genus. It produces artemisinin, a sesquiterpene lactone

Figure 1.63 *Artemisia absinthium.*

used to treat malaria (see malaria case study in the section titled "*Achillea millefolium*: Re-Exploring Herbal Remedies for Combating Malaria" in Chapter 4).

Other useful plants include common vegetables such as lettuce (*Lactuca sativa* L.), chicory (*Cichorium intybus* L.), endive (*Cichorium endivia* L.), artichokes (Cynara scolymus L.), and Jerusalem artichoke (*Helianthus tuberosus* L.). Sunflower (*Helianthus annuus* L.) and Safflower (*Carthamus tinctorius* L.) produce valuable oils used in foods and cosmetics. The shrub *Stevia rebaudiana* (Bertoni) Bertoni is a native of South America that is grown for its sweet-tasting leaves. It produces ent-kaurane-type diterpenes known as steviol glycosides, primarily stevioside and rebaudiosides A and C (Ceunen and Geuns 2013).

Arnica montana L., often called wolf's bane, is an old remedy for bruises and sore muscles. The sesquiterpene lactone helenalin is the presumed active component. It has anti-inflammatory activity and can be found in several other members of Asteraceae. Chicory (*C. intybus*) root contains bitter sesquiterpene lactones with anti-inflammatory activity. Chicory has been used as a coffee substitute, especially in times of hardship when coffee is unavailable. Chicory root also contains inulin, a polysaccharide used as a pre-biotic for digestive health. Inulin is present in many other plant species, but chicory is often used as the commercial source.

Apiaceae: Carrot Family

Apiaceae (Umbelliferae) is an important source of food crops, herbs, medicines, and poisons. Most members grow a large taproot, which is consumed as a vegetable in the case of carrot (*Daucus carota* L.), celeriac (*Apium graveolens* L.), and parsnips (*Pastinaca sativa* L.). Other vegetable crops like celery (*Apium graveolens* L.) and fennel (*Foeniculum vulgare* Mill.) are harvested for their fleshy stems. Dill (*Anethum graveolens* L.), parsley [*Petroselinum crispum* (Mill.) Fuss], cilantro (*Coriandrum sativum* L.), and culantro (*Eryngium foetidum* L.) are valued for the essential oils in their leaves and are used as culinary herbs. The seeds of coriander, cumin (*Cuminum cyminum* L.), caraway (*Carum carvi* L.), and fennel (*Foeniculum vulgare* Mill.) also contain essential oils and are used as spices. Poison hemlock (*Conium maculatum* L.) was notoriously used to carry out death sentences in ancient Greece including Socrates. The piperidine alkaloids coniine and γ-coniceine are primarily responsible for the toxic properties; *C. maculatum* can contain over 2% coniine w/w. The alkaloids are neurotoxins that bind to nicotinic acetylcholine receptors, causing respiratory failure. Apiaceae is also the source of furanocoumarins, such as psoralen and angelicin. Many furanocoumarins are photoactive; their activity is potentiated by UV light. They can bind to DNA, and interact with proteins and lipids. Wild parsnip (*Pastinaca sativa* L.) produces furanocoumarins as a defense mechanism against the parsnip webworm. When humans come into contact with furanocoumarins, it can cause skin blistering and hyperpigmentation (phytophotodermatitis). *Ferula marmarica* Asch. & Taub. ex Asch. & Schweinf is the source of ammoniac of Cyrenaica gum, a traditional medicine from Northern Africa.

Araliaceae: Ginseng Family

There are about 1500 species in Araliaceae; ginseng (*Panax* spp.) is the most notable from an ethnobotanical perspective. There are 11 species of *Panax*, which are found mostly in cooler temperate regions. *Panax vietnamensis* Ha & Grushv. grows in higher elevations of Vietnam, but this is the southernmost range for the genus. Most commercially available ginseng is either Asian (*P. ginseng* C.A.Mey., Figure 1.64) or American (*P. quinquefolius* L.) ginseng. They are both marketed for their adaptogenic properties, providing long life and vitality. Siberian ginseng [*Eleutherococcus senticosus* (Rupr. & Maxim.) Maxim.] is also in Araliaceae, but distantly related to the "true" ginsengs. The Asian (or Korean) species has a 5000 year history of medicinal use and is the most widely studied of all the species. Native Americans used the American species long before European settlers arrived. Traditionally, the rhizome is either peeled and dried (white ginseng), or steamed and dried without peeling (red ginseng). The root could be taken as a solid oral dose, liquid dose (extract), or prepared as a tea. The purported active principles are ginsenosides, triterpene saponins that have been the subject of intensive research for decades. There are over 150 ginsenosides isolated from every part of the plant. They are classified as dammarane- and oleanane-type based on the aglycone structure. Ginsenoside Rf is found in *P. ginseng* but not in *P. quinquefolius*, while 24(R)-pseudo-ginsenoside F_{11} is found in *P. quinquefolius* but only found in trace amounts in *P. ginseng*. Therefore, a high ratio of ginsenoside Rf to 24(R)-pseudo-ginsenoside F_{11} can be used to distinguish *P. ginseng* from *P. quinquefolius* after processing. Siberian

Figure 1.64 Ginseng (*Panax ginseng*) farm in China. © 2007 Isidor Byeongdeok Yu.

ginseng contains the structurally diverse eleutherosides instead of ginsenosides and can easily be distinguished from *Panax* spp. Eleutherosides have coumarin, lignan, phenyl-propanoid, polysaccharide, sterol, triterpene, and a few other miscellaneous structures.

References and Additional Reading

The Angiosperm Phylogeny Group (2009) An update of the Angiosperm Phylogeny Group classification for the orders and families of flowering plants: APG III. *Botanical Journal of the Linnean Society*, **161**, 105–121.

Bardil, A., Dantas de Almeida, J., Combes, M.C., Lashermes, P., and Bertrand, B. (2011) Genomic expression dominance in the natural allopolyploid *Coffea arabica* is massively affected by growth temperature. *New Phytologist*, **192**, 760–774.

Bechtold, T., and Mussak, R. (eds) (2009) *Handbook of Natural Colorants*. John Wiley & Sons, Ltd, Chichester.

Bemis, J., Curtis, L., Weber, C., and Berry, J. (1978) The feral buffalo gourd. *Cucurbita foeti. Economic Botany*, **1**, 87–95.

Borlinghaus, J., Albrecht, F., Gruhlke, M.C., Nwachukwu, I.D., and Slusarenko, A.J. (2014) Allicin: Chemistry and biological properties. *Molecules*, **19** (8), 12591–12618.

Bolsinger, C.L., and Jaramillo, A.E. (1990) *Taxus brevifolia* Nutt. Pacific yew, in *Silvics of North America Volume 1, Conifers*. (eds R.M. Burns and B.H. Honkala), Agriculture Handbook 654. United States Department of Agriculture Forest Service, Washington, DC.

Burns, R.M., and Honkala, B.H. (1990) *Silvics of North America Volume 1, Conifers*. Agriculture Handbook 654. United States Department of Agriculture Forest Service, Washington, DC.

Ceunen, S., and Geuns, J.M. (2013) Steviol glycosides: Chemical diversity, metabolism, and function. *Journal of Natural Products*, **76** (6), 1201–1228.

Christensen, L.P. (2009) Ginsenosides chemistry, biosynthesis, analysis, and potential health effects. *Advances in Food and Nutrition Research*, **55**, 1–99.

Cole, R.J., Dorner, J.W., Lansden, J.A. *et al.* (1977) Paspalum staggers: Isolation and identification of tremorgenic metabolites from sclerotia of *Claviceps paspali*. *Journal of Agricultural and Food Chemistry*, **25** (5), 1197–1201.

Cook, D., Gardner, D.R., and Pfister, J.A. (2014) Swainsonine-containing plants and their relationship to endophytic fungi. *Journal of Agricultural and Food Chemistry*, **62** (30), 7326–7334.

Dampc, A., and Luczkiewicz, M. (2015) Labrador tea – the aromatic beverage and spice: A review of origin, processing and safety. *Journal of Agricultural and Food Chemistry*, **95** (8), 1577–1583.

Eich, E. (ed) (2008). *Solanaceae and Convolvulaceae: Secondary Metabolites: Biosynthesis Secondary Metabolites Biosynthesis, Chemotaxonomy, Biological and Economic Significance (A Handbook)*. Springer, New York.

Finn, C.E., Retamales, J.B., Lobos, G.A., Hancock, J.F. (2013) The Chilean Strawberry (*Fragaria chiloensis*): Over 1000 Years of Domestication. *Hortscience*, **48** (4), 418–421.

Food and Agriculture Organization of the United Nations (2013). FAOSTAT Database Collections. Food and Agriculture Organization of the United Nations, Rome, http://faostat.fao.org (accessed 14 September, 2015).

Fraga, B.M. (1994) The trachylobane diterpenes. *Phytochemical Analysis*, **5**, 49–56.

Gaillard, Y., Krishnamoorthy, A., and Bevalota, F. (2004) *Cerbera odollam*: A "suicide tree" and cause of death in the state of Kerala, India. *Journal of Ethnopharmacology*, **95** (2–3), 123–126.

Gard, M. (2010) The toxicity of extracts of *Tephrosa virginiana* (Fabaceae) in Oklahoma. *Oklahoma Native Plant Record*, **10**, 54–64.

Gao, Q.H., Wu, C.S., and Wang, M. (2013) The jujube (*Ziziphus jujuba* Mill.) fruit: A review of current knowledge of fruit composition and health benefits. *Journal of Agricultural and Food Chemistry*, **61** (14), 3351–3363.

Giacosa, A., Guido, D., Grassi, M. *et al.* (2015a) The effect of ginger (*Zingiber officinalis*) and artichoke (*Cynara cardunculus*) extract supplementation on functional dyspepsia: A randomised, double-Blind, and placebo-controlled clinical trial. *Evidence Based Complementary and Alternative Medicine*, **2015**, 915087.

Giacosa, A., Morazzoni, P., Bombardelli, E., Riva, A., Bianchi Porro, G., and Rondanelli, M. (2015b) Can nausea and vomiting be treated with ginger extract? *European Review for Medical and Pharmacological Sciences*, **19** (7), 1291–1296.

Ginsberg, J. (2003) *Discovery of Camptothecin and Taxol*. American Chemical Society, Washington, DC.

Grobosch, T., Schwarze, B., Stoecklein, D., and Binscheck, T. (2012) Fatal poisoning with *Taxus baccata*: Quantification of paclitaxel (taxol A), 10-deacetyltaxol, baccatin III, 10-deacetylbaccatin III, cephalomannine (taxol B), and 3,5-dimethoxyphenol in body fluids by liquid chromatography-tandem mass spectrometry. *Journal of Analytical Toxicology*, **36** (1), 36–43.

Hasan, C.M., Healey, T.M., and Waterman, P.G. (1982) Kolavane and kaurane diterpenes from the stem bark of *Xylopia aethiopica*. *Phytochemistry*, **21** (6), 1365–1368.

Hughes Jr., C.L. (1988) Phytochemical mimicry of reproductive hormones and modulation of herbivore fertility by phytoestrogens. *Environmental Health Perspectives*, **78**, 171–174.

Ilic, N.M., Dey, M., Poulev, A.A., Logendra, S., Kuhn, P.E., and Raskin, I. (2014) Anti-inflammatory activity of grains of paradise (*Aframomum melegueta* Schum) extract. *Journal of Agricultural and Food Chemistry*, **62** (43), 10452–10457.

Janick, J., and Moore, J.N. (eds) (1996) *Fruit Breeding Volume I. Tree and Tropical Fruits*. John Wiley & Sons, Ltd, Chichester.

Katzer, G. (2015) Gernot Katzer's Spice Pages, http://gernot-katzers-spice-pages.com/engl/ Xylo_aet.html (accessed 17 December 2015).

Kim, Y., Goodner, K.L., Park, J.-D., Choi, J., and Talcott, S.T. (2011) Changes in antioxidant phytochemicals and volatile composition of *Camellia sinensis* by oxidation during tea fermentation. *Food Chemistry*, **129**, 1331–42.

Kiple, K.F., Ornelas, K.C. (eds) (2000) *Cambridge World History of Food*. Cambridge University Press, Cambridge.

Kumari, S., Pundhir, S., Priya, P. *et al.* (2014) EssOilDB: A database of essential oils reflecting terpene composition and variability in the plant kingdom, http://nipgr.res.in/ Essoildb/about.html (accessed April 8, 2016).

Latz, P. (1995) *Bushfires & Bushtucker. Aboriginal Plant Use in Central Australia*. IAD Press, Alice Springs.

Lee, E.K., Cibrian-Jaramillo, A., Kolokotronis, S.-O. *et al.* (2011) A functional phylogenomic view of the seed plants. *PLoS Genetics*, **7** (12), available from, http:// journals.plos.org/plosgenetics/article?id=10.1371/journal.pgen.1002411

Lee, J., Dossett, M., and Finn, C.E. (2014) Mistaken identity: Clarification of *Rubus coreanus* Miquel (Bokbunja). *Molecules*, **19**, 10524–10533.

Liu, Y., Heying, E., and Tanumihardjo, S.A. (2012) History, global distribution, and nutritional importance of citrus fruits. *Comprehensive Reviews in Food Science and Food Safety*, **11** (6), 530–545.

Markert, A., Steffan, N., Ploss, K. *et al.* (2008) Biosynthesis and accumulation of ergoline alkaloids in a mutualistic association between *Ipomoea asarifolia* (Convolvulaceae) and a clavicipitalean fungus. *Plant Physiology*, **147** (1), 296–305.

Martin, M.N. (1991) The latex of *Hevea brasiliensis* contains high levels of both chitinases and chitinases/lysozymes. *Plant Physiology*, **95** (2), 469–476.

McGovern, P.E. (2003) Georgia as homeland of winemaking and viticulture. In *National Treasures of Georgia*, Soltes, O. Z. (ed), Philip Wilson, London, pp. 58–59.

Moerman, D. (2015) University of Michigan Dearborn Native American Ethnobotany database, http://herb.umd.umich.edu (accessed October 24, 2015).

Morton, J.F. (2013) *Fruits of Warm Climates*. Echo Point Books & Media, Brattleboro.

Nambisan, B. (2011) Strategies for elimination of cyanogens from cassava for reducing toxicity and improving food safety. *Food and Chemical Toxicology*, **49** (3), 690–693.

Orwa, C., Mutua, A., Kindt, R., Jamnadass, R., and Anthony, S. (2009) Agroforestree Database: A Tree Reference and Selection Guide, version 4.0. World Agroforestry Centre, Kenya, http://www.worldagroforestry.org/resources/databases/agroforestree (accessed 14 October, 2014).

Ruprich, J., Rehurkova, I., Boon, P.E. *et al.* (2009) Probabilistic modelling of exposure doses and implications for health risk characterization: Glycoalkaloids from potatoes. *Food and Chemical Toxicology*, **47** (12), 2899–2905.

Singh, R.J. (2011). *Genetic Resources, Chromosome Engineering, and Crop Improvement: Medicinal Plants*, vol 6. CRC Press, Boca Raton.

Smith III, A.B., Kingery-Wood, J., Leenay T.L., Nolen E.G., and Sunazuka, T. (1992) Indole diterpene synthetic studies. 8. The total synthesis of (+)-paspalicine and (+)-paspalinine. *Journal of the American Chemical Society*, **114** (4), 1438–1449.

Neuwinger, H.D. (1996) *African Ethnobotany: Poisons and Drugs: Chemistry, Pharmacology, Toxicology*. CRC Press, Boca Raton.

Pridgeon, A.M., Cribb, P.J., Chase, M.W., and Rasmussen, F.N. (eds) (2014) *Genera Orchidacearum: Epidendroideae*, vol 6, part 3. Oxford University Press, Oxford.

United States Department of Justice Drug Enforcement Administration Office of Intelligence (1992) *Opium Poppy' Cultivation and Heroin Processing in Southeast Asia Heroin Manufacture*. DEA publication 92004, https://www.ncjrs.gov/pdffiles1/Digitization/141189NCJRS.pdf (accessed 21 December, 2014).

United States Department of Agriculture Forest Service (2004) Fire Effects Information System (FEIS), http://www.feis-crs.org/beta/(accessed 16 September, 2014).

University of California, Riverside (2015) Plants of Economic or Asthetic Importance, http://www.faculty.ucr.edu/~legneref/botany/botindx2.htm (accessed 2 October, 2014).

Wang, Y.H., Avula, B., Nanayakkara, N.P.D., Zhao, J., and Khan, I.A. (2013) Cassia cinnamon as a source of coumarin in cinnamon-flavored food and food supplements in the United States. *Journal of Agricultural and Food Chemistry*, **61** (18), 4470–4476.

Watson, J.T., Jones, R.C., Siston, A.M. *et al.* (2005) Outbreak of food-borne illness associated with plant material containing raphides. *Clinical Toxicology*, **43** (1), 17–21.

Watson, L., and Dallwitz, M.J. (2015) The Families of Flowering Plants: Descriptions, Illustrations, Identification, and Information Retrieval, http://delta-intkey.com (accessed 21 August 2014).

Wil, A. (2005) *Coffee: A Dark History*. W. W. Norton & Company, New York.

Plant Anatomy and Architecture: The Highlights

Identifying Plants: Basic Anatomy

There are several excellent plant biology textbooks that thoroughly cover basic plant structures, metabolism, genetics, growth, and life cycles such as Stern's *Introductory Plant Biology, Botany: An Introduction To Plant Biology* by Mauseth, or *Plant Biology* by Graham *et al*. This section will primarily focus on structures relevant for ethnobotanical plant identification.

Plants are complex organisms, made up of different cell types, tissues, vegetative organs, and reproductive organs. Leaves, roots, and stems are vegetative organs. Angiosperm reproductive organs are the flower, fruit, and seed; gymnosperms have seeds and cones but no fruit or flowers. Lower vascular plants such as pteridophytes have an entirely different system of reproduction. They reproduce by spores and have two different generations, the gametophyte and the sporophyte.

Angiosperms and gymnosperms may occur as woody or herbaceous perennials. A **woody perennial** grows for more than two years and has persistent woody stems and roots. A **herbaceous perennial** (Figure 1.65) does not have persistent woody stems, but lives for more than two years. Flowering plants may occur as perennials, annuals, or biennials. An **annual** plant completes its lifecycle in one season, from flowering to seed, then death. A **biennial** such as *Digitalis purpurea* L. (Figure 1.66) completes its lifecycle in two years. In the first year it grows only vegetative tissue. The second year it flowers, produces seeds, and dies.

Figure 1.65 Examples of herbaceous perennials, *Echinacea purpurea* (L.) Moench (foreground, Asteraceae), *Rudbeckia hirta* L. (Asteraceae), and *Crocosmia* sp. (Iridaceae)

Vegetative Tissues

Roots

Roots are water-absorbing organs that contain vascular tissue. They have three main functions: water and nutrient absorption, anchoring the plant to the ground, and storing food. During seed

Figure 1.66 **Example of a biennial,** *Digitalis purpurea* L. (Plantaginaceae).

germination, the radicle becomes the primary root of the plant. This primary root may become a thick central root called a **taproot**. A taproot may or may not have lateral roots. Examples of tap-roots include carrot (*Daucus carota* L.), beets (*Beta vulgaris* L.), and radish (*Raphanus sativus* L.). The primary root may instead give rise to many thin but tough **fiberous roots**. Pteridophytes lack vascular tissue and therefore do not have true roots. Instead, they have thread-like **rhizoids**. **Root hairs** in vascular plants are similar to rhizoids. They are out-growths of root epidermal cells that increase the surface area of the root, thereby increasing the uptake of water and nutrients. Root hairs should not be confused with **mycorrhizae**. In a mycor-rhizal association, a fungus associates with the roots of a vascular plant, form-ing mycelium that also aids in the uptake of water and nutrients.

Most plants have fibrous roots, a tap-root, or some combination of the two. Some plants have specialized roots that serve additional functions for the plant. **Aerial roots** (Figure 1.67) grow above the ground, as the name implies. They provide support, enable climbing plants and epiphytes to attach themselves to surfaces like rocks and tree bark, and may serve additional functions. **Pneumatophores** are porous, erect out-growths that aid in gas exchange for waterlogged roots. These can be seen in swamp and mangrove plants like black mangrove (*Avicennia germinans* L.). Although bald cypress [*Taxodium distichum* (L.) Rich.] trees have "knees" that look like pneumato-phores, there is currently no consensus about their true function. Parasitic plants like mistletoe (*Viscum album* L.) have **haustorial roots** that can attach to and penetrate host plant tissues for the purpose of absorbing nutrients. **Prop roots** begin as aerial roots growing from the base of a tree. They eventually grow into the soil and provide support. Corn (*Zea mays* L.) prop roots keep the cornstalk upright during windstorms. Tropical figs like banyan trees (*Ficus bengalensis* L.) produce large and numerous prop roots that look like many trunks (Figure 1.68). **Buttress roots** also provide support like prop roots, but are vertically flattened, like fins on the bottom of tree trunks.

Most roots store some level of food and water for the plant; however, some plants are adapted for storing large quantities of food or water. Members of Cucurbitaceae that live in arid regions such as the Southwest region of North America have large water storage roots. One *Cucurbita foetidissima* Kunth had a root weighing 72.12 kg. Members of the genus *Marah* also have huge storage roots, weighing around 30 kg. **Tuberous roots** are lateral roots that produce numerous parenchyma cells. These swollen areas store considerable amounts of starch and other carbohydrates. Sweet potato [*Ipomoea batatas* (L.) Lam.] is an example of tuberous roots.

Figure 1.67 Ariel roots extend down from tropical rainforest trees in Puerto Rico.

Figure 1.68 "Strangler fig" (*Ficus aurea* Nutt.) prop roots look like multiple trunks.

Roots that develop from something other than the primary root are called **adventitious roots**. They most often arise from part of the stem. Stem or leaf cuttings may be placed in soil to form adventitious roots as a means of plant propagation.

The Shoot System: Stems and Leaves

The shoot system of a typical flowering plant consists of the stem and leaves, along with buds, flowers, and fruits. A stem is the main ascending axis or stalk of the plant. In angiosperms, it arises from the epicotyl and is divided into nodes and internodes. The nodes bear buds, leaves, and/or flowers, and the internode is the part of the stem between the nodes. Stems have several important functions including structural support for the leaves, flowers, and fruit; transport of water, nutrients, and carbohydrates from photosynthesis; storage of water and starch; and production of new plant tissues.

Stems can grow in many directions. A **decumbent** stem grows in a reclining fashion near the ground; a **prostrate** stem grows across the ground; and a **creeping** stem grows on the ground and forms adventitious roots at the nodes. An **erect** stem grows perpendicular to the ground, while an **ascending** stem grows in a slanted upward direction. Scandent, or climbing plants, use **tendrils**, small twisting appendages that attach to other plants or objects.

An entire book could be written about the structure and function of **wood**. Wood is made up of secondary xylem. Variations in the chemical composition lead to a wide variation in the density and color of wood. When viewed in cross-section, trees growing in temperate climates have concentric annual rings, each representing one season of growth. On the basis of annual rings, bristlecone pines (*Pinus aristata* Engelm. *P. longaeva* D.K. Bailey, and *P. balfouriana* Balf.) and redwoods (Cupressaceae subfamily Sequoioideae) are some of the oldest trees on the planet, growing for thousands of years. The larger cells on the inner part of the annual ring that form in the springtime are called **springwood** or earlywood. Smaller cells that form later in the season are called **summerwood** or latewood. Different trees have different distribution patterns of large-diameter vessels. Some are found primarily in the springwood, while others are found throughout spring and summerwood. Often, wood in the center of the tree is darker than the periphery. The dark-colored wood is called **heartwood**, and the lighter wood on the periphery is called **sapwood** (Figure 1.69).

Sapwood contains functioning xylem cells that actively transport water and dissolved minerals. Heartwood xylem cells are clogged with terpene resins and gums that protect the tree from decay, pathogens, and insect attack. The resins are produced by the vascular cambium and transported to the center of the trunk or stem by medullary rays. The resins enter axial paratracheal parenchyma cells. As the tree ages, the parenchymal cells form thyloses. They grow into the vessels, forming a bubble, and deposit their resinous contents into dead xylem cells in the heartwood. Some gymnosperms have **resin ducts** or canals to produce and transport resins. The resins are actually produced by epithelial cells lining the ducts. Many types of plant resins are economically important. The volatile liquid fraction is called **turpentine**. **Rosin** is made from resin that has been heated to remove the turpentine fraction and hardened (see section titled "Taxonomy: Plant Families," Figure 1.10).

Laticifers are narrow secretory cells often found in the secondary phloem, but they do occur in other parts of the plant as well. They form a network of living vessels that produce latex. Latex is a white or colored mixture of gums, oils, proteins, sugars, and

Figure 1.69 Cross section of red cedar (*Juniperus virginiana* L.) showing the vibrant red heartwood and lighter sapwood. According to the growth rings, this tree was 35 years old.

isoprenes. When exposed to air, isoprenes polymerize to become rubber or latex (see section titled "Taxonomy: Plant Families," Figure 1.32). Euphorbiaceae contains many plants that have laticifers including rubber tree (*Hevea brasiliensis* Müll.Arg.), pointsettia (*Euphorbia pulcherrima* Willd. ex Klotzsch), crown of thorns (*Euphorbia milii* Des Moul.), cassava (*Manihot esculenta* Crantz), and castor bean (*Ricinus communis* L.). Rubber is harvested from rubber trees by cutting into the laticifer vessels through the bark. The rubber flows out of the trunk and into a bucket attached to the tree. A few other plants produce latex of economic value. The latex from opium poppy (*Papaver somniferum* L.) seedpods contain powerful alkaloids including morphine. *Manilkara* sp. trees produce chicle latex that is still used as a natural chewing gum.

Bark is the protective covering over the wood of a tree. It often contains secondary compounds to protect it from pathogens and herbivory including tannins, alkaloids, and pigments. The texture of bark is often used for identification purposes. Bark may be ringed, scaly, peeling, smooth, fissured, or cracked. Some bark remains green and photosynthetic, like palo verde trees (*Parkinsonia* sp.).

Sclerenchyma tissue fortifies plant stems and organs. It is composed of fiber or sclereid cells that have lignin-fortified secondary cell walls. **Sclereids** often occur in a sheet as in seed coats, but can occur individually as stone cells in pear fruit (*Pyrus* sp.). **Fiber cells** are elongated cells that occur in various formations, end to end, in aggregates, or individually. Angiosperm hardwoods contain dense clusters of fiber cells highly fortified with lignin. But gymnosperm wood lacks the same dense clusters, thereby making it a softer, lighter wood that is more easily cut and damaged. Fiber cells are also the basis of paper, cordage, and many textiles.

Some plant families have **cladodes** (also called phylloclades or cladophylls), stems that look more like leaves, as they are flattened and photosynthetic (Figure 1.70). Some well-known examples include members of Asparagaceae *(Asparagus)*, Ruscaceae (*Ruscus aculeatus*, Butcher's Broom), Euphorbiaceae (*Phyllanthus*), and Cactaceae (*Opuntia* spp., Prickly Pear; *Schlumbergera truncata*, Christmas Cactus). Cactaceae have thick cladodes covered in helically arranged nodes. A narrow leaf and short shoot called an areole grow at each node. The leaf falls off, while the areoles produce leaf spines in the form of detachable **glochids**, a cluster of rigid spines, or both. Areoles are a distinguishing feature of the family Cactaceae.

Most stems are conspicuous aboveground structures; however, there are several types of modified stems that creep along the ground or grow underground. **Stolens**, also called runners, are creeping aboveground stems produced by mother plants as a means of vegetative reproduction. Stolens can form adventitious roots at the nodes and grow daughter plants. Strawberries (*Fragaria*) are a classic example of a stoleniferous plant. **Rhizomes** are horizontal underground stems with reduced scale-like leaves. Many members of the *Iris* genus are rhizomatous. Other fleshy underground stems include the following: **bulb**, a short basal underground stem surrounded by fleshy leaves; **tuber**, an enlarged fleshy root tip; and **corm**, a fleshy bulb-like stem without fleshy leaves covered instead with papery thin dry leaves. Some root crops are actually modified, underground stems. For example, onions (*Allium cepa* L.) are bulbs, Irish potatoes (*Solanum tuberosum* L.) and the Andean crop oca (*Oxalis tuberosa* Molina) are tubers (Figure 1.71), taro [*Colocasia esculenta* (L.) Schott] is a corm, and ginger (*Zingiber officinale* Roscoe) is a rhizome.

Figure 1.70 Dragon fruit (*Hylocereus* sp.) and many other species in Cactaceae are excellent examples of cladodes.

Figure 1.71 Women selling the Andean tuber oca (*Oxalis tuberosa* Molina, foreground) at a market in the Urubamba Valley, Peru.

The surfaces of stems often have features that may be useful for identification. **Lenticels** are raised patches of parenchymatous tissue of various shapes that may be found on bark, tubers, or fruit. Lenticels provide gas exchange. Many members of Rosaceae have conspicuous lenticels. Roses (*Rosa* sp.) have **prickles**, which are sharp, pointed extensions of the cortex and epidermis. **Thorns** are modified, sharp stems that arise from a bud and can produce a leaf or be branched. Hawthorn (*Crataegus* spp.) produces true thorns.

Leaves are usually flat, thin, and phyotosynthetic. They come in an astounding array of shapes, sizes, and colors, making them ideal features for plant identification. Leaves grow on stems in alternate, opposite, whorled, or rosette arrangements (Figure 1.72). They can be simple, with one leaf blade per petiole, or compound, with many leaflets per petiole. Compound leaflets attach to the petiole without individual bud nodes. They can be arranged pinnately, palmately, or double pinnately (Figure 1.72). The shape of the leaf, leaf apex, leaf base, and leaf margin are all important for plant identification. The most common leaf shapes are depicted in Figure 1.73. Leaves are often temporary in nature, usually related to seasonal changes in temperature, day length, or rainfall.

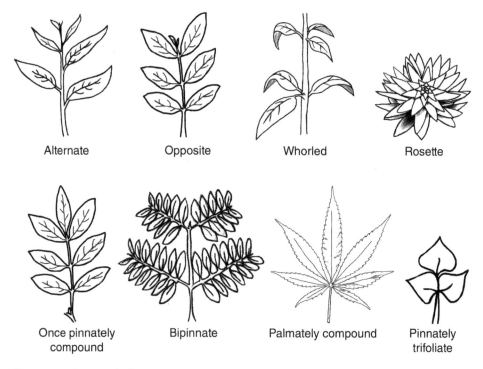

Alternate Opposite Whorled Rosette

Once pinnately compound Bipinnate Palmately compound Pinnately trifoliate

Figure 1.72 Common leaf arrangements.

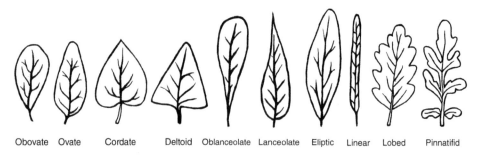

Obovate Ovate Cordate Deltoid Oblanceolate Lanceolate Eliptic Linear Lobed Pinnatifid

Figure 1.73 Common types of leaf margins.

Deciduous leaves fall at the end of the growing season, while evergreen leaves are persistent. An excellent example is the mixed temperate forests of northern latitudes. In autumn, most woody dicots form a zone of abscission at the base of the petiole. The leaf moves its remaining nutrients into the stem, its color fades, and it falls to the ground. Gymnosperms, on the other hand, do not undergo the same process. They remain "evergreen" through all seasons.

There are several types of modified leaves that have different roles compared to the typical leaf. A **spine** is a modified leaf, stipule (a projection at the base of a leaf), or leaf part that typically contains vascular tissue made of cells with hard cell walls. Members of Cactaceae produce spines. They should not be confused with thorns or prickles, which are modified stems. **Bulb scales** are actually thick storage leaves; the bulb itself is a modified stem.

Flowers

Flowers are quite intricate in design and therefore may be described in several different ways including symmetry, reproductive capability, arrangement or position of floral parts, and their arrangement on the plant. For a full understanding of flower morphology, consult Jones and Luchsinger, 1986. The generalized form of a flower is depicted in Figure 1.74. A **complete flower** contains sepals, petals, stamens, and carpel (ovary, ovules, and pistil). If it lacks one of these structures, it is an **incomplete flower**. Most plants have bisexual or perfect flowers that contain both stamen and pistil. A unisexual or imperfect flower has one or the other and is referred to as **staminate** or **pistillate**. A **monoecious** plant has staminate and pistillate flowers on the same plant. Examples include oaks (*Quercus* spp.), pines (*Pinus* spp.), and corn (*Zea mays* L.). While most unisexual plants are monoecious, there are some **dioecious** plants with staminate and pistillate flowers occurring on separate plants. Examples include holly (*Ilex* spp.), juniper (*Juniperus* spp.), mulberry (*Morus* spp.), willow (*Salix* spp.), and ginkgo (*Ginkgo biloba* L.).

If a flower is bisected through the perianth, its type of symmetry can be determined. Most flowers have regular, radial symmetry; they can be bisected in different ways to create mirror images. These flowers are called actinomorphic. An **actinomorphic** flower is often star shaped. **Zygomorphic** or irregular flowers only have a single line of symmetry, creating two mirror images. Orchids (Orchidaceae; Figure 1.75) and snapdragons are classic examples of zygomorphic flowers.

Inflorescence is the arrangement of flowers on a plant. Some plants have single, **solitary** flowers. But others have aggregations of flowers that may be simple or difficult to characterize. Figure 1.76 shows some common types of inflorescences.

Fruit and Seeds

Fruit and seeds are an important source of food for humans and animals alike. A fruit is the mature ovary of a flower containing seeds. Seeds are the mature ovules. Fruits protect the seeds as well as aiding in their dispersal. Like leaves and flowers, they are important for plant classification.

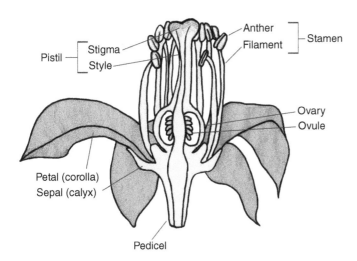

Figure 1.74 Diagram of a complete flower.

Figure 1.75 *Phalaenopsis* sp. orchid, an example of a zygomorphic flower.

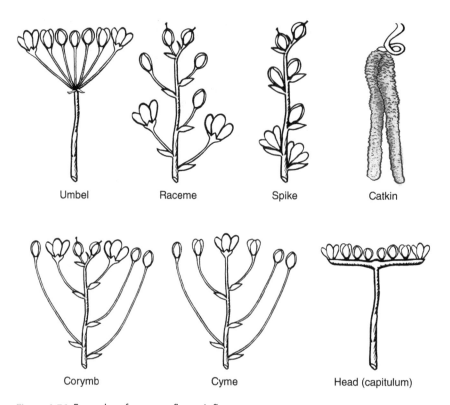

Umbel   Raceme   Spike   Catkin

Corymb   Cyme   Head (capitulum)

Figure 1.76 Examples of common flower inflorescences.

Vernacular use of the term *fruit* typically refers to a fleshy, sweet, edible structure like an apple (*Malus* spp.), mango (*Mangifera* spp.), or strawberry (*Fragaria* spp.). Other fruits like cucumbers (*Cucumis sativus* L.), eggplant (*Solanum melongena* L.), peppers (*Capsicum* spp.), and tomatoes (*Solanum lycopersicum* L.) are often called vegetables.

In fact, in 1893 the United States Supreme Court ruled that tomato is a *vegetable* for customs purposes, favoring the common use of the term over botanical correctness. The EU classifies rhubarb stalks and carrots to be fruits for the purposes of jam making, so the confusion over botanical classification of economic plant products goes both ways. Botanically speaking, any mature ovary is a fruit including seedless or **partheno-carpic** fruit and cereal grains like wheat (*Triticum* spp.) or barley (*Hordeum vulgare* L.). A fruit may also contain accessory organs that developed along with the ovary. A mature fruit typically has three regions collectively known as the **pericarp**. The **exocarp** is the outer skin layer, while the **endocarp** is the area surrounding the seed chamber. This may be papery like an apple core or hard like the pit of stone fruit. The **mesocarp** is the area between the exocarp and endocarp, which is often fleshy. Dry fruit like grain or legumes do not have a fleshy mesocarp.

It should be noted, that there is considerable disagreement among fruit classification scholars regarding the definition of fruit categories and which fruits fall into which category. The often-cited 1994 publication by Richard Spjut, *A Systematic Treatment of Fruit Types*, is extensive and well respected. However, it uses complex terminology not practical for the average botanist. For example, the term *berry* is replaced with *bacca*. Bananas and papayas are classified as pepos rather than berries, and coconuts and almonds are called "nucleaniums." The following is a simplified fruit classification schematic based on works by Camelbeke (2008) and Jones and Luchsinger (1986). Fruit can be classified by several different features including flower structure and any accessory organs involved in development; type of pericarp; whether the pericarp splits open at maturity or not and the manner in which it splits; number of seeds; type and number of ovaries; and number of carpels. Using these features, fruit can be divided into four main groups:

1) Dry, indehiscent, one-seeded fruit

   Achene: Fruit with one carpel and one seed. Seed nearly fills firm pericarp but does not adhere to it; pericarp easily separated from seed coat, for example, sunflower (*Helianthus* spp.).

   Grain (caryopsis): Pericarp is fused with the seed and often enclosed in a chaff at maturity, for example, grass seed (Poaceae; Figure 1.77).

   Nuts and nutlets: Fruit developed from a pistil with more than one carpel, having a hard woody coat, for example, oak acorns (*Quercus* spp.).

   Samara: An achene or nutlet with a wing, for example, one-seeded elm (*Ulmus* spp.) or two-seeded maple (*Acer* spp.).

   Schizocarp: A compound ovary that splits along the midline when mature, forming two indehiscent halves with one seed each, called mericarps, for example, Apiaceae.

   Utricle: One-seeded fruit with thin, bladder-like wall, for example, *Amaranthus* spp.

2) Dry, dehiscent, with many seeds

   Capsule: Fruit originating from two or more carpels, with various forms of dehiscence including:

   a) Valves (loculicidal), dehiscent lengthwise down the middle of the carpels (e.g., *Iris*).

   b) Pores, dehiscent by pores on top (e.g., Papaveraceae).

   c) Lid (pyxis), dehiscent in a circumscissile manner (e.g., *Plantago*).

   d) Septicidal, dehiscent lengthwise at the junction of the carpels (e.g., *Yucca*).

Figure 1.77 Grasses produce caryopses, dry indehiscent fruit in which the pericarp is fused with the seed and enclosed in a chaff at maturity.

Follicle: Fruit develops from a single pistil, dehiscent along one margin (e.g., *Delphinium*).

Legume: One-locular fruit splits along two seams (e.g., Fabaceae).

Silique and Silicle: Two locular fruit splits along two seams; siliques are elongated, and silicles are less than three times long as they are wide (e.g., Brassicaceae).

3) Fleshy, simple

Berry: Soft fleshy pericarp, usually from a compound ovary with more than one seed, for example, tomato (*Solanum lycopersicum* L.), and grapes (*Vitis* spp.). Some berries are from inferior ovaries and have remnants of flower parts at their tip, for example, blueberries and cranberries (*Vaccinium* spp.), pomegranates (*Punica granatum* L.), and bananas (*Musa* spp.). Bananas are berries even though their exocarp is tough and leathery and they are parthenocarpic. Avocados (*Persea Americana* Mill.) are berries with a single seed. The endocarp is the fleshy white covering over the hard seed coat. Pomegranates (*Punica granatum* L.) are berries with seeds covered by a fleshy, juicy aril.

Drupe: Fleshy mesocarp with hard stony endocarp around a single seed, for example, plums, apricots, peaches, cherries, and almonds (*Prunus* spp.), olives (*Olea europaea* L.), mangos (*Mangifera* spp.), coffee (*Coffea arabica* L.), and coconuts (*Cocos nucifera* L.). Both coconuts and almonds have a fibrous husk (mesocarp and exocarp) that is removed before marketing. The endocarp is cracked open to reveal the edible seed inside. Some authors recognize multi-seeded drupes (up to five stony seeds) and move fruit like *Vitis*, *Ilex*, *Sambucus*, and *Coffea* into this category.

Hesperidium: A berry with a tough leathery skin, for example, citrus from Rutaceae.

Hip: Aggregation of achenes surrounded by accessory tissue, the receptacle, and hypanthium, for example, rose (*Rosa* spp.).

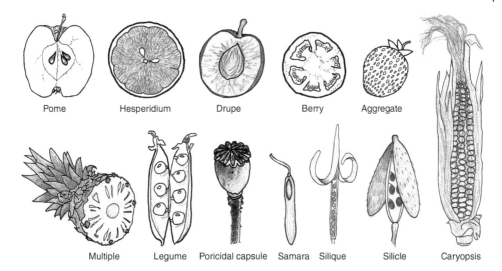

| Pome | Hesperidium | Drupe | Berry | Aggregate |

| Multiple | Legume | Poricidal capsule | Samara | Silique | Silicle | Caryopsis |

Figure 1.78 Common types of fruit. © 2016 M. T. Nguyen.

Pepo: One locular fruit developed from an inferior ovary; typically has a thick rind, for example, Cucurbitaceae.

Pomes: Simple fleshy fruit from compound inferior ovary with seeds encased in a papery wall, for example, apple (*Malus domestica* Borkh.). The mesocarp is derived from the enlarged floral tube that grows around the ovary.

1) Aggregate and Multiple ("False Fruits")

Aggregate: Cluster of fruits from a single flower with many pistils but common receptacle. In blackberry (*Rubus* spp.), the individual fruits are tiny drupelets, each with one hard seed inside. Strawberry (*Fragaria* spp.) fruits are tiny achenes embedded in the fleshy enlarged receptacle. A rose hip may also be called an aggregate fruit, since it consists of achenes surrounded by a fleshy receptacle.

Multiple: On a common axis, but derived from the ovaries of several flowers on one inflorescence that fuse together into a single, larger fruit. For example, mulberries (*Morus* spp.) are a cluster of drupelets; figs (*Ficus carica* L.) are a collection of drupelets inside a fleshy inflorescence; osage orange [*Maclura pomifera* (Raf.) C.K.Schneid.] is a collection of fused druplets that are often seedless; and pineapple [*Ananas comosus* (L.) Merr.] is a bunch of berries embedded in a fleshy stem, each with a jagged bract.

The most common types of fruit are illustrated in Figure 1.78.

References and Additional Reading

Camelbeke, K. (2008) A typology of fruits. *International Dendrology Society Yearbook*, http://www.dendrology.org/site/en/tree-info/-114-a-typology-of-fruits-koen-camelbeke-2008/(accessed 15 December, 2015).

Jones, S.B., and Luchsinger, A.E. (1986) *Plant Systematics*, 2nd edition. McGraw-Hill Publishing Co., New York.

Spjut, R.W. (1994) A systematic treatment of fruit types. *Memoirs of the New York Botanical Garden*, **70**, 1–182.

Keys for Plant Identification

Knowledge of plant anatomy is important for identifying plant species. Without it, a plant may be misidentified, with potentially serious consequences. Plant identification keys are useful tools if you have enough knowledge of plant anatomy to properly use them. A dichotomous key provides the user with two choices at each step, whereas a polyclave key may provide several choices at each step. Electronic polyclaves typically allow the user to input several known characteristics of the plant, thereby eliminating many species in the key. It may also provide the likelihood or probability that the remaining species are the correct choice and may prompt the user to input other characteristics to eliminate more species.

There are a few basic rules to creating a dichotomous key. The entries should begin with a couplet that has identical first words (e.g., Leaves opposite or Leaves alternate) but are contradictory statements. A leaf cannot be both alternate and opposite, so the decision is clear. To avoid confusion, do not have multiple entries in a row that begin with the same word, and do not use overlapping ranges of measurement (e.g., 1–5 mm or 4–8 mm). Negative statements (e.g., Leaves not opposite) should also be avoided. Couplets can be numbered, lettered, or a combination of both. Sometimes indented keys use no numbers or letters at all.

Below is a simple dichotomous key to identify genera of neotropical Juglandaceae (Milliken 2009):

> 1. Leaves opposite ... go to 2
> 1. Leaves alternate ... go to 3
>
> 2. Fruits with samaroid wings ... *Oreomunnea*
> 2. Fruits without samaroid wings ... *Alfaroa*
>
> 3. Perianth present in flower ... *Juglans*
> 3. Perianth absent in flower ... *Carya*

Although this key is only a few lines, it contains terms that require specialized knowledge of plant anatomy. If you do not know what "samaroid wings" or "perianth" are, the key is useless.

Field guides often contain keys and illustrations to aid in plant identification. They typically focus on a limited geographic region and are compact enough to carry in the field. A field guide allows the user to compare the unknown plant with known plants that grow in the region. Even with a good field guide and key, it is often difficult to identify a plant down to the species level, especially if it is not flowering or fruiting. If possible, a botanist will collect a specimen to take back to the lab for a more thorough examination.

Herbaria, Plant Collection, and Voucher Specimens

A **herbarium** (plural: herbaria) is a collection of dried plant specimens for research purposes. The main functions of a herbarium are to provide reference materials for identification of newly collected plants, serve as a resource for botanists and botany courses, documenting the presence of a species in a certain geographic region, resolving classification issues, and housing type and voucher specimens. A **type specimen** is the exact specimen on which the name of a taxon is based. They are highly valuable and may be housed separately or mounted onto different colored paper to avoid loss or damage. A **voucher specimen** serves as the basis for a scientific study. It is a reliable method to verify the exact identity of the plant used for the study. Without voucher specimens, if the identity of the plant material comes into question, there is no way to assure others the plant was properly identified in the first place.

A wise plant collector will not collect rare or endangered plants or dig up the only remaining plant in a given area. (S)he is mindful of local laws and obtains any required permits. Landowner permission should be obtained to collect plants on private land. Botanists typically identify a plant only after it is pressed and dried, but some prefer to identify material directly in the field, especially if the situation prohibits collection. To prepare a plant for deposition in a herbarium, plants are typically collected in the field and immediately pressed between newspaper using a field press constructed of wood, cardboard, blotting paper, newspaper, and strapping material. Sticky plants may be pressed between waxed paper. The arrangement on the newspaper is rather important; once a specimen is dry, it is difficult to rearrange. The plant should be arranged to show all the important features including lower and upper leaf, flower heads, and so on. Large plants can be folded or cut into sections. Bulky fruit can be cut in half; large cones can be tagged to be stored in a box. Plants should be dried as quickly as possible; forced air dryers are sometimes used (Figure 1.79).

Information about the plant's geographic location, GPS coordinates, date of collection, and collection number should be recorded at the time of collection in a field notebook or electronically. This information is required to prepare the label for the herbarium specimen sheet. The collection number should be written on the margin of the sheet of newspaper. Once the plant is properly identified, the name can be written alongside the collection number. Herbarium specimen labels are approximately 2.75 × 4.25 in. (7 × 11 cm) (Figure 1.80). They are arranged in various ways, but typically contain a heading with state or province, county or parish, country, and name of institution associated with the specimen. The scientific name (genus, species, author), specifics about the locality such as geographic features or distance from nearest town or landmark, and habitat details (soil, elevation, moisture, etc.) follow. The date of collection, name of the collector, and collection number are also important information. The label is placed at the bottom-right-hand corner of a 11.5 × 16.5 inch herbarium mounting sheet and glued fast (Figure 1.81). Glue is also applied to the back of the plant specimen, and it is arranged on the mounting sheet. Small weights may be placed on top of the plant until the glue is dry. Fragments of plant material such as seeds or pollen can be placed in a small folded pocket, which is also glued fast to the sheet. Once dry, the herbarium stamps the sheet with its name and assigns it an accession number. It is then

Figure 1.79 Plant material drying in a forced air drier at the New York Botanical Garden Herbarium.

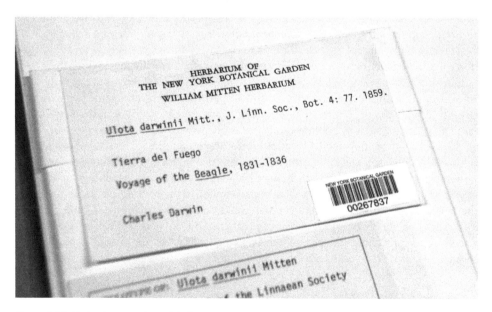

Figure 1.80 Herbarium specimen label.

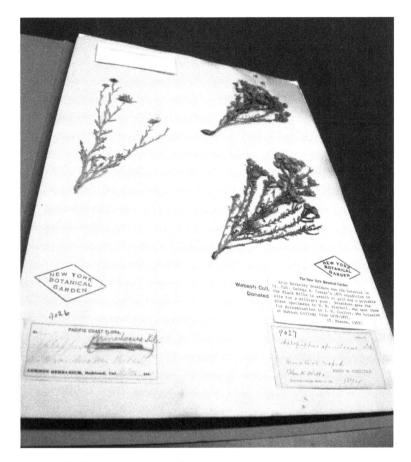

Figure 1.81 Herbarium specimen sheet.

filed in the herbarium cabinets that are typically organized alphabetically according to plant family, genus, and species. Some herbaria use numerical organization according to the Adolf Engler system of classification.

References and Additional Reading

Jones, S.B., and Luchsinger, A.E. (1986) *Plant Systematics*, 2nd edition. McGraw-Hill Publishing Co., New York.

Judd, W.S., Campbell, C.S., Kellog, E.A., Stevens, P.F., and Donoghue, M.J. (2015) *Plant Systematics: A Phylogenetic Approach*, 4th edition. Sinauer Associates, Inc., Sunderland.

Milliken, W. (2009) Neotropical Juglandaceae. In Neotropikey—Interactive Key and Information Resources for Flowering Plants of the Neotropics. Milliken, W., Klitgård, B., and Baracat, A. (eds), available at, http://www.kew.org/science/tropamerica/neotropikey/families/Juglandaceae.htm

2

Phytochemistry

D. M. Klaser Cheng

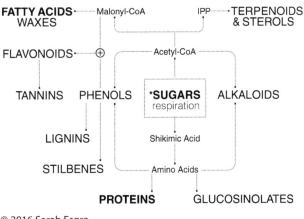

PLANT METABOLISM

© 2016 Sarah Esgro

Primary and Secondary Metabolites

Phytochemicals are separated into two classes: primary metabolites and secondary metabolites. Primary compounds have direct roles in plant growth and development, and structure. They are essential for processes such as photosynthesis, citric acid cycle, and the electron transport chain. Primary metabolites include intermediates of catabolic and anabolic pathways, which occur in all plants, have the same metabolic functions, and are necessary for plant survival. Compounds produced in these cycles include amino acids, sugars, and fatty acids. These small-molecular-weight compounds may assemble into long chains to form macromolecules of proteins, carbohydrates, lipids, and nucleic acids. Secondary metabolites are produced downstream from primary metabolic pathways including those of photosynthesis, the Calvin cycle, citric acid cycle

Ethnobotany: A Phytochemical Perspective, First Edition. Edited by B. M. Schmidt and D. M. Klaser Cheng.
© 2017 John Wiley & Sons Ltd. Published 2017 by John Wiley & Sons Ltd.

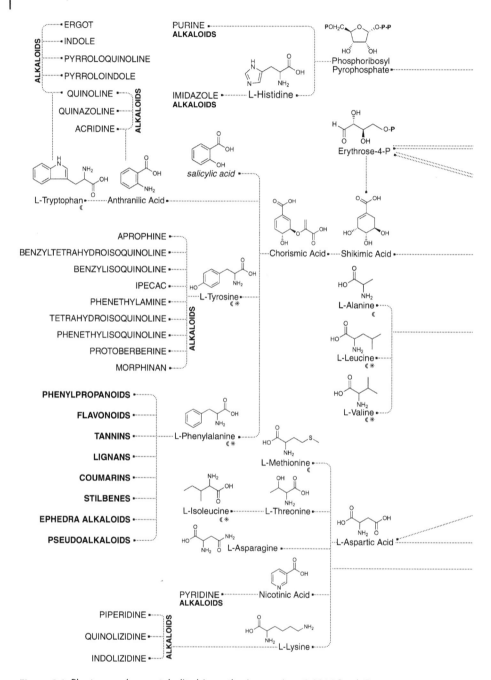

Figure 2.1 Plant secondary metabolite biosynthesis overview. © 2016 Sarah Esgro.

(aka. tricarboxylic acid cycle or Krebs cycle), glycolysis, and the pentose phosphate pathway (Figure 2.1).

Plant secondary metabolites were initially thought to be waste products, as their function for the plant was not known. However, many phytochemicals have now been shown to play pivotal roles in plant defense and resistance against pests, diseases, as

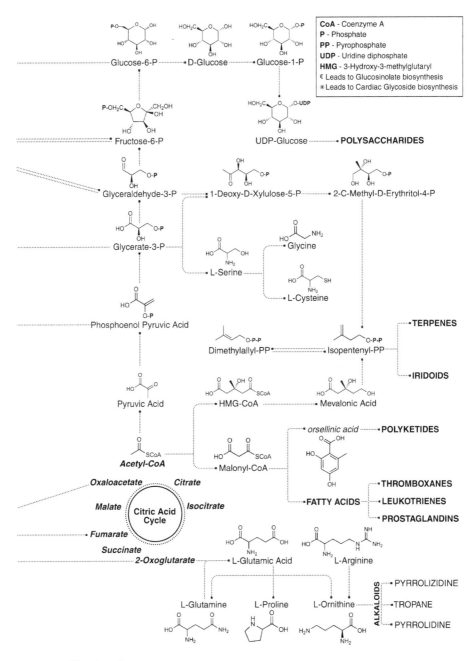

Figure 2.1 (Continued)

well as biological and environmental stresses. Plant secondary metabolites, also referred to as plant secondary compounds or natural products, often exclusively occur in certain plant families or accumulate in specific tissues, as opposed to the universal occurrence of primary metabolites. As such, many extraordinary functions of a plant extract, such as anti-microbial, anti-inflammatory, and anti-cancer effects, as well as flavors, fragrances, and dyes properties are attributed to plant secondary metabolites rather than primary metabolites.

There are different ways of organizing plant compounds: biological function, chemical structure, and biochemical pathways of synthesis. Most often, and in this chapter, plant secondary metabolites are classified according their pathways of synthesis. Secondary metabolites can be organized into three major groups: terpenoids, phenylpropanoids, and nitrogen-containing compounds. The major plant secondary metabolic pathways are the shikimic acid pathway, which leads into the phenylpropanoid biosynthesis; the mevalonic acid and MEP/DOXP (2-C-methyl-D-erythritol 4-phosphate/1-deoxy-D-xylulose 5-phosphate) pathways that lead to terpenoid biosynthesis; and the acetate pathway leading to fatty acid and polyketide biosynthesis. Alkaloids, cyanogenic glycosides, and glucosinolates are biosynthesized from amino acids; however, multiple pathways may also be involved in their synthesis.

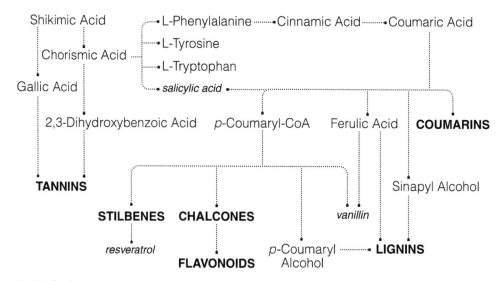

© 2016 Sarah Esgro

Phenylpropanoids are phenolic compounds, meaning they contain a phenol group (a hydroxyl group on an aromatic ring). Polyphenols contain more than one phenolic ring in the structure (not two or more hydroxyls on the same ring). There are two routes of phenylpropanoid biosynthesis: the shikimic acid pathway, which occurs in most plants (but not in animals), and the malonic acid (propanedioic acid) pathway, which occurs in fungi and bacteria. In the shikimic acid pathway, simple carbohydrates (from glycolysis and the pentose phosphate pathway) are converted to aromatic amino acids (Figure 2.1 and 2.2). The enzyme phenylalanine ammonia lyase (PAL) is considered the branch point between primary and secondary metabolism. Simple phenylpropanoids have a C6-C3 structure, denoting a six-carbon aromatic ring with a three-carbon chain. They include cinnamic acid, *p*-coumaric acid, caffeic acid, and ferulic acid. Phenylpropanoid lactones have a C6-C1 structure and include coumarins such as umbelliferone, psoraalen, and furanocoumarins. Benzoic acid derivatives are C6-C1 compounds and include vanillin and salicylic acid. Stilbenoids, such as resveratrol, follow a C6-C2-C6 structure.

Flavonoids have a C6-C3-C6 structure of two six-carbon aromatic rings connected by a three-carbon bridge. Flavonoids are further classified into the following groups on the basis of variations in double bond and/or hydroxyl group location: anthocyanins, catechins, flavonols, and isoflavones (Figure 2.2).

Tannins are either polymers of flavonoid units (condensed tannins) or heterogeneous polymers containing phenolic acids (esp. gallic acid) and sugars (hydrolyzable tannins).

Lignins are highly branched polymers of C6-C3 phenylpropanoids, generally made of three phenylpropanoid alcohols including coniferyl alcohol, coumaryl alcohol, and sinapyl alcohol.

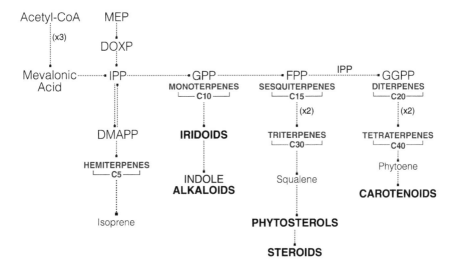

© 2016 Sarah Esgro

Terpenoids (or terpenes) are biosynthesized from acetyl-CoA or glycolytic intermediates (Figure 2.3). They are also called isoprenoids because they are formed from the fusion of five carbon isoprene units. Some scientists use the term *terpenoid* only when referring to terpenes with oxygen functional groups, but often the two terms are used interchangeably. Terpenoid biosynthesis from primary metabolites occurs by two pathways: the mevalonic acid (or HMG-CoA reductase) pathway in the cytosol and the 2-C-methyl-D-erythritol 4-phosphate/1-deoxy-D-xylulose 5-phosphate (MEP/DOXP) pathway in plastids. The mevalonic acid pathway is the main pathway that occurs in the cytosol and leads to biosynthesis of triterpenes, sterols, and many sesquiterpenes. The non-mevalonic MEP/DOXP pathway occurs in plastids and leads to biosynthesis of monoterpenes, some sesquiterpenes, diterpenes, and carotenoids. Three molecules of acetyl-CoA are joined to form mevalonic acid, a six-carbon intermediate that is pyrophosphorylated, then decarboxylated, and finally dehydrated to form isopentenyl diphosphate (IPP). Hemiterpenoids are five-carbon isoprenoids. Monoterpenoids are ten-carbon terpenoids formed by two isoprene units. Examples include essential oil components such as limonene and menthol. Sesquiterpenoids such as caryophyllene and zingiberene contain 15 carbons formed by three isoprene units. Diterpenoids contain twenty carbon terpenes formed by four isoprene units. The plant hormone gibberellic acid is an example of a diterpenoid.

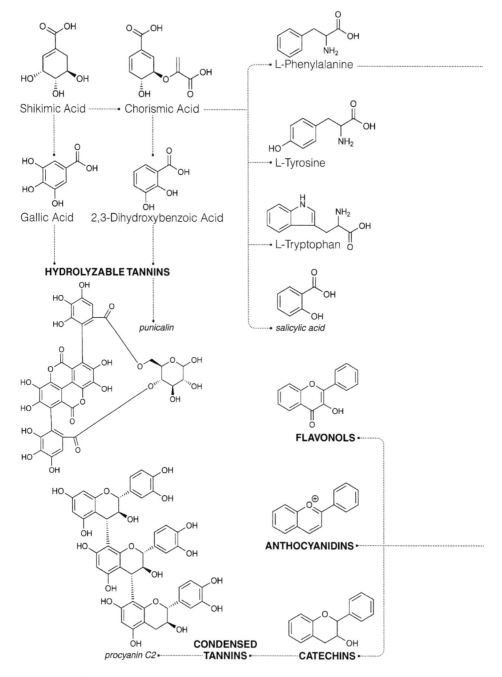

Figure 2.2 **Phenylpropanoid biosynthesis.** © 2016 Sarah Esgro.

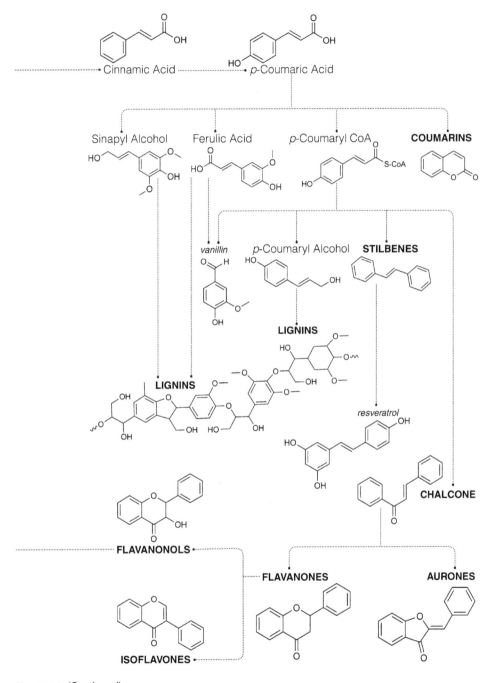

Cinnamic Acid ⋯⋯ p-Coumaric Acid

Sinapyl Alcohol   Ferulic Acid   p-Coumaryl CoA   **COUMARINS**

vanillin   p-Coumaryl Alcohol   **STILBENES**

LIGNINS

LIGNINS

resveratrol

LIGNINS

CHALCONE

FLAVANONOLS

FLAVANONES   **AURONES**

ISOFLAVONES

Figure 2.2 (Continued)

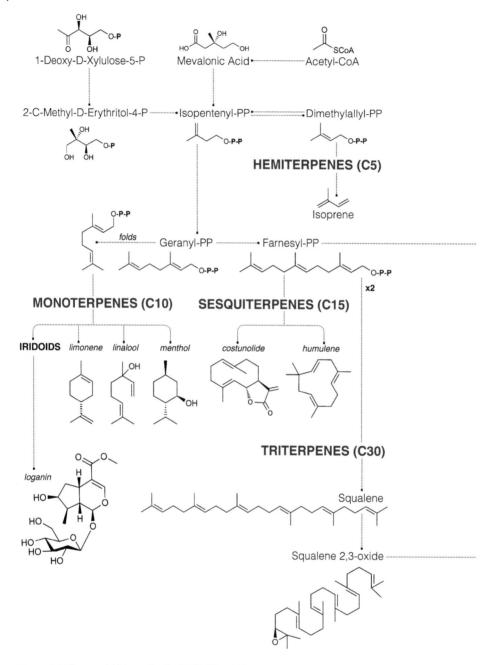

Figure 2.3 Terpenoid biosynthesis. © 2016 Sarah Esgro.

Triterpenoids contain 30 carbons formed by two sesquiterpenoids to form the backbone of phytosterols. Examples of triterpenoids include brassinosteroids, cardenolides, saponins, phytoecdysteroids, and limonoids. Tetraterpenoids contain 40 carbons from two 20-carbon units. The only examples of naturally occurring tetraterpenoids are the

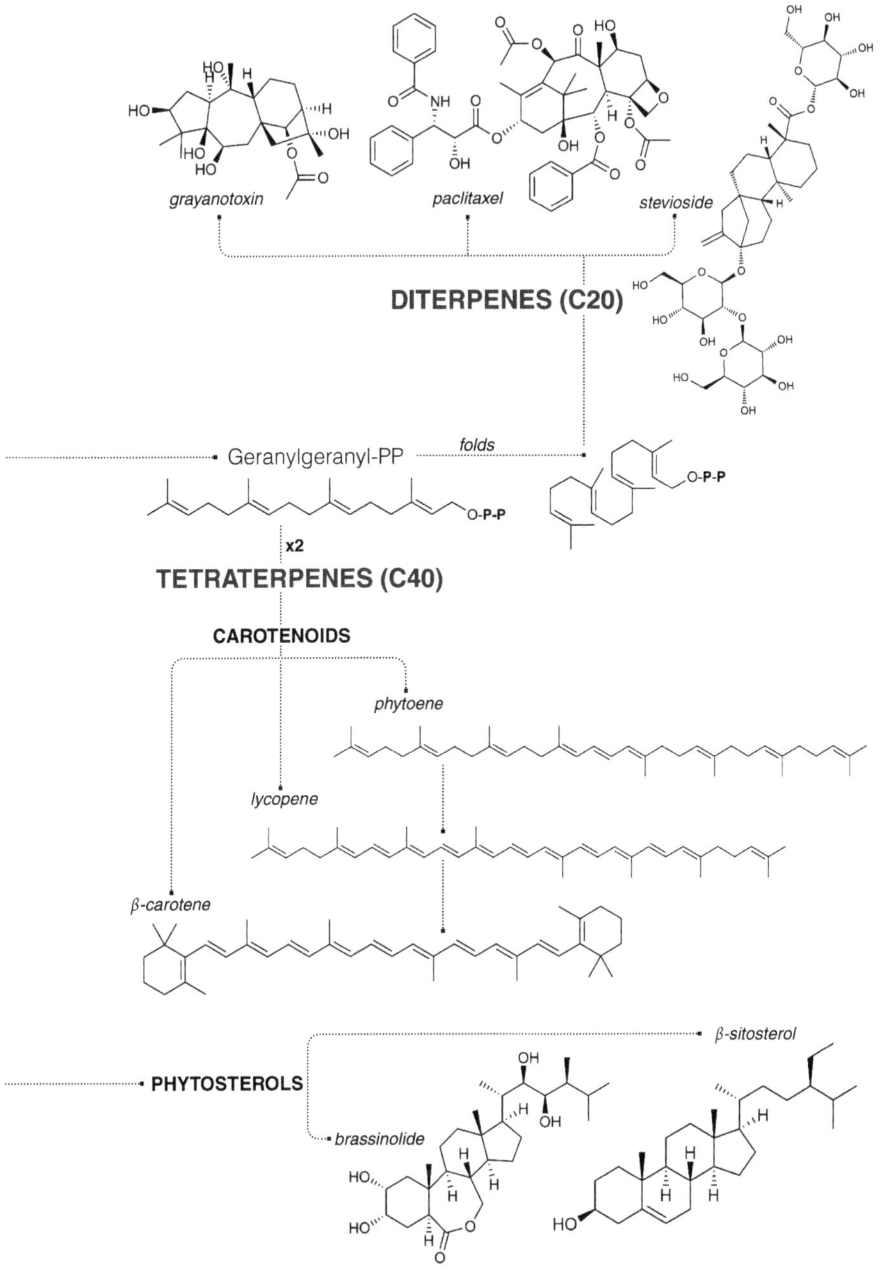

Figure 2.3 (Continued)

carotenoids, which include lutein and zeaxanthin. Terpenoids with greater than 40 carbons, formed from more than eight isoprene units, are polyterpenoids.

Nitrogen-containing compounds comprise the third large phytochemical group and can be organized into alkaloids, cyanogenic glycosides, glucosinolates, and non-protein amino acids. Many of them are poisons, stimulants, narcotics, and medicines.

Alkaloids contain one or more nitrogen atoms as part of a heterocyclic ring of nitrogen(s) and carbons. They are typically alkaline; therefore, they have a positive charge at physiological pH and are stored in protonated form mostly in the acidic environment of the vacuole. Alkaloids are synthesized from a few common amino acids while their carbon skeleton is often formed through the terpenoid pathway, though they can also be formed from other biosynthetic pathways (Figure 2.1).

Alkaloids derived from terpenoid biosynthesis include terpene alkaloids and steroidal alkaloids (veratrum alkaloids, solanum alkaloids). Piperidine alkaloids are derived from fatty acids or polyketides from the acetate pathway (Figure 2.4).

Many alkaloids are derived from aromatic amino acids tryptophan, phenylalanine, tyrosine, and anthranilic acid from the shikimate pathway (Figure 2.1). From tryptophan, indole (vinblastine and catharanthine), pyrroloindole, and ergot alkaloids (ergotamine) are derived as well as pyrroloquinoline and quinoline (quinine) alkaloids. Ephedra alkaloids are derived from phenylalanine. Isoquinoline and thiazole alkaloids are derived from tyrosine. Isoquinoline alkaloids include berberine, morphine, papaverine, emetine, secologanin, and thebaine. Acridine, quinoline, and quinazoline alkaloids are derived from anthranilic acid (Figure 2.5). Betalains are indole alkaloid pigments (not related to flavonoids) synthesized from tyrosine. They impart a red color to plants of the Caryophyllales order and are often mistaken for the more commonly occurring anthocyanin.

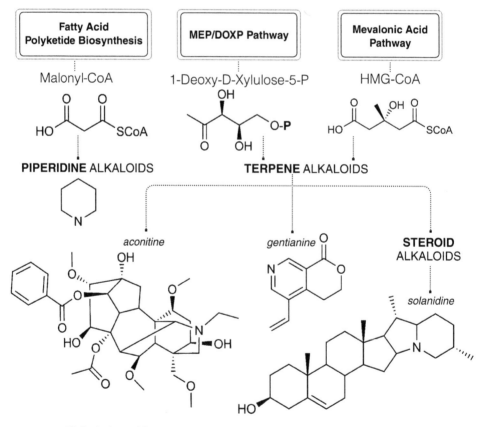

Figure 2.4 Alkaloids derived from terpenes, steroids, and the acetate pathway. © 2016 Sarah Esgro.

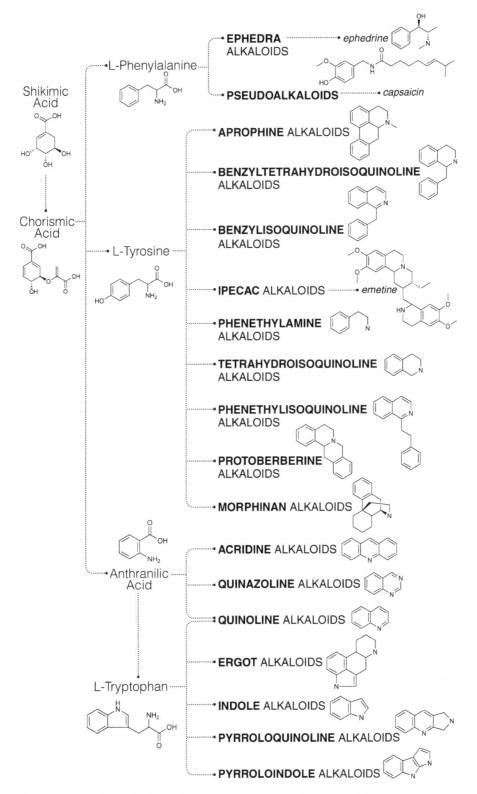

Figure 2.5 Alkaloids derived from the shikimic acid pathway. © 2016 Sarah Esgro.

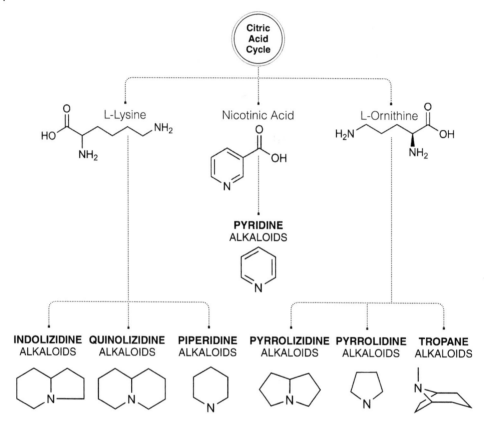

Figure 2.6 Alkaloids derived from ornithine, nicotinic acid, and lysine. © 2016 Sarah Esgro.

Alkaloids derived from ornithine, lysine, and nicotinic acid include indolizidine, quinolizidine, pyridine, pyrrolizidine, pyrrolidine, and tropane alkaloids (Figure 2.6). Examples of tropane alkaloids include cocaine and scopolamine.

Alkaloids derived from histidine and purine include imidazole alkaloids (ex. histamine) and purine alkaloids (ex. caffeine and theobromine) (Figure 2.7).

Cyanogenic glycosides are composed of an α-hydroxynitrile or cyanohydrin and a glycoside. They are typically sequestered within the vacuoles, separated from enzymes that can remove the cyano group (Vetter 2000). Cyanogenic glycosides may be derived from phenylalanine, tyrosine, valine, isoleucine, leucine, or non-protein amino acids (Figure 2.1).

Glucosinolates are nitrogen- and sulfur-containing compounds known for their pungent smell. When plants with glucosinolates are crushed, thioglucosidase or myrosinase are released, cleaving glucose from sulfur compounds. This leads to a chemical rearrangement, producing mustard-smelling volatiles. Glucosinolates may be derived from methionine, leucine, isoleucine, valine, alanine, phenylalanine, tyrosine, or tryptophan (Figure 2.1).

Plants also produce non-protein amino acids, amino acids that are not one of the 20 standard amino acids used by plants and animals as protein building blocks. Examples include theanine and γ-aminobutyric acid.

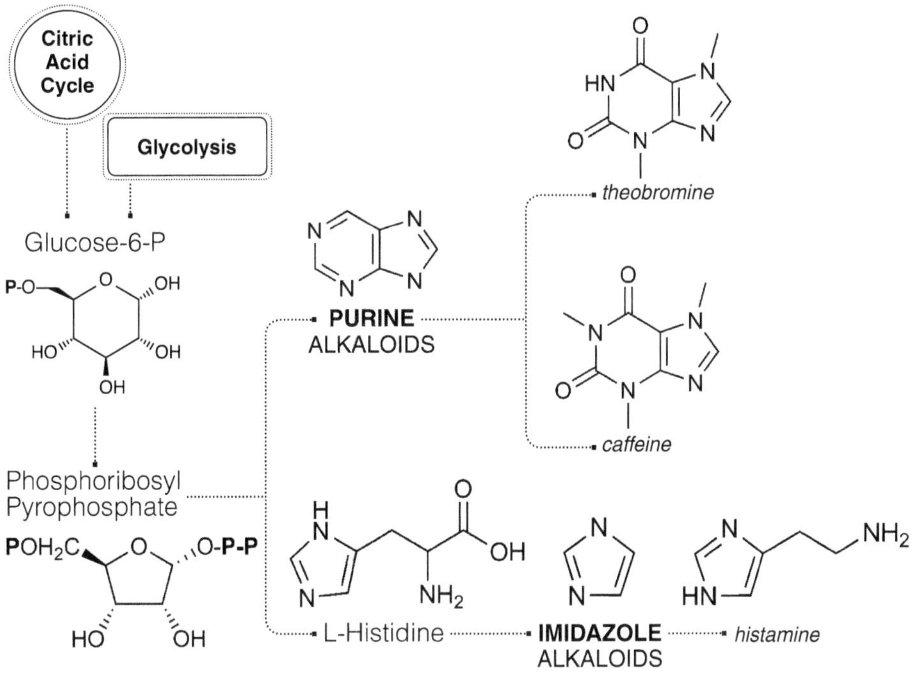

Figure 2.7 Alkaloids derived from histidine and purine. © 2016 Sarah Esgro.

The acetate pathway initiates through carboxylation of acetyl-Coenzyme A (CoA) with CO_2 using ATP and the coenzyme biotin to form malonyl-CoA (Figure 2.8). The pathway for fatty acid and aromatic polyketide biosynthesis then diverge. For aromatic polyketides, the chain extension process continues to form poly-β-keto esters that are stabilized by enzyme association to allow for addition of malonyl groups and then cyclization. Cyclization of the poly-β-keto esters leads to the formation of simple phenolics. An example is phloroglucinol, a trihydroxybenzene. Polymerization of phloroglucinol units leads to formation of phlorotannins. Anthraquinones, such as emodin and endocrocin, are polyketides formed from eight two-carbon units. Dianthrones, such as hypericin, can be formed from oxidative coupling of two anthrone systems.

For fatty acids, the carbonyl groups are reduced before attachment of additional malonyl groups. Synthesis of fatty acids involves fatty acid synthase (FAS). Acetyl-CoA and malonyl-CoA are converted to thioesters bound to FAS and then form acetoacetyl-ACP. Reduction, dehydration, and reduction reactions follow to generate an acyl-ACP that has increased in length by two carbons compared to the starting acetyl-CoA. This fatty acid can feed back into the system and subsequently elongate by two carbons through successive reduction, dehydration, and reduction steps (Dewick 2001). In addition to saturated and unsaturated fatty acids, prostaglandins (modified C_{20} fatty acids), thromboxanes, and leukotrienes are also form through fatty acid biosynthesis.

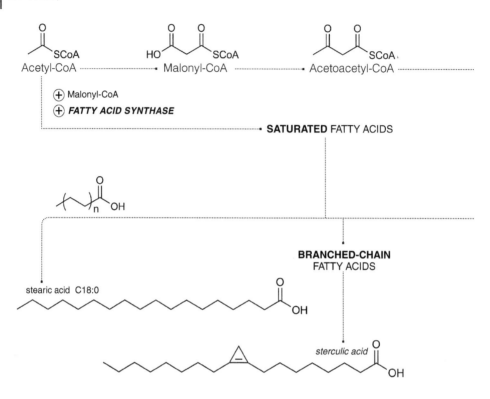

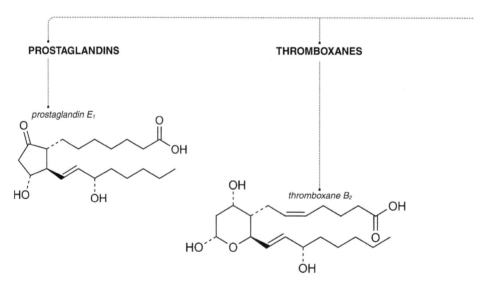

Figure 2.8 **Acetate pathway:** Fatty acids and polyketide biosynthesis. © 2016 Sarah Esgro.

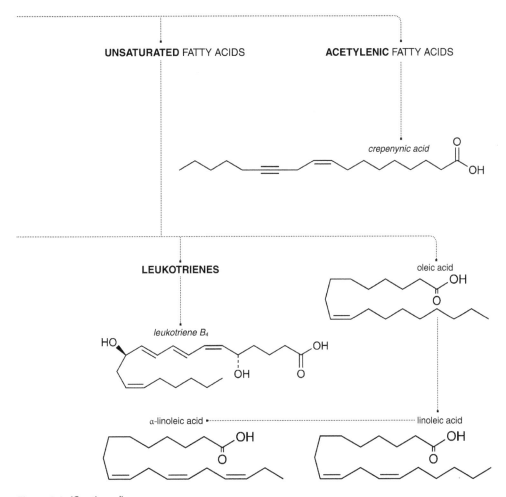

orsellinic acid

POLYKETIDES •••••• emodin •••••••••• hypericin

UNSATURATED FATTY ACIDS   ACETYLENIC FATTY ACIDS

crepenynic acid

LEUKOTRIENES   oleic acid

leukotriene B₄

α-linoleic acid ••••••   linoleic acid

Figure 2.8 (Continued)

Databases

Several databases for plant secondary metabolic pathways, biological activities, chemical structures, ethnobotanical uses, content in foods, and pharmacology are open source and queryable (Scalbert *et al.* 2011). A few useful databases are listed below and may continue to serve as updated sources of information.

KEGG (Kyoto Encyclopedia of Genes and Genomes)
Pathways drawn from molecular interactions and reaction networks.
http://www.genome.jp/kegg/pathway.html

Plant Metabolic Network
Network of plant metabolic pathway databases with curated information about genes, enzymes, compounds, reactions, and pathways involved in primary and secondary metabolism in plants.
http://www.plantcyc.org/

MetaCYC
Curated database of experimentally elucidated metabolic pathways containing 2260 pathways from 2600 different organisms.
http://metacyc.org/

Dictionary of Natural Compounds
Chemical substances including descriptive and numerical data on chemical, physical, and biological properties of compounds; systematic and common names of compounds; literature references; structure diagrams and their associated connection tables.
http://dnp.chemnetbase.com

Dr. Dukes's Phytochemical and Ethnobotanical Databases
Searchable databases of chemicals, activities, plant contents, and ethnobotanical uses.
http://www.ars-grin.gov/duke/

PubChem
Bioactivity information of small molecules and structure similarity search tool.
http://pubchem.ncbi.nlm.nih.gov/

USDA databases
Phytonutrient content in foods databases on flavonoid, carotenoid, proanthocyanidin, and isoflavones.
http://fnic.nal.usda.gov/food-composition/phytonutrients

Phenol-Explorer
Database of polyphenol content foods, polyphenol metabolism, and effects of food processing and cooking.
http://phenol-explorer.eu

NAPRALERT
Natural products chemistry, pharmacology, and biochemical activity.
http://www.napralert.org/about.aspx

References and Additional Reading

Dewick, P.M. (2001) *Medicinal Natural Products*, 2nd edition. John Wiley & Sons, Chichester.

Heldt, H., and Heldt, F. (eds) (2005) *Plant Biochemistry*, 3rd edition. Academic Press, Burlington.

Hopkins, W.G., and Huner, N.P.A. (eds) (2003) *Introduction to Plant Physiology*, 3rd edition. John Wiley & Sons, USA.

Scalbert, A., Andres-Lacueva, C. Masanori, A. *et al.* (2011) Databases on food phytochemicals and their health-promoting effects. *Journal of Agricultural and Food Chemistry*, **59** (9), 4331–4348.

Vetter, J. (2000) Plant cyanogenic glycosides. *Toxicon*, **38** (1), 11–36.

Extraction and Chromatographic Techniques

Ethnobotany encompasses a broad span of botanical applications including their use in medicine, supplements, cosmetics, and food. Some of the most important applications and areas of research relate to maintenance and promotion of human health. Deciphering active constituents, modes of action(s), and efficacy begins with phytochemical extraction.

Extraction

There are several traditional methods of preparation of plant material for healing applications. Traditional methods of extraction include decoctions, tinctures, and poultices. Starting plant material may be freshly harvested material or, more commonly, dried plant material such as tea leaves. Decoctions are extracts made by boiling the plant material in hot water, as opposed to teas, which are merely steeped. The boiling of the material helps to extract more phytochemicals, but prolonged heat also degrades temperature-sensitive compounds. Tinctures are concentrated alcoholic extracts of plant material. Poultices are made from maceration of plant material typically applied to the skin or on a wound.

For scientific investigation of phytochemicals, modern methods of extraction use organic solvents to increase efficiency of phytochemical extraction compared to traditional methods of preparation. While solvents are typically removed prior to application by rotary evaporators (Figure 2.9) or spray dryers, choice of solvents may be limited according to the safety profile of solvents. For example, the use and level of solvent residues are regulated in food and cosmetic applications. Common solvents used for phytochemical extraction include water, ethanol, methanol, acetone, and hexane. Solvent to plant material extraction ratios are typically written as a ratio, 1:5, noting one part plant material and five parts solvent, either weight by weight (w/w) or weight to volume (w/v). Plant material is typically ground by mortar and pestle, milled, or by other means of homogenization to reduce plant particle size prior to mixing with solvents. This process increases the surface area and disrupts cellular structures to increase extraction yields. The mixture may be extracted for a short time to several days sitting idle or with shaking, stirring, or sonication. Some methods like percolation involve running continuous flows of solvent over plant material.

Figure 2.9 Rotary evaporators are used to remove solvents from plant extracts. The extract is heated under vacuum; volatile solvents condense on chilled coils and drip into a collection flask.

Figure 2.10 Example of a drum manifold lyophilizer. After solvents are removed, plant extracts are frozen in round-bottomed flasks, and then placed on the lyophilizer until water is removed via sublimation.

While fresh, frozen, or dry plant material may be used, drying of plant material normalizes plant dry mass and minimizes differences in extraction efficiency due to water content. Lyophilization, or freeze-drying (Figure 2.10), of plant material helps to preserve phytochemicals in the state of harvest by minimizing enzyme activity and the phytochemical degradation that occurs in cases of air drying or heat-assisted drying (oven or dehydrator). However, changes in phytochemical levels and composition may or may not be significant, and in some cases may be desired.

Hydrodistillation is a conventional method to extract essential oils from herbs, spices, and flowers used for flavors, fragrances, and aromatherapy. In Clevenger distillation, the plant material to be extracted is immersed in water. In steam distillation, the plant material sits on a surface to allow steam to pass through. By heating the water, non-water soluble compounds are volatilized with the water and form an azeotropic mixture. The vapors are condensed, and liquid is collected.

Supercritical fluid (SCF) extraction controls the pressure and temperature of a solvent to reach above its critical point, the point when the liquid and gaseous state of a fluid cannot be distinguished and solvation power is increased. SCFs have lower viscosity and higher diffusivity than liquids, allowing for better penetration into porous materials (Herrero *et al.* 2006). While several solvents can be used, CO_2 is one of the most commonly used solvents because of its selectivity, reusability, and easy separation from extracts through depressurization. Selectivity and separation efficiency can be adjusted by changing temperature and pressure conditions (Lang and Wai 2001).

Separation Methods

The process of compound identification involves methods of fractionating the complex mixtures present in crude extracts. One of the first steps is liquid-liquid partitioning, the separation of compounds based on their chemical preference in one of two or more immiscible solvents. Column chromatography is a technique to isolate or separate compounds from mixtures using the properties of compound interaction

Figure 2.11 Vacuum column chromatography using a hydroxylated methacrylic polymer as the stationary phase.

between a solid (stationary) phase and a liquid (mobile) phase. A solid adsorbant, such as silica gel or alumina, is packed into a column to act as the stationary phase, and a sample extract is loaded at the top and eluted through the matrix with organic solvents (the mobile phase) (Figure 2.11). The rate at which a compound is eluted from the column is based largely on polarity. The polarity of silica gel attracts polar compounds, and thus nonpolar compounds will elute first. Silica gel chromatography is normal phase chromatography. In reverse phase chromatography, hydrophobic C_{18} or C_8 groups are attached to silica gel, allowing more polar compounds to be eluted first. Ion-exchange chromatography separates compounds based on charge affinity; in addition, compound charge can be modified by pH adjustment. Size-exclusion (gel permeation or gel filtration) chromatography is used to separate compounds based on their size/molecular weight in solution. Mixed molecular weight compounds are applied to the column, and smaller compounds diffuse into the pores of the gel, slowing their elution from the column, while larger compounds are excluded and flow through the column quickly. Gel permeation is typically used to separate polymers such as polysaccharides and proteins.

Affinity chromatography uses ligands that recognize and bind to specific structural motifs, thus increasing specificity for compound purification. Affinity chromatography has also been applied in the evaluation of an isolated compound as well as complex mixture devoid of a targeted compound. Termed *knockout extract*, this mixture minus compound X, is evaluated for biological activity as a means of determining the importance of phytochemical interactions.

Ideal compound separation is achieved by further selection and optimization of solvent systems. Thin layer chromatography (TLC) uses the same principles as column chromatography, but the solid phase is affixed to a solid flat surface such as a glass or aluminum plate. Dissolved extracts are spotted on TLC plates, and the mobile phase is drawn up the solid phase, pulling along compounds up the plate. When the solvent reaches the top of the plate, the plates are dried, and colored bands are either visualized directly (for colored components) or the plate is sprayed with a developing solution for visualization (Figure 2.12). TLC is typically performed prior to column chromatography to select an ideal solvent system for compound separation. In addition, TLC can be used as a quick, inexpensive method for qualitative evaluation of compound presence. Increasingly, high-resolution TLC methods are used for natural product authentication (Tistaert *et al.* 2011).

Counter current separation, which includes counter current chromatography and centrifugal partition chromatography, is a liquid chromatography technique that uses

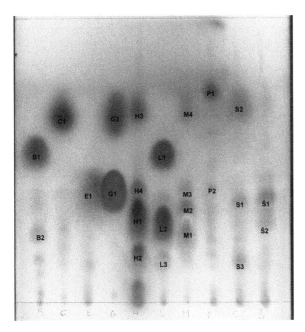

Figure 2.12 Silica gel TLC plate of essential oils developed with mobile phase toluene–ethyl acetate (93:7 v/v), sprayed with vanillin in H_2SO_4, and heated.

only liquids for both the mobile and stationary phases to achieve compound separation. Similar to liquid–liquid partitioning, counter current separation makes use of differential partitioning of analytes in a two-phase solvent system (Ito 2005; Pauli *et al.* 2015). Counter current separation is a preparative technique and can be optimized for compound separation and purification. As there is no adsorptive solid support, there is no irreversible binding of compounds. In addition to complete recovery of the loading sample, other benefits include high sample input and mild separation conditions.

Analytical Methods

Natural products are identified and quantified by a combination of analytical techniques. Typically, unknown compounds are compared to standards using (ultra) high performance liquid chromatography [(U)HPLC] for non-volatiles and gas chromatography (GC) for volatiles. These separation technologies are usually coupled with ultraviolet (UV) absorbance or diode array detection (DAD) for UV absorbing compounds (Figure 2.13). For non-UV absorbing compounds, evaporative light scattering detection (ELSD), chemiluminescence detection, and/or mass spectrometry (MS; Figure 2.14) may be used for natural products detection (Tistaert *et al.* 2011). In high-resolution MS, the mass-to-charge ratio of a molecule can be determined to the ten thousandth unit to limit the number of possible molecular formulas as an exact mass corresponds to limited possibilities of elemental compositions. Tandem MS or MS-MS, MS^n, can be used to ionize specific chromatographic peaks to help identify compounds according to their fractionation patterns. The gold standard for compound elucidation is nuclear magnetic resonance (NMR) spectroscopy: ^{1}H and ^{13}C NMR (Figure 2.15). NMR spectroscopy is a based on the resonance of nuclei from the absorption of electromagnetic

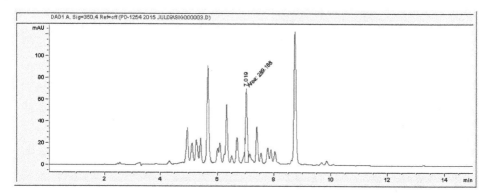

Figure 2.13 HPLC chromatogram using a diode array detector (DAD).

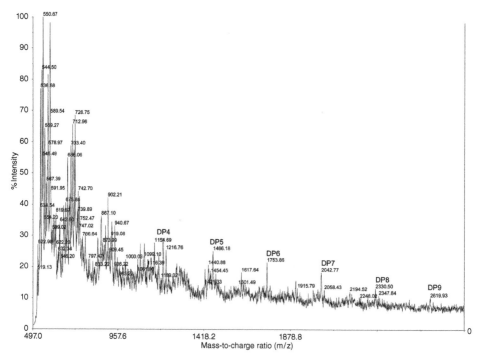

Figure 2.14 Mass spectrum of a blueberry fraction showing increasing degrees of oligomeric proanthocyanidin polymerization.

radiation in the radio frequency region when a sample is placed in a magnetic field. It is a non-destructive, reproducible technique that allows for identification and quantification of isolated purified molecules. More recently, NMR has also been used to identify and quantify components of mixtures for extract authentication and species differentiation.

Commercial botanical extracts typically consist of complex mixtures. As such, chromatographic fingerprints (developed using HPLC or TLC) and marker compounds

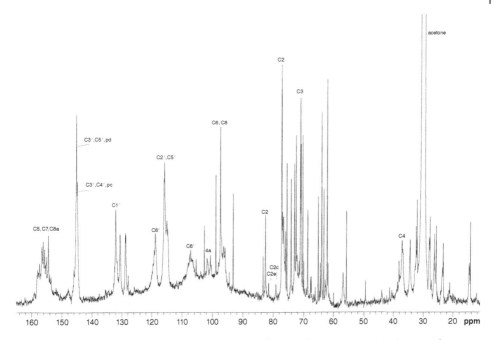

Figure 2.15 NMR absorption spectrum of a blueberry fraction showing increasing degrees of oligomeric proanthocyanidin polymerization.

are typically used to standardize extracts. However, in cases where active constituents cannot be determined, extracts have been standardized using relevant bioassays. In addition, the advent of omics technologies, such as metabolomics, allows for complex phytochemical profiling as well as evaluation of metabolic changes. Principal component analysis may be used to handle robust data for classification of botanicals including searching for analytical markers and phytochemical groupings (Tistaert *et al.* 2011).

References

Herrero, M., Cifuentes, A., and Ibañez, E. (2006) Sub- and supercritical fluid extraction of functional ingredients from different natural sources: Plants, food-by-products, algae and microalgae: A review. *Food Chemistry*, **98** (1), 136–148.

Ito, Y. (2005) Golden rules and pitfalls in selecting optimum conditions for high-speed counter-current chromatography. *Journal of Chromatography, A*, **1065** (2), 145–168.

Lang, Q., and Wai, C.M. (2001) Supercritical fluid extraction in herbal and natural product studies—a practical review. *Talanta*, **53** (4), 771–782.

Pauli, G.F., Pro, S.M., Chadwick, L.R. *et al.* (2015) Real-time volumetric phase monitoring: Advancing chemical analysis by countercurrent separation. *Analytical Chemistry*, **87** (14), 7418–7425.

Tistaert, C., Dejaegher, B., and Heyden, Y.V. (2011) Chromatographic separation techniques and data handling methods for herbal fingerprints: A review. *Analytica Chimica Acta*, **690**, 148–161.

Evaluation of Biological Activities

Phytochemical Screening

The well-established use of plants as sources of medicine and nutritional enhancement is long-standing and continues to endure as an important part of personal health care. In the 1990s, natural products research greatly expanded with advances in isolation and high-throughput screening technologies. Later, progress in rational drug design and synthetic medicinal chemistry for pharmaceutical drug discovery shifted the focus away from natural products programs (David *et al.* 2015). The number of drugs derived from natural products or analogs of natural products decreased from about 80% in 1990 to about 50% by 2007 (Li and Vederas 2009). However, the constant drive for novel chemical structures, coupled with the limited coverage of synthetic chemical libraries relative to chemicals projected in natural biodiversity, reinforces the value of natural resources. Botanicals are still largely under-explored chemical resources with only about 15% of plant species investigated phytochemically and 6% investigated pharmacologically (Atanasov *et al.* 2015). In plant species that are phytochemically characterized, typically the focus is on measuring notable marker compounds or classes of compounds that have potential biological value as it is nearly impossible to identify all compounds. Thus, many phytochemicals may be overlooked, but at the same time offer opportunities for reinvestigation. Developments in genomics, transcriptomics, proteomics, and metabolomics technologies will help unravel the multifactorial and systematic effects of natural products, which are major challenges for natural products research.

Evaluations of phytochemical activities usually take a reductionist approach where one or a few compounds are isolated through iterative fractionation, evaluated for targeted biological function(s) in vitro, and validated in vivo. In what is also known as reverse pharmacology, numerous compounds isolated from plants are screened through bioassays based on promising pharmacological targets. Bioassays are developed for biologically relevant markers that are sensitive, selective, reproducible, reliable, and ideally have medium- to high-throughput capabilities.

Forward pharmacology is the discovery of drug function through evaluation of phenotypic changes first, followed by identification of molecular action(s) of extracts or phytochemical(s) (Atanasov *et al.* 2015). By and large, the forward pharmacology approach has been performed by humans for thousands of years through trial and error. However, modern research requires empirical evidence of potential effectiveness prior to phenotypic evaluation in humans or animals. Thus, in vitro bioassays are commonly employed for initial evaluations.

Beyond random screening of chemical libraries, several knowledge-based approaches are used to focus and direct phytochemical screening programs. Through ethnobotany, plants with a history of use and/or cross-cultural use patterns related to a pharmacological target can be prioritized for fractionation, isolation, and bioactivity screening. As related plants evolved similar phytochemical pathways, additional plant leads may be selected according to chemosystematic or phylogenetic relationships. In addition to identifying bioactive compounds, a screening database provides information for further development and improvement. New compounds may be synthesized and/or modified in order to optimize the biological functional impact. Small modifications of chemical structures can be made to evaluate structure activity relationships and develop pharmacophore models for in silico screens. Taken together, in vitro screening hits from any one or a combination of these approaches may serve as leads for in vivo validation.

High-throughput screening (HTS) programs should lead to functional or clinical relevance. Assays for screens may be based on chemical reactions, enzyme-catalyzed reactions, protein (receptor, transporter, enzyme) activity, or cellular assays. Cellular assays may be derived from primary cell lines or immortalized cell lines. While immortalized cell lines are easier to culture and serve as an approachable initial bioassay, it is important to keep in mind that they may not have relevant responses due to their often cancerous origin. Primary cell line evaluations provide further evidence of biological function but require freshly harvested tissue for cell culture. In cell culture, changes in intracellular signaling can be measured by several molecular techniques including immuno-cytochemistry, fluorescence-activated cell sorting analysis, real-time PCR, Western blotting, immunoprecipitation, or omics techniques of genomics, transcriptomics, proteomics, and metabolomics. However, in vitro activity (binding or inhibition of protein activity) may not always lead to function in more complex biological systems (animals) or clinical experiments. Still, chemical and in vitro bioassays are necessary to determine potential modes of action and limit compounds to be tested in vivo. Where appropriate, testing of molecules in ex vivo conditions, where intact organs are externalized, may better simulate effects on biological tissue such as in dermal applications.

Whole cell organisms have been explored for HTS, including *Caenorhabditis elegans*, yeast, and zebrafish. The zebrafish model has been of particular interest as zebrafish have high genetic, physiological, pharmacologic similarity with humans, and many bioassays have been developed for biomedical research including cancer, cardiovascular, skeletal, neurological, and metabolic disorders (Crawford *et al.* 2008). Their small size (1–5 mm), rapid development, high fecundity, and requirement of only microgram quantities of test compounds makes them compatible for medium and HTS. However, the predictive value of zebrafish is yet to be known, as clinical relevance of molecules identified in the zebra model platform have not yet been assessed (Crawford *et al.* 2011). Bioassays that utilize whole cell organisms may provide an informative bridge to in vivo testing.

There are many issues to consider in order to avoid misinterpretation of results when performing in vitro experiments. These include chemical, physical, and toxicoloigical properties of phytochemical targets. As most bioassays require dissolution of compounds or extracts into the medium, it is important to determine that compounds do not aggregate, precipitate, or decompose over the duration of the experiment. In addition, test compounds and extracts should not interfere with detection or create artifacts. For example, phytochemical compounds may quench fluorescence or be fluorescent, thus interfering with colorimetric endpoints of measure. Extract complexity may make it nearly impossible to evaluate every compound in a mixture. A de-tannization step may help avoid false effects in a bioassay as tannins are promiscuous protein binders (Kingston 2011). Often blank or control samples are included to account for extract interference; however, solutions are assay dependent.

Of primary importance in cell-based assays is the determination of cellular toxicity. Except in certain cases such as cancer cells or pathogens, toxicity is an undesired biological outcome (Atanasov *et al.* 2015). Another major issue with bioassays for evaluation of biological function is the dose. Typically, phytochemicals are treated in micromolar to millimolar levels in bioassays, which often preclude their physiological relevance. Yet, they are tested and reported at these levels because this is the level at which an effect can be shown. In vitro bioassays are imperfect systems, and while a reductionist approach may suffer from critical issues, including a focus on nonspecific cellular phenomena, questionable dose levels, and neglect of pharmacokinetics and pharmacodynamics aspects relevant to clinical conditions, bioassays can be an efficient and relevant first step in evaluation of bioactivity. Extensions of claims of biological effect can only be made after proper evaluation; in the case of health claims, further evaluation would involve animal systems and/or clinical trials.

Polypharmacology and synergy are important and often challenging aspects of botanical-extract-based research. The concept of phytochemical synergy is when there is a greater than additive effect of two or more compounds. Synergy is often used explain the importance of mixture effects. For example, bioassay guided fractionation is designed to lead the isolation process toward an active secondary metabolite; however, in some cases fractionation results in loss of biological activity through inadvertent separation of synergistic compounds. Synergism may be due to enhanced solubility, dissolution, absorption, and/or metabolism of active components as well as multiple target effects and/or reversing antagonism. A method to evaluate synergy, antagonism, or additive effects is by designing experiments with several dose combinations to plot and evaluate isobolograms (Figure 2.16; David *et al.* 2015).

Compounds that interact and produce a greater than additive effect are considered synergistic or potentiating. Compounds applied together that lead to less than an additive biological effect would be considered antagonists, where one compound may act as an inhibitor. Beyond biological assays, computational algorithms have been designed to account for pharmacodynamics and calculate a combination index to measure synergy (further reading: [Chou 2006, 2010]).

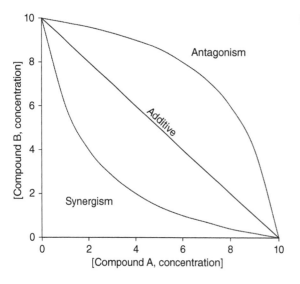

Figure 2.16 Isobologram.

To better understand mixture effects, newer modes of evaluation have been developed including knock-out and DESIGNER extracts. The use of knock-out extracts are another approach to explore, measure, and confirm bioactivity contributed in a phytochemical mixture. Similar to genetic knock-outs, where a gene is deleted or silenced to evaluate biological function, knock-out extracts involve the removal of one or multiple chemical components from a mixture in order to further characterize the biological effect of the target compound(s). Deplete and Enrich Select Ingredients to Generate Normalized Extract Resources (DESIGNER) extracts takes a step further by including both deletion and augmentation of target compound(s) in an extract as well as integration of chemical characterization (Ramos Alvarenga *et al.* 2014). These methods help address the limitation of single-compound characterization from the traditional reductionist approach, which does not account for the entirety of bioactivity from a complex extract.

Botanical Extract Standardization

Quality, safety, efficacy, and consistency are required for botanical supplements and medicines. Standardization and reproduction of a newly discovered bioactive extract may be achieved by several means. An extract may be standardized to one or more phytochemicals if they are readily available and of good quality. In mixtures where multiple reference standards are difficult to come by but have similar chemical properties for quantification (absorption, ionization, reactivity, etc.), phytochemical extracts may be characterized by using a single pharmacopoeial quality reference standard to determine multiple components, such as using quercetin to quantify flavonols or cyanidin glucoside for anthocyanins. Biochemical assays for total polyphenol content and other phytochemical subclasses such as anthocyanidins, proanthocyanidins, and flavonols are frequently used for standardization but lack specificity compared to HPLC, (U)HPLC-MS, NMR, and GCMS quantification. In addition to analytical fingerprints, multicomponent bioassays may be used to establish batch-to-batch consistency of biological effect (Guo *et al.* 2015). This provides a checkpoint for extract bioactivity, which in some instances may be the only form of quality evaluation that is reproducible.

Manufacturing controls also help ensure small molecule and extract quality. Good Agricultural and Collection Practices (GACP), Good Laboratory Practices (GLP), Good Manufacturing Practices (GMP), and Good Clinical Practices (GCP) may all be needed to achieve a high standard of botanical extract quality. DNA barcoding, in conjunction with analytical fingerprinting, may be used to ensure authentic plant material. In some cases, control over botanical material quality may be accomplished by only using plant raw materials from the same cultivars and grown in the same regions as the clinically investigated materials.

In Vivo and Clinical Evaluations

After sufficient evidence of bioactive potential is identified in vitro, botanicals move to preclinical, or in vivo, models requiring larger quantities of test compounds and extracts. Rodents are the primary models for in vivo testing where numerous disease models are available. For human clinical trials, double blind placebo (vehicle) controlled clinical studies are considered the gold standard. While many botanicals have demonstrated effectiveness in animal models, biological efficacies in humans are few and rather inconsistent. For claims related to maintaining and promoting good health, botanicals must be evaluated in healthy populations. Health claims related to disease must follow a drug route requiring FDA (Food and Drug Administration) approval in the United States and similar routes in other countries.

Regulations

The use of botanical therapies is not only long-standing, but worldwide with established systems of practice in many regions: In Asia, traditional Chinese medicine (TCM), Japanese medicine (Kampo), Korean medicine, Jamu (Indonesia), and Ayurvedic medicine (India), and in Europe, phytotherapy (Itokawa *et al.* 2008). In general, medicinal products are used to treat symptoms of an illness, a therapeutic effect, whereas food products support the physiology of healthy consumers, for health maintenance (Chinou *et al.* 2014).

Modern regulations are designed to improve safe use and availability of herbal medicinal products (HMPs). In Europe, use of botanicals for health applications beyond basic nutrition is regulated according to Directive 2004/24/EC, which harmonizes the use of HMPs among the 28 member states of the European Union (EU, Directive 2004/24/EC). HMPs must have defined manufacturing processes at all stages and quality tests that follow specifications to ensure reproducibility (Chinou *et al.* 2014). HMPs fall under two categories: well-established use HMPs and traditional HMPs.

Well-established HMPs can be granted marketing authorizations. Well-established HMPs have adequate clinical data, which include at least one controlled clinical study with pharmacological support showing efficacy in specific therapeutic indications and data demonstrating a positive risk benefit profile. Well-established HMPs also require well-documented medicinal use for 10 years in the EU. Traditional HMPS can be granted a registration and have no existing proof of clinical efficacy. They are eligible for registration if they have been in use for 30 years and at least 15 years in an EU country. These directives aim to protect public health while allowing for free movement of herbal medicinal products within the EU.

In the United States, botanical therapies are within alternative, complementary, or integrative medicine, which also include other traditional remedies (meditation, massage, homeopathy, etc.). Botanical extracts taken orally are marketed as dietary supplements in the United States. According to the Dietary Supplement Health and Education Act (DSHEA) of 1994, dietary supplements fall under the category of foods and are required to be labeled as dietary supplements. While dietary supplements cannot make disease claims to diagnose, cure, mitigate, treat, or prevent disease, claims regarding structure-function and mechanisms of action in humans are allowed to be labeled. In contrast, a botanical drug product is intended for use in the diagnosis, cure, mitigation, treatment, or prevention of disease in humans (FDA 2004). In 2003, the Center for Drug Evaluation and Research (CDER) established a botanical review team and published *Guidance for Industry: Botanical Drug Products* in 2004. Development of botanical drugs involves an extensive regulatory process through the FDA requiring filing of an investigative new drug (IND) application and can be marketed only after approval under a new drug application (NDA) (FDA 2004).

There are currently only two FDA-approved botanical drugs on the market: Veregan™ (sinecatechins) approved in 2006 and Fulyzaq (crofelemer) approved in 2012. Veregan ointment contains 15% w/w sinecatechins purified from *Camellia sinensis* (L.) Kuntze

(Theaceae). It was approved for topical treatment of external genital warts and perianal warts. Sinecatechins is a mixture containing 85–95% w/w catechins and other green tea components. The safety and efficacy of Veregan were established in two randomized, double-blind, vehicle controlled studies (Center for Drug Evaluation and Research 2006; Lee *et al.* 2015). Two main points were reviewed to ensure quality and consistency: (1) botanical raw material consistency and (2) efficacy and dose range. In order to reduce chemical variability at plant and raw material levels but still have natural variability, cultivation sites were required to follow GACP. Botanical raw materials intended for future marketed batches were limited to cultivars used in the clinical studies. For quality control of efficacy, clinical results showed no significant differences between two doses (10% and 15%), which suggested variation in this range would not have an impact on therapeutic efficacy (Center for Drug Evaluation and Research 2006; Lee *et al.* 2015).

The other botanical drug, Fulyzaq (crofelemer), was approved for symptomatic relief of noninfectious diarrhea in patients with HIV/AIDS on antiretroviral therapy. A tablet contains 125 mg of the botanical drug crofelemer, a mixture of oligomeric proanthocyanins derived from the red latex of *Croton lechleri* Müll. Arg, (Euphorbiaceae). The red latex is also known as dragon's blood and has been used to treat diarrhea in South America (Gupta *et al.* 2008). Due to the difficulties in chemical characterization of the oligomeric mixture of crofelemer, FDA required a clinically relevant bioassay to evaluate drug function and quality. In this case, a bioassay was developed on the basis of the mode of action of crofelemer. Crofelemer is an inhibitor of both the cAMP stimulated cystic fibrosis transmembrane conductance regulator (CFTR) chloride channel and the calcium-activated chloride channels (CaCC). This inhibition blocks chloride secretion and high volume water loss that occurs with diarrhea, and normalizes chloride and water flow in the gastrointestinal tract (Center for Drug Evaluation and Research 2012). Thus, in addition to restricted harvesting, GACP, and clinical data, a bioassay based on crofelemer's mechanism of action was included for quality assurance of the botanical drug (Center for Drug Evaluation and Research 2012; Lee *et al.* 2015).

Human health is a dominant area of natural products research; however, similar approaches of phytochemical characterization, bioactive or function discovery, and evaluation strategies are taken in pharmaceutical, food, cosmetic, textile, and agricultural industries. The following case studies highlight the numerous uses of botanicals from around the world. Phytochemical research revealed the compounds responsible for their function, including sesquiterpene lactones for anti-malarial therapy, betalains for textile dyes, arecatannins for teeth blackening, vanillin and vanilla extract for flavors and fragrances, and iridoids for food preservation. It is clear from these examples that there are a multitude of uses of phytochemicals which continue to grow; as such, so does the importance of phytochemical research.

References

Atanasov, A.G., Waltenberger, B., Pferschy-Wenzig, E. *et al.* (2015) Discovery and resupply of pharmacologically active plant-derived natural products: A review. *Biotechnology Advances*, **33** (8), 1582–1614.

Center for Drug Evaluation and Research (2012) Approved Labeling for Fulyzaq (NDA 202292), http://www.accessdata.fda.gov/drugsatfda_docs/nda/2012/202292Orig1s000Lbl.pdf (accessed 11 October 2015).

Center for Drug Evaluation and Research (2006) Approved Labeling for Veregen (NDA 021902), http://www.accessdata.fda.gov/drugsatfda_docs/nda/2006/021902s000_prntlbl.pdf (accessed 11 October 2015).

Chinou, I., Knoess, W., and Calapai, G. (2014) Regulation of herbal medicinal products in the EU: An up-to-date scientific review. *Phytochemistry Reviews*, **13**, 539–545.

Chou, T. (2010) Drug combination studies and their synergy quantification using the Chou-Talalay method. *Cancer Research*, **70** (2), 440–446.

Chou, T. (2006) Theoretical basis, experimental design, and computerized simulation of synergism and antagonism in drug combination studies. *Pharmacological Reviews*, **58** (3), 621–681.

Crawford, A.D., Esguerra, C.V., and de Witte, P.A.M. (2008) Fishing for drugs from nature: Zebrafish as a technology platform for natural product discovery. *Planta Medica*, **74** (6), 624–632.

Crawford, A.D., Liekens, S., Kamuhabwa, A.R. *et al.* (2011) Zebrafish bioassay-guided natural product discovery: Isolation of angiogenesis inhibitors from East African medicinal plants. *PLoS ONE*, **6** (2), e14694.

David, B., Wolfender, J., and Dias, D. (2015) The pharmaceutical industry and natural products: Historical status and new trends. *Phytochemistry Reviews*, **14** (2), 299–315.

FDA (2004) Guidance for Industry: Botanical Drug Products, http://www.fda.gov/downloads/Drugs/GuidanceComplianceRegulatoryInformation/Guidances/UCM070491.pdf (accessed 11 October 2015).

Guo, D., Wu, W.Y., Ye, M., Liu, X., and Cordell, G.A. (2015) A holistic approach to the quality control of traditional Chinese medicines. *Science*, **347** (6219 Suppl), S29–S31.

Gupta, D., Bleakley, B., and Gupta, R.K. (2008) Dragon's blood: Botany, chemistry and therapeutic uses. *Journal of Ethnopharmacology*, **115** (3), 361–380.

Itokawa, H., Morris-Natschke, S.L., Akiyama, T., and Lee, K.H. (2008) Plant-derived natural product research aimed at new drug discovery. *Journal of Natural Medicine*, **62** (3), 263–280.

Kingston, D.G.I. (2011) Modern natural products drug discovery and its relevance to biodiversity conservation. *Journal of Natural Products*, **74** (3), 496–511.

Lee, S.L., Dou, J., Agarwal, R. *et al.* (2015) Evolution of traditional medicines to botanical drugs. *Science*, **347** (6219 Suppl), S32–S34.

Li, J.W.H., and Vederas, J.C. (2009) Drug discovery and natural products: End of an era or an endless frontier? *Science*, **325** (5937), 161–165.

Ramos Alvarenga, R.F., Friesen, J.B., Nikolić, D. *et al.* (2014) K-Targeted metabolomic analysis extends chemical subtraction to DESIGNER extracts: Selective depletion of extracts of hops (*Humulus lupulus*). *Journal of Natural Products*, **77** (12), 2595–2604.

Part II

Case Studies

3

Introduction

Relating phytochemistry to a plant's traditional use provides a deeper scientific under-standing of ethnobotany, answering questions about how a plant works and why people use it. This understanding can promote creativity and new discoveries. The following section contains case studies of ethnobotanical plants from around the world. They are organized alphabetically by region: Africa, Americas, Asia, Europe, and Oceania. The regional introductions set the stage for the case studies, providing general information on the geography, botany, native peoples, cultures, and history of the region. Many of the plants and ethnobotanical uses transcend geographic borders and were difficult to cate-gorize into one region. Some plants are used more universally than others, and there are instances where a plant is native to one region, but people in an entirely different region developed its ethnobotanical use. Quite often, new discoveries challenge currently held beliefs about a plant's origin and the way it has been used throughout history.

   Contributing authors submitted case studies in or close to their field of expertise, fol-lowing a fairly uniform format describing the ethnobotanical use of one or several plants, phytochemicals responsible for the observed activity or usefulness, followed by modern use. Sometimes modern use is only theoretical, such as ethnomedicines with drug potential. Or modern use may be nearly the same as traditional use, such as using indigo to dye fabric. Ethnomedicine often dominates the field when phytochemistry is considered. But here we present a range of topics in an effort to highlight some of the diverse ways people have used plants that are intricately linked to phytochemistry.

Ethnobotanical Use	Chapter
Ethnomedicine	
Malaria	4
HIV	4
Maqui	5
Artemisia	6
Jaundice	6
Lavender	7
Banana	8

Ethnobotany: A Phytochemical Perspective, First Edition. Edited by B. M. Schmidt and D. M. Klaser Cheng.
© 2017 John Wiley & Sons Ltd. Published 2017 by John Wiley & Sons Ltd.

Ethnobotanical Use	Chapter
Intoxicants/Psychoactives	
Artemisia	6
Mad Honey	7
Kava	8
Dyes and Industrial	
Biofuels	4
Betalains	5
Teeth Blackening	6
Indigo	7
Insecticides	
Neem	6
Essential Oils	7
Religious/Cultural	
Dai cake	6
Teeth Blackening	6
Sacred trees	6
Fibers	
Agave	5
Food Science/Nutrition	
Vanilla	4
Seaweed	5
Quinoa	5
Dai cake	6

4

Africa

M. H. Grace, P. J. Smith, I. Raskin, M. A. Lila, B. M. Schmidt, and P. Chu

© 2002 Ilya Raskin

Introduction

Africa is a vast continent (Figure 4.1), rich in botanical and cultural diversity. The majority of Africa's residents are still indigenous, but many areas have been heavily influenced by the cultures of trade partners, immigrants, and colonial powers. Climate change, especially desertification of the Sahara, also had a large influence on African ethnobotany.

Ethnobotany: A Phytochemical Perspective, First Edition. Edited by B. M. Schmidt and D. M. Klaser Cheng.
© 2017 John Wiley & Sons Ltd. Published 2017 by John Wiley & Sons Ltd.

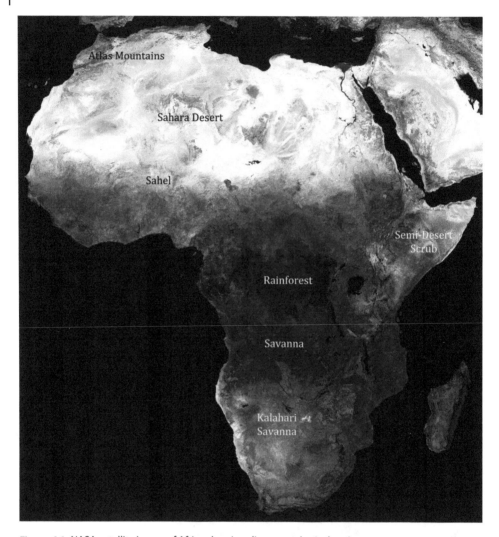

Figure 4.1 NASA satellite image of Africa showing diverse ecological regions.

African contributions to ethnobotany have been largely underreported and underappreciated. Since tribal ethnobotany was typically an oral tradition, there is little documentation until the arrival of Arabic botanists during the medieval period. The following cursory review mentions only a tiny fraction of the thousands of ethnic groups, each with their own language, culture, and unique knowledge of plants. For more in-depth understanding of African ethnobotany, see Carney and Rosomoff (2011) and other publications in the reference section.

Africa has a long and intricate history, beginning with the appearance of the first humans, domestication of animals, and the emergence of agriculture and metalworking. During prehistoric times, local indigenous tribes inhabited most of Africa. Savanna tribes were subsistence hunter-gatherer societies, raising domestic cattle, sheep, and goats, while collecting wild grasses and other food plants. Some believe agriculture first developed around 5000 BCE in the Nile Valley and developed much later in the rest of Africa (c. 3000 BCE) (van der Veen 1999). However, there is evidence of sorghum [*Sorghum bicolor* (Linn.)Moench] cultivation by 7000–6000 BCE and cotton (*Gossypium herbaceum* L.) domestication by 5000 BCE in Sudan (Carney and Rosomoff 2011). Furthermore, the ancient grain teff [*Eragrostis tef* (Zucc.) Trotter] (Figure 4.2) was domesticated between 8000 and 5000 BCE in Ethiopia, supporting the theory that plant domestication likely occurred over a broad region of Sub-Saharan Africa. Other crops domesticated around this time include pearl millet [*Pennisetum glaucum (L.) R.Br.*], melons (Cucurbitaceae spp.), cowpea [*Vigna unguiculata* (L.) Walp.], groundnut [*Vigna subterranea* (L.) Verdc.], finger millet (*Eleusine coracana* Gaertn.), noog [*Guizotia abyssinica* (L.f.) Cass.], yam (*Dioscorea* spp.), and most likely bottle gourd [*Lagenaria siceraria* (Molina) Standl., Figure 4.3]. Even after the development of domesticated species, wild plants were still collected to supplement the food and medicine supply, especially in areas with unpredictable rainfall (Carney and Rosomoff 2011). Some of the later domesticated crops play a major role in today's economy, like coffee (*Coffea arabica* L. and *Coffea canephora* Pierre ex A.Froehner, syn. *C. robusta*), oil palm (*Elaeis guineensis* Jacq.), kola nut (*Cola* spp.), shea nut (*Vitellaria paradoxa* C.F. Gaertn.), gum arabic [*Acacia senegal* (L.) Willd., *A. seyal* Delile], locust bean [*Parkia biglobosa* (Jacq.) R. Br. ex G. Don], khat [*Catha edulis* (Vahl) Endl., Figure 4.4], tamarind (*Tamarindus indica* L.), okra [*Abelmoschus esculentus* (L.) Moench], and African rice (*Oryza glaberrima* Steud.).

There are many different ways to organize a summary of African ethnobotany. Special consideration could be given to West Africa, as this was the center of domestication for many economic plants. Trade with the Arabian Peninsula and the Far East influenced the Horn of Africa and Swahili coast before many regions had any contact with the

Figure 4.2 Teff (*Eragrostis tef*) harvest in Northern Ethiopia. © 2007 A. Davey Coogan.

Figure 4.3 Calabash gourds (*Lagenaria siceraria*) hanging from a support.

Figure 4.4 Chat (*Catha edulis*, qatt, jaad, or khat) market near Harar, in southeastern Ethiopia. Chat users prefer to chew the newest leaves from freshly harvested stems. For this reason, fresh chat stems are bought and sold at daily markets. © 2007 A. Davey Coogan.

outside world, making their ethnobotany quite unique compared to tribes from West Africa or the interior forests. But for simplicity, the following section separates the continent into north and south, roughly along the Sahara desert (Figure 4.1). The Sahara desert was a geographical barrier, separating the northern and southern tribes of Africa for millennia. Islam, trade with Europe, and invading armies heavily influenced North African culture. However, indigenous tribes south of the Sahara (Sub-Saharan Africa) historically had less contact with outsiders, especially those living in the central rainforests. So many of the tribes and cultures remain even today. One exception is the Horn of Africa on the northeast coast. While technically part of Sub-Saharan Africa, it was introduced to Islam early on, due to its proximity to the Arabian Peninsula.

Northern Africa

For the purposes of this section, Northern Africa is a relatively small strip of land north of and including the Sahara Desert (Figure 4.1). It is dominated by the Atlas Mountains and steppes, with some fertile farmland along the Mediterranean, Nile River, and some mountain valleys. North African recorded history begins with Ancient Egypt, Nubia, Maghreb (formerly Barbery Coast in Northwest Africa), and the Horn of Africa (peninsular Northeast Africa). The ancient Egyptians brought Cedars of Lebanon (*Cedrus libani* A.Rich.) to Egypt as early as 3000 BCE, using the resin for mummification. Archeological evidence shows the Egyptians grew a number of Mediterranean crops such as grapes (*Vitis vinifera* L.), garlic (*Allium sativum* L.), onions (*Allium* spp.), lettuce (*Lactuca sativa* L.), leeks (*Allium ampeloprasum* L.), and chufa (*Cyperus esculentus* L.). They also grew some native plants for medicine, spice, and oil such as senna (*Senna alexandrina* Mill.), sesame (*Sesamum indicum* L.), licorice (*Glycyrrhiza glabra* L.), cumin (*Cuminum cyminum* L.), black cumin (*Nigella sativa* L.), coriander (*Coriandrum sativum* L.), and ajwain (*Trachyspermum ammi* Sprague) (Table 4.1). Khat [*Catha edulis* (Vahl) Endl.] (Figure 4.4) is a stimulant plant native to the Horn of Africa. It contains

Table 4.1 Crops that originated in Northern Africa.

Medicine and stimulants		Spices, oils, and resins	
Bishop's weed	*Ammi majus* L., Apiaceae	Argan seed oil	*Argania spinosa* (L.) Skeels, Sapotaceae
Khat	*Catha edulis* (Vahl) Endl., Celastraceae	Black cumin	*Nigella sativa* L., Ranunculaceae
Licorice	*Glycyrrhiza glabra* L., Fabaceae	Coriander	*Coriandrum sativum* L., Apiaceae
Senna	*Senna alexandrina* Mill., Fabaceae	Cumin	*Cuminum cyminum* L., Apiaceae
Latex and resins		Sesame	*Sesamum indicum* L., Pedaliaceae
Ammoniac gum resin	*Ferula marmarica* Asch. & Taub. ex Asch. & Schweinf., Apiaceae		
Frankincense	*Boswellia* spp., Burseraceae		
Lagos silk	*Funtumia elastica* (Preuss) Stapf, Apocynaceae		
Myrrh	*Commiphora* spp., Burseraceae		

the phenylalanine-derived alkaloid cathinone, which is structurally similar to amphetamines. The practice of chewing fresh khat leaves in Egypt and other North African societies is thought to predate the use of coffee as a stimulant. The Egyptians used bishop's weed (*Ammi majus* L.) to treat vitiligo, a skin condition characterized by progressive depigmentation. Today the active compound, β-methoxypsoralen, is used for to treat several skin disorders.

The Greeks and Phoenicians arrived in Northern Africa around 900–600 BCE. The seafaring Phoenicians built ships from Cedar of Lebanon, maintaining maritime trade along the north coast of Africa and with cities along the Mediterranean and Red Sea. They were known for their valuable Tyrian purple dye made from murex sea snail shells. The Horn of Africa, particularly the city of Punt, supplied aromatic wood, myrrh (resins from *Commiphora* spp.), frankincense (resins from *Boswellia* spp.) (Figure 4.5), and luxury spices to Ancient Egypt, the Phoenicians, and other civilizations in the region. They traveled by sea in ships made of wood either from local *Acacia* spp. or Cedar of Lebanon obtained from Egypt and the Phoenicians. The Egyptians were not a seafaring people, but they did construct reed boats in the city of Saqqara. Egyptian and Phoenician demand for raw materials led to trade with Sub-Saharan Africa, which had developed rather

Figure 4.5 Common incense resins and powders (from top down, left to right) Makko powder (*Machilus thunbergii* Siebold & Zucc., Lauraceae), Borneol camphor [*Dryobalanops sumatrensis* (J.F.Gmel.) Kosterm., Dipterocarpaceae], Sumatra benzoin (*Styrax benzoin* Dryand., Styracaceae), Omani frankincense (*Boswellia sacra* Flueck., Burseraceae), guggul [*Commiphora wightii* (Arn.) Bhandari, Burseraceae], golden frankincense [*Boswellia papyrifera* (Caill. ex Delile) Hochst., Burseraceae], Tolu balsam [*Myroxylon balsamum* (L.) Harms, Fabaceae], Somalian myrrh [*Commiphora myrrha* (Nees) Engl., Burseraceae], labdanum (*Cistus creticus* L., Cistaceae), opoponax [*Commiphora erythraea* (Ehrenb.) Engl., Burseraceae], and white Indian Sandalwood powder (*Santalum album* L., Santalaceae). © 2007 S. J. S. Chen.

independently until this point (c. 500 BCE). The Phoenicians sailed down the eastern coast, trading beads and jewelry for raw materials like antimony and tin in port towns as far south as present day South Africa. They may have even planted crops along the coast to keep themselves fed during the long voyage. The Garamantes of the central Sudan became a major power by planting Mediterranean crops like wheat (*Triticum* spp.) and barley (*Hordeum vulgare* L.) and adopting irrigation systems originally developed by the Persians (Carney and Rosomoff 2011). They also imported agricultural slaves from the Sahel (transition zone south of the Sahara). The slave farmers introduced two important crops domesticated in the Sahel: pearl millet and sorghum. Since these two crops were adapted to hot, dry weather, kingdoms on the north coast could now grow Mediterranean vegetables and cereal grains (e.g., wheat and barley) in the winter and sorghum and millet in the summer. This increase in agricultural productivity was critical for the growth of their civilizations. Later in this chapter, the ethnobotanical contribution of African slaves across the world (the African Diaspora) will be explored further.

Sometime during the sixth century CE, Muslims from the Arabian Peninsula overtook Persia, Western Asia, and much of Northern Africa, extending down the Swahili coast. Their trading networks with the Far East brought citrus (Rutaceae), sugarcane (*Saccharum* spp.), and Asian rice (*Oryza sativa* L.) to the region, but they also exported African slaves and several Sub-Saharan African crops like sorghum, millet, melons, coffee, kola nut, grains of paradise (*Aframomum melegueta* K.Schum.), and gum arabic to their widespread kingdom. Even after the Arab conquest in North Africa, Berber tribes remained functionally in control of the northwest region. Many native Berbers learned Arabic and shifted their ethnic identity from Berber to Arab after converting to Islam. Some migrated to Muslim-occupied southern Spain, introducing sorghum to the region. Berber goat herders used the argan tree [*Argania spinosa* (L.) Skeels] (Figure 4.6) leaves and fruit as a source of forage for their goats, leading to widespread damage to the species. Today, argan trees are protected for their valuable seed oil. Argan oil is an important source of income for women in modern-day Morocco.

From the sixteenth through the nineteen centuries, European powers fought for control of Northern Africa. In 1912, the Italian Alessandro Trotter was hired to be the agronomist on an expedition to Italian-controlled Tripolitania, Libya. He documented local species along with information on plant uses from the local people, knowledge that could be useful to new Italian settlers (De Natale and Pollio 2012). His collection reflects the history of trade in the area. There were plants from Mediterranean (e.g., Greece, Turkey, Morocco), Asiatic, and European origin. Ironically, few drugs were made from indigenous plants. One exception was ammoniac of Cyrenaica gum resin from *Ferula marmarica* Asch. & Taub. ex Asch. & Schweinf. Comparing Trotter's collection with other ethnobotanical studies across North Africa reveals a shared traditional medicine along the coast, which is likely a reflection of shared climate, culture, and language (De Natale and Pollio 2012).

Southern Africa

There are a wide range of climates and ecosystems south of the Sahara Desert including hot savannas and deserts, tropical rainforests, and broadleaf forests. Archeological evidence reveals that costal communities maintained trade relations with the Phoenicians,

Figure 4.6 Goats in an argan tree (*Argania spinosa*), Morocco. © 2014 Mikel Santamaria

Figure 4.7 Tamarind comes from the sticky pulp of a legume.

India, and China more than a millennium before the arrival of Arab traders. As early as 3000 BCE, people from the Ethiopian Highlands and East African Highlands brought their goods to the Horn of Africa and the Swahili coast. From there, seafaring merchants used the monsoon winds to traverse the Indian Ocean to trade with South Asia. They introduced sorghum, pearl millet, finger millet, cowpea, and watermelon [*Citrullus lanatus* (Thunb.) Matsum. & Nakai] to Asia between 2000 and 1200 BCE (Carney and Rosomoff 2011). Quite possibly they also introduced pigeon pea [*Cajanus cajan* (L.) Millsp.] and castor bean (*Ricinus communis* L.), but their origins are still uncertain (i.e., Africa versus India or both). Introduction of tamarind (Figure 4.7) and okra soon followed. In exchange, African traders brought back taro [*Colocasia esculenta* (L.) Schott] and banana [*Musa* spp.; see case study titled "Banana (*Musa* spp.) as a Traditional Treatment for Diarrhea" in Chapter 8]. Banana became an incredibly important crop across Africa. It was easy to grow, reproduced rapidly, and produced edible starchy stems and fruit. At least 180 cultivars were developed, providing farmers in tropical inland regions like the Congo Basin and the Great Lakes with a reliable crop to trade with the rest of the continent.

Numerous forest-dwelling Pygmy, Bantu, and Khoisan tribes have inhabited most of Southern Africa since prehistoric times. The kingdom of Ghana (c. 300–1235 CE), located in present-day Mali and Mauritania in West Africa, was the first to gain control over the southern end of the Sahara. They used domesticated camels to establish gold trade routes to the north coast. Slaves from Lake Chad, ivory, ostrich feathers, salt, and kola nuts were also sent up the Sahara trade routes to be sold to the Arabs. The kola tree [*Cola nitida* (Vent.) Schott & Endl.], native to West African rainforests, produces caffeine-containing kola seeds or "nuts" (Figure 4.8). Kola nuts have been chewed as a stimulant across West Africa in social and ceremonial settings for centuries. Extracts from kola seeds eventually become famous worldwide as a component of Coca-Cola.

West Africa contains a savanna-to-forest transition zone in Cameroon and Nigeria that is thought to contain the richest plant diversity on the continent (Carney and Rosomoff 2011). It is the origin of several protein-rich crops including pigeon pea and cowpea, as well as coffee, okra, tamarind, oil palm, grains of paradise, and yams. There is evidence that oil palm (Figure 4.9) was used as early as 3000 BCE. Egyptian tombs contained casks of palm oil, a reflection of its importance in their society. Found throughout West Africa, the oil palm provided communities with not only oil for cooking, fuel, and personal care products, but also the leaves and trunk were used for building materials, the roots were used for medicine, and the fruit and sap were fermented into wine. Today, palm oil is a huge industry, embroiled in many environmental and social controversies. In most West African countries, it remains an important source of income for women, as they are the ones in charge of the processing and sale of the oil (Carrere 2013).

Madagascar was not settled until c. 600 CE by people from Southeast Asia and eastern Africa. The people from Southeast Asia brought banana and Asian rice cultivation to the island along with several other new crops. As agriculture expanded in Madagascar and southeastern Africa, there was an increase in population and migration of people to the region. Madagascar

Figure 4.8 Fruits of *Cola nitida* contain stimulating kola nuts. © 2012 Michael Hermann.

Figure 4.9 Fruit of the oil palm (*Elaeis guineensis*). © 2015 Tafilah Yusof.

Figure 4.10 Madagascar periwinkle (*Catharanthus roseus*) flower. © 2013 Bishnu Sarangi.

periwinkle [*Catharanthus roseus* (L.) G.Don] (Figure 4.10) is a native ornamental that has been used in traditional medicine for a wide range of ailments. Today it is the source of the antihypertensive monoindole alkaloids ajmalicine and serpentine and the antileukemic bisindole alkaloids vincristine and vinblastine. The latter are antimitotics that bind to tubulin, blocking the formation of microtubules required for cell division. All of the alkaloids are too complex for chemical synthesis, so they are still derived from the plant. It is worth noting that Madagascar periwinkle can be highly neurotoxic if taken in sufficiently high doses.

By the end of the Middle Ages, interior forest kingdoms were trading forest products (e.g., shea nuts and butter, kola nuts, and gum arabic) through established trade routes. Gold mines in southeast Africa brought wealth to the region, dominated by grazing cattle and cotton fields. By the sixteenth century, Portuguese explorers were looking for a route around Africa to reach the spices of India and Southeast Asia. They established colonies in Guinea and Angola in the west, to take advantage of the slave trade. In the east, they explored the region around Mozambique in hopes of finding gold. Their settlements were far from the goldmines, but they remained an unchallenged presence in the region until the arrival of the British and Dutch in the seventeenth century. The British and Dutch were also on a mission to find a trade route around Africa, but they required ports to restock their ships on the long voyages. The British East India Company set up a settlement on the island of Saint Helena in 1659; it remains in British control today. The Dutch East India Company made a permanent settlement at the southern tip of Africa at the Cape of Good Hope below Table Mountain in 1652, known as Cape Town today. They established trade with the Khoikhoi tribes, acquiring cattle and eventually all of their grazing land to the northeast through warfare and smallpox outbreaks.

Herbal medicine has been used extensively throughout Africa since prehistoric times. Some of the most common remedies are listed in Table 4.2. Two of the Africa case studies provide an in-depth look at traditional African medicines for treating malaria and HIV. Combined, these two diseases claim at least 1.5 million African lives each year, creating urgency for effective treatments. Unlike many of the traditional medicines listed in Table 4.2, the treatments discussed in the case studies are not well known and in most cases, not well understood. The third case study profiles vanilla, a flavoring and

Table 4.2 Common medicinal plants from Southern Africa.

Plant or medicine	Latin name	Primary traditional use
Aloe	*Aloe* spp., Xanthorrhoeaceae	Laxative, topical for burns and wounds
Artemisia	*Artemisia herba-alba* Asso, Asteraceae	Antiseptic
Bitter melon	*Momordica charantia* L., Cucurbitaceae	Diabetes
Devil's claw	*Harpagophytum procumbens* (Burch.) DC., Pedaliaceae	Anti-inflammatory and analgesic
Gotu kola	*Centella asiatica* (L.) Urb., Apiaceae	Skin disorders
Gum arabic	*Acacia senegal* (L.) Willd., *A. seyal* Delile, Fabaceae	Infections
Honeybush tea	*Cyclopia genistoides* (L.) Vent., Fabaceae	Urinary and stomach health
Madagascar periwinkle	*Catharanthus roseus* (L.) G.Don, Apocynaceae	Bitter tonic, promotes lactation, emetic
Rooibos tea	*Aspalathus linearis* (Brum.f.) R. Dahlgren, Fabaceae	Soothing for colicky babies; general health tea
Umckaloabo tubers	*Pelargonium sidoides* DC, Geraniaceae	Respiratory infections
Yohimbe	*Pausinystalia johimbe* (K.Schum.)Pierre ex Beille, Rubiaceae	Aphrodisiac

Table 4.3 Crops that originated in Southern Africa.

Fruits and vegetables		Grains and legumes	
Bottle gourd	*Lagenaria siceraria* (Molina) Standl, Cucurbitaceae	African rice	*Oryza glaberrima* Steud., Poaceae
Ensete	*Ensete ventricosum* (Welw.) Cheesman, Musaceae	Cowpea	*Vigna unguiculata* (L.) Walp, Fabaceae
Natal plum	*Carissa macrocarpa* (Eckl.) A.DC. Apocynaceae	Finger millet	*Eleusine coracana* Gaertn., Poaceae
Okra	*Abelmoschus esculentus* (L.) Moench, Malvaceae	Groundnut	*Vigna subterranea* (L.) Verdc., Poaceae
Sugar plum	*Uapaca kirkiana* Müll.Arg., Phyllanthaceae	Locust bean	*Parkia biglobosa* (Jacq.) R. Br. ex G. Don, Fabaceae
Tamarind	*Tamarindus indica* L., Fabaceae	Pearl millet	*Pennisetum glaucum* (L.) R.Br., Poaceae
Watermelon	*Citrullus lanatus* (Thunb.) Matsum. & Nakai, Cucurbitaceae	Sorghum	*Sorghum bicolor* (Linn.) Moench, Poaceae
Yam	*Dioscorea* spp., Dioscoreaceae	Teff	*Eragrostis tef* (Zucc.) Trotter, Poaceae
Fibers and oils		**Stimulants**	
Cotton	*Gossypium herbaceum* L., Malvaceae	Coffee	*Coffea arabica* L., *C. canephora* Pierre ex A. Froehner, Rubiaceae
Jute	*Corchorus olitorius* L., Malvaceae	Khat	*Catha edulis* (Vahl) Endl., Celastraceae
Lagos silk	*Funtumia elastica* (Preuss) Stapf, Apocynaceae	Kola	*Cola nitida* (Vent.) Schott & Endl., Malvaceae
Palm oil	*Elaeis guineensis* Jacq., Arecaceae	**Gums and spices**	
Roselle	*Hibiscus sabdariffa* L., Malvaceae	Grains of paradise	*Aframomum melegueta* K.Schum.
Shea butter	*Vitellaria paradoxa* C.F.Gaertn., Sapotaceae	Gum arabic	*Acacia senegal* (L.) Willd., *A. seyal* Delile, Fabaceae
Kenaf	*Hibiscus cannabinus* L., Malvaceae		

fragrance from the vanilla orchid. Although there are several traditional African remedies derived from orchids (see Pant 2013 for a review), sales of vanilla beans and extracts eclipses them economically. Table 4.3 lists other economically important crops that, unlike vanilla, originated in Southern Africa.

The African Diaspora

The African people were exploited as slaves for centuries, transported from their homes to other parts of Africa, the Arabian Peninsula, Europe, and the Americas. This forced migration led to the African Diaspora, the cultural legacy of African communities dispersed throughout the world. Slaves brought sorghum and millet to Northern Africa from their homelands south of the Sahara. Slaves forced to work in the sugarcane fields of Mesopotamia introduced sorghum, millet, and melons. Just as European explorer's ships carried botanical specimens from exotic lands home to Europe, slave ships unintentionally carried an African botanical legacy to the New World. African plants were not typically transported for the purpose of trade or botanical research, but to be sustaining food and medicine for the millions of West and Central African slaves transported across the Atlantic. Typical rations consisted of African legumes mixed in a starchy gruel made of yams or manioc. Tamarind and kola nut were used to improve the palatability of stagnant water (Carney and Rosomoff 2011). It was the African slaves themselves that took the initiative to plant their native crops in the New World including eggplant, okra, yams, kola nut, tamarind, cowpea, and other African legumes. In the New World tropics, they used their ethnobotanical knowledge and African crops to support whole European settlements unfamiliar with tropical farming (see Carney and Rosomoff 2011, ch. 6, for further reading). And while enslaved Native Americans were also successful at raising their staples in tropical America, only the African slaves had dual knowledge of tropical crop production and raising livestock introduced to the continent by the Europeans.

Much of the American success story can indeed be credited to African slaves. During the late seventeenth century, Africans from the Upper Guinea coast brought along African rice (*Oryza glaberrima* Steud.) and the knowledge of how to successfully grow rice to the New World. Upper Guinean slaves with rice-growing skills were a minority of all slaves bound for the New World (about 1 in 5) during the early period of rice cultivation (Eltis *et al.* 2006). After 1750, they were preferentially shipped to the Carolina low country rice-growing region and fetched a higher price than ordinary slaves because of their specialized ethnobotanical knowledge. They planted a cultivar from West Africa that was eventually named Carolina Gold in the swamps of the Carolinas. Carolina Gold quickly became a high-value export crop to Europe.

European settlers that landed in tropical regions of the New World did not know how to grow tropical crops. Their food supply was highly dependent on African slaves and their knowledge of African crops (Figure 4.11). Sometimes slave owners allowed their slaves to grow a garden with their own traditional crops to supplement the food supply. Marooned or runaway slaves established their own farms raising African crops like cowpeas, okra (Figure 4.12), rice, and jute mallow (leaves from *Corchorus* spp.). Sometimes these plants were used in ceremonies to honor ancestors or as an offering to gods (Carney and Rosomoff 2011). In Suriname, rice was their primary food source. Forward-thinking women would smuggle rice in their hair when they were taken from Africa or escaped from a slave plantation (Eltis *et al.* 2005) to ensure their community had access to the best rice cultivars. Slaves taken from various cultures throughout Africa grew plants and prepared the dishes from their individual homelands, sharing their culinary and ethnobotanical knowledge with their fellow slaves. The combined knowledge of many individual African ethnic groups (e.g., Wolof, Mandinka, Serer, Jola, Balanta, and Fula people) was a reinvention of traditional African ethnobotany. This amalgamation of knowledge became known collectively as "African" (i.e., African plants, African cuisine, African traditional medicine) in the Americas. Traditional African dishes shaped the cuisine from Brazil through the Caribbean to the US south. Gumbo, a popular dish in Louisiana, comes from the Bantu word for okra (Baudry des Lozières

Figure 4.11 African American men and women planting sweet potatoes [*Ipomoea batatas* (L.) Lam.] or yams (*Dioscorea* spp.) at James Hopkinson's Plantation circa 1862/63 (photo by Henry P. Moore).

Figure 4.12 A row of okra (*Abelmoschus esculentus*).

1803). The word *yam* is derived from a West African Wolof word *nyami* meaning to eat or taste (Eisnach and Covey 2009). Yams were frequently provided as sustenance to slaves on the Atlantic crossing. They are a high-yielding nutritious crop with special spiritual and cultural significance to some West African tribes (Eisnach and Covey 2009; Bird 2010). The Igbo and Igala people of Nigeria display yams to honor Ela, the Earth Mother. African communities in southwest Lousiana ritualistically cut sweet potatoes and use them as amulets during the Yambilee festival (Bird 2010). In the *Isu Egbegbe* or "mysterious yam ritual," a yam is cut and placed in a bottle of water hung above the ground. A pregnant woman is supposed to drink the yam water for three days to induce vomiting and purge the body of toxins and disease (Bird 2010). In the Americas, cowpeas (*Vigna unguiculata*), also called black-eyed peas, are the foundation of dishes like Hoppin' John (US South), beans and rice (throughout the Caribbean and South America), *akara* (Brazil and throughout the Caribbean), and *buñuelos de fríjol de cabecita negra* (Colombian Atlantic coast).

References and Additional Reading

Baudry Des Lozières, L.-N. (1803) *Second Voyage à la Louisiane*. Chez Charles, Paris.

Bird, S.R. (2010) *The Big Book of Soul: The Ultimate Guide to the African American Spirit*. Hampton Roads Publishing, Newburyport.

Carney, J.A., and Rosomoff, R.N. (2011) *In the Shadow of Slavery: Africa's Botanical Legacy in the Atlantic World*. University of California Press, Berkeley.

Carrere, R. (2013) *Oil Palm in Africa, Past Present and Future Scenarios*. World Rainforest Movement, Montevideo.

De Natale, A., and Pollio, A. (2012) A forgotten collection: the Libyan ethnobotanical exhibits (1912–14) by A. Trotter at the Museum O. Comes at the University Federico II in Naples, Italy. *Journal of Ethnobiology and Ethnomedicine*, **8**, 4.

Eisnach, D., and Covey, H.C. (2009) *What the Slaves Ate: Recollections of African American Foods and Foodways from the Slave Narratives*. Greenwood Press, Santa Barbara.

Eltis, D., Morgan, P., and Richardson, D. (2006) Slave prices, the African slave trade, and productivity in eighteenth-century South Carolina: A reassessment. *The Journal of Economic History*, **66** (04), 1054–1065.

Eltis, D., Morgan, P., and Richardson, D. (EMR) (2005) The African contribution to rice cultivation in the Americas. In *Georgia Workshop in Early American History and Culture*. University of Georgia, Athens, GA.

Gurib-Fakim, A. (2006) Medicinal plants: Traditions of yesterday and drugs of tomorrow. *Molecular Aspects of Medicine*, **27** (1), 1–93.

Juliani, H.R., Simon, J.E., and Ho, C.-T. (2009) *African Natural Plant Products: New Discoveries and Challenges in Chemistry and Quality*. American Chemical Society, Washington, DC.

Mahomoodally, F.M. (2013) Traditional medicines in Africa: An appraisal of ten potent African medicinal plants. *Evidence-Based Complementary and Alternative Medicine*. Available from: <http://www.hindawi.com/journals/ecam/2013/617459/> [11 November, 2015].

Neuwinger, H.D. (1996) *African Ethnobotany: Poisons and Drugs*. Chapman & Hall, London.

Pant, B. (2013) Medicinal orchids and their uses: Tissue culture a potential alternative for conservation. *African Journal of Plant Science*, **7** (10), 448–467.

Van der Veen, M. (1999) *The Exploitation of Plant Resources in Ancient Africa*. Springer, New York.

Voeks, R., and Rashford, J. (2012) *African Ethnobotany in the Americas*. Springer, New York.

Wild, A. (2004). *Coffee: A Dark History*. W. W. Norton & Company, New York.

Zohary, D., Hopf, M., and Weiss, E. (2012) *Domestication of Plants in the Old World: The origin and spread of domesticated plants in Southwest Asia, Europe, and the Mediterranean Basin*, 4th edition. Oxford University Press, Oxford.

Achillea millefolium: Re-Exploring Herbal Remedies for Combating Malaria

(M. H. Grace, P. J. Smith, I. Raskin, M. A. Lila)

Ethnobotany and Ethnopharmacology

Globally, an estimated 3.2 billion people are at risk of being infected with malaria, and 1.2 billion are at high risk of developing the disease. According to the latest World Health Organization (WHO) estimates, 212 million cases of malaria occurred globally in 2015, and the disease led to 429,000 deaths. The burden is heaviest in Africa, where an estimated 92% of all malaria deaths occur, and 78% of all deaths occur in children less than 5 years old (WHO 2016). The very high prevalence of this disease and the increased resistance of the dangerous parasite (*Plasmodium falciparum*) to currently used treatments have motivated a continuous search for new antimalarial compounds. Plants have been historically used as a first-line remedy for the infection in malaria-endemic areas (Bourdy *et al.* 2008), and many have served as a source of novel antimalarial compounds (Kaur *et al.* 2009), as most evidenced by the antimalarial agents quinine and artemisinin, initially isolated from *Cinchona* spp. (Smith 1976), and *Artemisia annua* (Antimalaria studies on Qinghaosu 1979), respectively. As resistance develops to existing drugs, new ones need to be introduced. Artemisinin-based combination therapy (ACT) which relies on mixtures of compounds with alternative modes of action is the most recommended current natural product treatment for *P. falciparum* malaria.

During World War II, a project was initiated by Merck & Co. in response to US government appeals to discover new plant-based antimalarial compounds. Quinine, the phytoactive compound most commonly used to treat malaria at the time, was in dwindling supply, and combat soldiers in malaria-prone regions of Southeast Asia were at risk. A small team of top Merck scientists partnered with New York Botanical Garden botanists and succeeded in describing nearly 80 plant species (of more than 600 tested), mainly from the United States, active in vivo against avian malaria (Spencer *et al.* 1947). However, as the war ended and quinine supplies returned to normal, the funding for this ambitious research project was abandoned, and the promising leads were merely archived and never advanced to late stages of drug development. Over 50 years later, after reading some reports about the natural products research being conducted in our research teams, one of the original Merck researchers, Christine Malanga Wilson, offered to share the archived prioritized leads with Dr. Ilya Raskin, Rutgers University. This profoundly significant work, particularly in the context of emerging drug resistance to current antimalarials, won the financial support of The Medicines for Malaria Venture (MMV) organization. In collaboration with researchers at the

University of Cape Town, South Africa, we were able to follow up on these significant and largely forgotten antimalarial screening efforts to discover novel antimalarial compounds, but now, with the advantage of state-of-the-art analytical instrumentation (high field nuclear magnetic resonance spectroscopy (950 MH$_Z$), liquid chromatography-ion trap-top of flight mass spectrometry, and centrifugal partition chromatography) and streamlined antiplasmodial bioassays to facilitate the investigations. Many plants from the original WWII-era screening were re-evaluated in vitro for their antiplasmodial activity, and the extracts which showed most activity were further investigated, through bioassay-guided fractionation, to isolate active compounds. Promising antimalarial leads identified from this project to date include sesquiterpenes [*Canella winterana* (Grace *et al.* 2010) and *Liriodendron tulipifera* (Graziose *et al.* 2011)], *N*-alkylamides [*Spilanthes acmella* (Mbeunkui *et al.* 2011)], alkaloids [*Liriodendron tulipifera* (Graziose *et al.* 2012) and *Geissospermum vellosii* (Mbeunkui *et al.* 2012)], triterpenes [*Cornus florida* (Graziose *et al.* 2012)], glycolipids [*Cassia fistula* (Grace *et al.* 2012)], and cucurbitacin glycosides [*Datisca glomerata* (Graziose *et al.* 2013)].

We report here on one typical case study based on the MMV-sponsored follow-up to the original WWII era screening trials. *Achillea millefolium* L. (Asteraceae) (Figure 4.13), known commonly as yarrow, is one of the most widespread medicinal plants in the world. It grows wild all around Europe, Asia, North Africa, and North America. *A. millefolium* has been used in folk medicine for ages for treatment of wounds and infectious diseases as well as many other ailments. Recent studies reported

Figure 4.13 *Achillea millefolium* L. © 2016 Mary Grace.

antimicrobial, antiphlogistic, hepatoprotective, antispasmodic, and calcium antagonist activities of its polar extracts (Stojanovic *et al.* 2005; Yaeesh *et al.* 2006) and a protective effect of its infusions against H_2O_2-induced oxidative damage in human erythrocytes and leucocytes (Konyalioglu and Karamenderes 2005). The species has been used to heal malaria in the folk medicine of Europe since the sixteenth century (Adams *et al.* 2011), especially in Southern Italy (Calabria), as evidenced in writings produced between the nineteenth and twentieth centuries (Tagarelli *et al.* 2010). The same species was also dispersed to the Pacific, and is now used as an antimalarial in Indonesia (Murnigsih *et al.* 2005). *A. millefolium* was among plants used by the local inhabitants of Pakistan to treat malaria (Shah and Khan 2006; Shah *et al.* 2014). Despite ample ethnobotanical history of use, few modern studies have isolated or identified the critical phytoactive constituents responsible for some of the historically observed activities. *Achillea* spp. are rich in secondary metabolites, which may explain their wide variety of medicinal uses. As it belongs to the Asteraceae, *A. millefolium*, the subject of this case study, is rich in sesquiterpenes. However, the antiplasmodial activity of its sesquiterpenes has never been evaluated, and only a single previous study described the antimalarial activity of flavonoids from this plant source (Vitalini *et al.* 2011).

Chemistry and Bioactivity

Many of the pharmacological activities of *Achillea millefolium* can be attributed to its sequiterpene components, which work in concert with other phytoactive compounds. Since many sesquiterpene lactones have been found to be active against *Plasmodium* parasites (Francois *et al.* 1996; Francois *et al.* 2004; Woerdenbag *et al.* 1994), the antiplasmodial evaluation focused on the lipophilic extract from *A. millefolium*. The dried aerial parts were extracted with dichloromethane, organic solvent evaporated, and the dried residue was fractionated over a silica gel column using a solvent gradient of increasing ethyl acetate in hexanes. Similar fractions were pooled together according to their thin layer chromatography profile to afford 10 fractions. The obtained fractions were tested for their inhibitory activity against the chloroquine-sensitive strain of *Plasmodium falciparum* (D10). Active fractions ($IC_{50} < 10$ µg/mL) were further fractionated and monitored for their antiplasmodial activity. Purification was, however, focused on those fractions showing significant enrichment of antiplasmodial activity upon fractionation so as to isolate the compounds primarily responsible for the observed biological activity. Four active compounds were isolated and identified as sesquiterpene lactones of the germacranolide carbon skeleton **1–4** (Figure 4.14). Structural elucidation was facilitated by spectroscopic data (NMR spectroscopy and mass spectrometry) and by direct comparison with published spectral data.

Sesquiterpene lactones are characteristic compounds not only for the genus *Achillea* but for the whole family Asteraceae. Among sesquiterpene lactones, germacranolide compounds, millefin, achillifolin, dihydroparthenolide, and balchanolide are mentioned most frequently in this genus (Wichtl 1997).

Isolated compounds **1–4** were evaluated for their antiplasmodial activity using the chloroquine-sensitive *Plasmodium* strain (D10) and cytotoxicity using the Chinese Hamster Ovarian (CHO) cell line. Results indicated moderate activity, with some toxicity (Table 4.4). Compound **2** showed the highest potency against the parasite, but also was most toxic to CHO cells, which indicates that the antiplasmodial activity was correlated with its general cytotoxicity. Compounds **1**, **2**, and **3** were significantly more active than

costunolide (1)

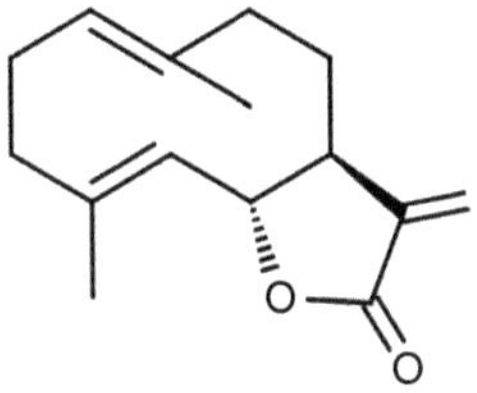

tulipinolide (2)

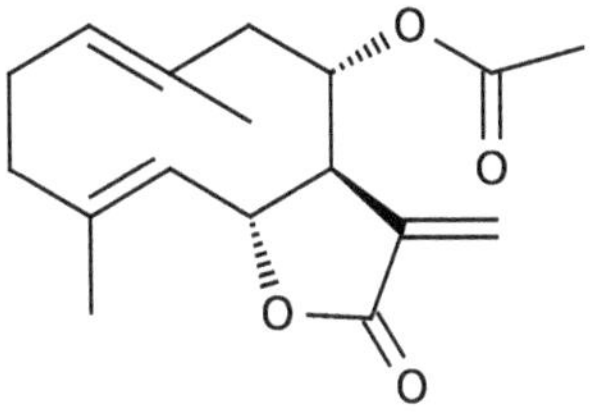

balchanolide acetate (3)

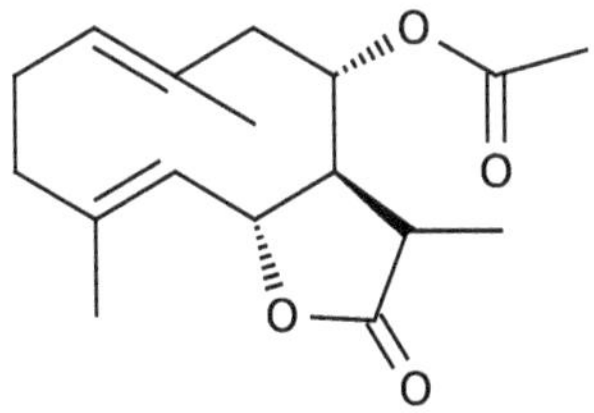

chihuahuin acetate (4)

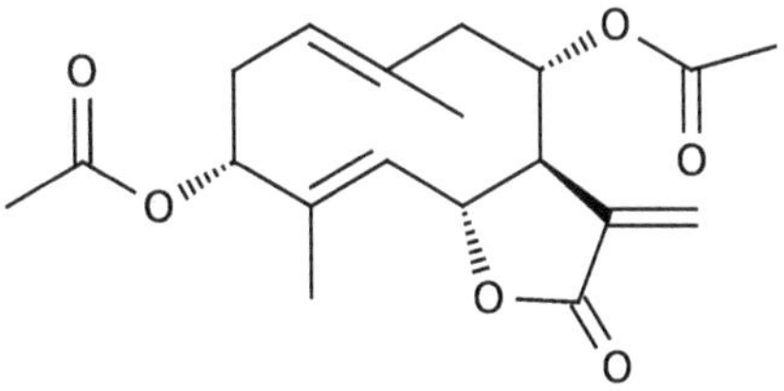

Figure 4.14 Chemical structures of sesquiterpene lactones isolated from *Achillea millefolium*.

Table 4.4 Antiplasmodial activity and cytotoxicity of sesquiterpene lactones **1-4** from *Achillea millefolium*.

Compounds	Plasmodium falciparum D10 (IC$_{50}$)		CHO cell line (IC$_{50}$)		SI
	µg/mL	µM	µg/mL	µM	
1	5.86 ± 1.0	25.48 ± 4.3	16.62 ± 2.2	72.26 ± 9.6	2.8
2	5.15 ± 1.1	17.76 ± 3.8	2.10 ± 0.8	7.24 ± 2.8	0.4
3	15.51 ± 2.0	53.12 ± 6.9	17.97 ± 2.0	61.96 ± 6.8	1.7
4	8.95 ± 2.0	25.7 ± 5.7	5.18 ± 1.5	14.88 ± 4.3	0.6
Chloroquine	0.1 ± 0.01	0.03 ± 0.03	—	—	—
Emetine	—	—	0.2 ± 0.1	0.40 ± 0.2	—

All experiments were performed in triplicate. Chloroquine was used as a positive control for antiplasmodial activity, and emetine for cytotoxicity in Chinese Ovarian hamster (CHO) cells. Selectivity Index (SI) = IC$_{50}$ CHO/IC$_{50}$ D10.

compound **3** alone, which suggests that the C-11-13 exocyclic double bond is responsible for activity. A previous report on the antiplasmodial activity of closely related germacranolide sesquiterpenes from *Oncosiphon piluliferum* (Asteraseae) indicated that reduction in the C-11-13 resulted in a significant decrease in activity (Pillay *et al.* 2007). A previous report about the growth inhibitory properties of sesquiterpene lactones indicated that the presence of an exocyclic methylene conjugated to a γ-lactone is a principal requirement for biological activity (Rodriguez *et al.* 1976). Although many sesquiterpene lactones have previously been tested for their antiplasmodial activity, this is the first report that has reported compounds **1–4** in terms of this activity. This research supports the traditional use of this plant in some parts of the world to treat malaria and provides, for the first time, direct evidence linking its phytoactive constituents to antimalarial activities. The isolated compounds could also be used as scaffolds to generate leads with enhanced antiplasmodial activity and reduced cytotoxicity.

Current and Future Status of Natural Products Research for Treatment of Malaria

Today, malaria continues to be a threat, with multiple risk areas identified around the world, including many regions currently reporting high incidence of anti-malarial drug resistance. Malaria is estimated to account for half of the absenteeism in African schools, and can contribute to lasting learning disorders.

Just within the past decade, progress in development of a vaccine against malaria seemed to be accelerating, thanks to advances in the technology. Increased funding and research is driving the discovery of new antigens and vaccine technologies, and many more malaria vaccine candidates are moving through the development pipeline. As recently as 2012, GlaxoSmithKline completed Phase 3 clinical development on the first candidate malarial vaccine. Enthusiasm about the potential for eradicating this disease through vaccine development actually caused a dramatic shift in research funding, away from natural product leads and into candidate vaccines.

This radical realignment of funding has had unfortunate consequences for sustained malaria control and treatment. Despite intensive work and funding, the development of effective malaria vaccines has not been as successful or rapid as anticipated. Multiple hurdles to vaccine introduction include (1) the lack of biomarkers for immunity, meaning that candidate vaccines can only be evaluated via expensive and time-intensive clinical trials, (2) few new antigens or adjuvants have been discovered over the past decade, and (3) correlative bioassays need to be developed to accelerate progress with vaccines.

Meanwhile, natural products remain a primary source for lead compounds with new/different mechanisms of action, and there are still a wealth of promising candidate plants and other natural products that need to be more thoroughly evaluated using the advanced laboratory tools available today. Rather than abandoning the established natural products leads in favor of vaccine development, a more prudent course of action would be to continue funding both lines of defense, as a better strategy for eventually overcoming this deadly disease.

References

Adams, M., Alther, W., Kessler, M., Kluge, M., and Hamburger, M. (2011) Malaria in the Renaissance: remedies from European herbals from the 16th and 17th century. *Journal of Ethnopharmacology*, **133**, 278–288.

Antimalaria studies on Qinghaosu (1979). *Chinese Medical Journal*, **92**, 811–816.

Bourdy, G., Willcox, M.L., Ginsburg, H., Rasoanaivo, P., Graz, B., and Deharo, E. (2008) Ethnopharmacology and malaria: new hypothetical leads or old efficient antimalarials? *International Journal for Parasitology*, **38**, 33–41.

Francois, G., and Passreiter, C.M. (2004) Pseudoguaianolide sesquiterpene lactones with high activities against the human malaria parasite *Plasmodium falciparum*. *Phytotherapy Research*, **18**, 184–186.

Francois, G., Passreiter, C.M., Woerdenbag, H.J., and Van Looveren, M. (1996) Antiplasmodial activities and cytotoxic effects of aqueous extracts and sesquiterpene lactones from *Neuralaena lobata*. *Planta Medica*, **62**, 126–129.

Grace, M.H., Lategan, C., Graziose, R., Smith, P.J., Raskin, I., and Lila, M.A. (2012) Antiplasmodial activity of the ethnobotanical plant *Cassia fistula*. *Natural Product Communications*, **7**, 1263–1266.

Grace, M.H., Lategan, C., Mbeunkui, F. *et al.* (2010) Antiplasmodial and cytotoxic activities of drimane sesquiterpenes from *Canella winterana*. *Natural Product Communications*, **5**, 1869–1872.

Graziose, R., Grace, M.H., Rathinasabapathy, T. *et al.* (2013) Antiplasmodial activity of cucurbitacin glycosides from *Datisca glomerata* (C. Presl) Baill. *Phytochemistry*, **87**, 78–85.

Graziose, R., Rathinasabapathy, T., Lategan, C. *et al.* (2011) Antiplasmodial activity of aporphine alkaloids and sesquiterpene lactones from *Liriodendron tulipifera* L. *Journal of Ethnopharmacology*, **133**, 26–30.

Graziose, R., Rojas-Silva, P., Rathinasabapathy, T. *et al.* (2012) Antiparasitic compounds from *Cornus florida* L. with activities against *Plasmodium falciparum* and *Leishmania tarentolae*. *Journal of Ethnopharmacology*, **142**, 456–461.

Kaur, K., Jain, M., Kaur, T., and Jain, R. (2009) Antimalarials from nature. *Bioorganic & Medicinal Chemistry*, **17**, 3229–3256.

Konyalioglu, S., and Karamenderes, C. (2005) The protective effects of *Achillea* L. species native in Turkey against H(2)O(2)-induced oxidative damage in human erythrocytes and leucocytes. *Journal of Ethnopharmacology*, **102**, 221–227.

Mbeunkui, F., Grace, M.H., Lategan, C., Smith, P.J., Raskin, I., and Lila, M.A. (2011) Isolation and identification of antiplasmodial N-alkylamides from *Spilanthes acmella* flowers using centrifugal partition chromatography and ESI-IT-TOF-MS. *Journal of Chromatography B*, **879**, 1886–1892.

Mbeunkui, F., Grace, M.H., Lategan, C., Smith, P.J., Raskin, I., and Lila, M.A. (2012) In vitro antiplasmodial activity of indole alkaloids from the stem bark of *Geissospermum vellosii*. *Journal of Ethnopharmacology*, **139**, 471–477.

Murnigsih, T., Subeki, Matsuura, H. *et al.* (2005) Evaluation of the inhibitory activities of the extracts of Indonesian traditional medicinal plants against *Plasmodium falciparum* and *Babesia gibsoni*. *Journal of Veterinary Medical Science*, **67**, 829–831.

Pillay, P., Vleggaar, R., Maharaj, V.J., Smith, P.J., and Lategan, C.A. (2007) Isolation and identification of antiplasmodial sesquiterpene lactones from *Oncosiphon piluliferum*. *Journal of Ethnopharmacology*, **112**, 71–76.

Rodriguez, E., Towers, G.H.N., and Mitchell, J.C. (1976) Biological activities of sesquiterpene lactones. *Phytochemistry*, **15**, 1573–1580.

Shah, M.G., Abbasi, A.M., Khan, N. *et al.* (2014) Traditional uses of medicinal plants against malarial disease by the tribal communities of Lesser Himalayas-Pakistan. *Journal of Ethnopharmacology*, **155**, 450–462.

Shah, G.M., and Khan, M.A. (2006) Common medicinal folk recipes of Siran Valley, Mansehra, Pakistan. *Ethnobotanical Leaflets*, **10**, 49–62.

Smith, D.C. (1976) Quinine and fever: The development of the effective dosage. *Journal of the History of Medicine and Allied Sciences*, **31**, 343–367.

Spencer, C.F., Koniuszy, F.R., Rogers, E.F. *et al.* (1947) Survey of plants for antimalarial activity. *Lloydia*, **10**, 145–174.

Stojanovic, G., Radulovic, N., Hashimoto, T., and Palic, R. (2005) In vitro antimicrobial activity of extracts of four *Achillea* species: the composition of *Achillea clavennae* L. (Asteraceae) extract. *Journal of Ethnopharmacology*, **101**, 185–190.

Tagarelli, G., Tagarelli, A., and Piro, A. (2010) Folk medicine used to heal malaria in Calabria (southern Italy). *Journal of Ethnobiology and Ethnomedicine*, **6**, 1–16.

Vitalini, S., Beretta, G., Iriti, M. *et al.* (2011) Phenolic compounds from *Achillea millefolium* L. and their bioactivity. *Acta Biochimica Polonica*, **58**, 203–209.

Wichtl, M. (1997) *Teedrogen und Phytopharmaka Ein handbuch fur die Praxis auf wissenschaftlicher Grundlage*, 3rd edition. Wissenschaftliche Verlagsgesellschaft GmbH, Stuttgart.

Woerdenbag, H.J., Pras, N., van Uden, W., Wallaart, T.E., Beekman, A.C., and Lugt, C.B. (1994) Progress in the research of artemisinin-related antimalarials: an update. *Pharmacy World & Science*, **16**, 169–180.

World Health Organization (2016) *World Malaria Report.* World Health Organization, Geneva.

Yaeesh, S., Jamal, Q., Khan, A.U., and Gilani, A.H. (2006) Studies on hepatoprotective, antispasmodic and calcium antagonist activities of the aqueous-methanol extract of *Achillea millefolium. Phytotherapy Research*, **20**, 546–551.

Vanilla: Madagascar's Orchid Economy

(B. M. Schmidt)

Ethnobotany

There are roughly 110 species of vanilla distributed in tropical regions around the world. Vanilla was originally produced from capsules of *Vanilla planifolia* Jacks. ex Andrews, an orchid native to Mexico and other parts of Central America. *Vanilla* is the only genus of orchid grown commercially for reasons other than its ornamental flowers. *V. planifolia* flowers (Figure 4.15) are naturally pollinated by small bees (*Melipona, Euglossa, Eulaema* spp.) native to Mexico; however, their pollination rate tends to be quite low. Therefore, hand pollination is required for higher yields. The seeds require the presence of specific mycorrhizal fungi, so the plants are most often vegetatively propagated.

There are several vanilla orchid cultivation methods, but all typically use small live trees called tutors to support the orchid vines (Figure 4.16). Vanilla requires deep shade, as direct sunlight will burn their leaves. Tutors are pruned in the shape of an umbrella, providing both shade and support. Vanilla cuttings are planted at the base of the tutor, and the vine gradually grows up the tree.

Vanilla "beans" (botanically a capsule) are individually hand-harvested about 6–9 months

Figure 4.15 Vanilla orchid (*Vanilla planifolia*) flower and pods. © 2015 Gustav Svensson.

Figure 4.16 Vanilla cultivation on *Dracaena reflexa* tutors, Réunion Island. © 2003 Bruno Navez.

after flower pollination. Immature beans are dark green, gradually turning light green as they ripen. When the capsule turns yellow at the flower end and starts to split along the line of dehiscence, it is ready to harvest. If left on the plant, the fruit will eventually begin to senesce and turn black. Green vanilla beans do not contain the classic vanilla aroma; they must first undergo a conditioning process. This process varies considerably by region. In the traditional process, fresh green beans are graded by length; they are washed, then submerged in hot (62–70°C) water for 1–5 minutes, depending on the grade of bean. Alternatively, this "killing" or "wilting" step can also be achieved in an oven or by sun exposure. Heat-treated beans are transferred to fabric-lined wooden boxes where they "sweat" at an initial temperature of 48–50°C for about two days. After sweating, beans are sun-dried in sheets and returned to sweating boxes. This process takes an additional 5–14 days, depending on the grade of bean. The beans are now a brown color and have lost much of their fresh weight. They also begin to develop a vanilla aroma. Next, the beans are rack-dried indoors for one week up to three months. Finally, the dried beans are classified according to color (gourmet black or extraction-quality red), moisture content, pod integrity (split or intact), texture, and aroma (Figure 4.17).

Historically, the use of vanilla as a flavoring started with the Totonaco Indians of present-day Vera Cruz, Mexico. According to their legend, the vanilla orchid was born from the spilled blood of Princess Xanat and her forbidden lover. The Aztecs conquered the Totonacos in the fifteenth century and gained the knowledge of vanilla. The Totonacos offered vanilla to Aztec nobility, and the Aztecs began using vanilla as a flavoring for chocolate beverage called *choclatl*. When the Spanish arrived, they observed Montezuma drinking *choclatl* flavored with *Tlilxochitl*, the Aztec word for cured vanilla beans. After Hernán Cortés conquered the Aztecs, he sent vanilla beans and plants to Europe along with the recipe for *choclatl*. Vanilla plants were eventually

Figure 4.17 Man sorting dried vanilla pods, Réunion Island. © 2004 Josélito Tirados.

shared with botanical gardens in France and England. Mexico (i.e., Spain, as they controlled Mexico during this time) held a monopoly in vanilla production until the mid-19th century because nobody could come up with a way to pollinate the flowers. In 1822, the French began growing vanilla on Réunion and Mauritius islands. They hoped to solve the pollination problem and start major production in these areas. The solution did not come from the best and brightest botanists of the time, but from a 12-year- old slave from Réunion named Edmond Albius. Edmond realized the male and female flower parts were separated by a rostellum, a flap that prevents orchid self-fertilization. By using a tool to lift the flap, he could easily transfer pollen to the stigma. This discovery facilitated an economic boom in Réunion vanilla production and broke the Spanish monopoly (Ecott 2004). The name Bourbon vanilla is derived from the former name of Réunion, Île Bourbon. Around 1850, vanilla was brought to Madagascar, where it became a major cash crop.

Phytochemistry

Cured vanilla beans contain an array of compounds that contribute to the aroma and flavor. Vanillin, 4-hydroxybenzaldehyde, 4-hydroxybenzoic acid, and vanillic acid (Figure 4.18A–D) are the four primary flavor/aroma constituents. An analysis of the ratios of these compounds can be used to identify the beans' geographic origin (John and Jamin 2004; Rao and Ravishankar 2000). Beans from Madagascar produce the highest-quality vanilla, with 2–3.4% vanillin content, whereas Mexican beans contain closer to 1.1–1.8% (Ruiz-Teran *et al.* 2001). There are about 200 volatile constituents in

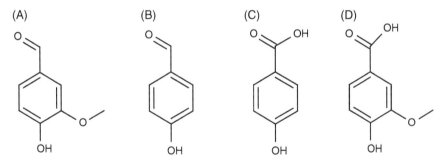

Figure 4.18 Chemical structures of the four major aroma constituents in vanilla: (A) vanillin, (B) 4-hydroxybenzaldehyde, (C) 4-hydroxybenzoic acid, and (D) vanillic acid (Takahashi *et al.* 2013).

cured vanilla beans that also contribute to the aroma, but most occur below 1 ppm (De La Cruz Medina *et al.* 2009).

Vanillin only occurs naturally in fully ripened or cured vanilla beans. Green vanilla pods contain glycosidic precursors to the aroma compounds, but the glycosides have no aroma. They must be enzymatically cleaved to produce the aromatic aglycones. For example, the enzyme β-glycosidase hydrolyzes the *O*-glycosyl linkage in glucovanillin found in green vanilla beans to produce vanillin. Both glucovanillin and β-glycosidase are found in the central placental tissue of the fruit (Odoux *et al.* 2003), but they are separated into different cellular structures. β-glucosidase has been observed in the cytoplasm and the apoplast, whereas glucovanillin seems to be localized in the vacuole or possibly the extracellular space (Odoux *et al.* 2003; Odoux *et al.* 2006). During fruit senescence, this cellular compartmentalization begins to break down, producing the reaction between enzyme and substrate. Post-harvest treatments on green vanilla beans can have the same effect. Freezing and heat treatments and the "sweating" stage of traditional curing serve to disrupt the cellular architecture, permitting hydrolysis of glucovanillin and other glycosidic aroma precursors (Odoux *et al.* 2006). The reaction still occurs even though the activity of β-glucosidase is greatly reduced by these treatments. Recent studies have shown that adding an enzymatic treatment step to the curing process can increase yields of vanillin and other aroma compounds by hydrolyzing any remaining glycosides (Ruiz-Terán *et al.* 2001).

Vanilla production is labor intensive, making it one of the most expensive spices in the world. Since the nineteenth century, world demand for natural vanilla quickly outstripped the supply. Not surprisingly, synthetic methods were developed to produce vanillin, the primary aroma component in vanilla. Nicolas-Theodore Gobley first isolated vanillin in 1858 in crystals from vanilla extract. Two German scientists, Ferdinand Tiemann and Wilhelm Haarmann, described its structure in 1874 and developed the first synthesis from coniferin, a glycoside found in pine bark. Two years later, Tiemann and Nagajosi Nagai developed a method to synthesize vanillin from eugenol found in clove oil. In the early twentieth century, wood pulp waste was used to provide lignin as the precursor. Today, synthetic vanillin is produced from petroleum-derived guaiacol using the Reimer–Tiemann reaction. Pure vanilla bean extract is made by soaking 100 g vanilla beans in 1 L of 35% ethanol. Synthetic vanillin is more widely used than natural vanilla extract; however, it lacks the same depth of aroma and flavor imparted by the numerous compounds found in natural vanilla extract.

Tonka bean adulteration has been a concern for the natural vanilla extract industry for some time. This type of adulteration not only cheats the consumer out of a higher-quality product, but tonka beans contain coumarin, a compound that causes hepatotoxicity. The US FDA has banned the use of tonka bean extract for human consumption, but it is still sometimes detected in natural vanilla bean extracts. Sophisticated methods of adding synthetic vanillin to natural vanilla extract have appeared in the last 20 years. Orchid-derived vanillin has a greater ratio of carbon-13 to carbon-12 than synthetic vanillin, on account of its unique biosynthetic pathway. ^{13}C-enriched vanillin, specific ^{13}C enrichment, and deuterium enrichment have all been utilized in an attempt to match the carbon signature of naturally derived vanillin (Remaud *et al.* 1997). However, modern analytical methods such as site-specific natural isotope fractionation–nuclear magnetic resonance (SNIF–NMR) can detect this type of adulteration, since the addition of ^{13}C to synthetic vanillin tends to be concentrated in the aldehyde and methyl groups (Remaud *et al.* 1997).

Modern Uses

Of the 15,000 tons of vanillin produced per year, less than 0.5% is derived from vanilla pods (Priefert *et al.* 2001; Zamzuri and Abd-Aziz 2013). The remaining majority is synthetic, from lignin or guaiacol. Madagascar and Indonesia are currently the largest producers of vanilla beans, producing around 90% of the world's vanilla; Mexico is a distant third (FAO 2013). Madagascar specializes in Bourbon vanilla, a type of *V. panifolia* grown only on the Bourbon Islands–Madagascar, Réunion, Comoros, and Seychelles. Many consider Madagascar Bourbon vanilla the highest-quality vanilla. Other African countries with significant (>5 tons per year) vanilla production include Uganda, Malawi, Kenya, and Zimbabwe (FAO 2013). French Polynesia and Papua New Guinea produce Tahitian vanilla from *V. tahitiensis*. This species is believed to be a cross between *V. odorata* and *V. panifolia*. Mexico produces vanilla from its native *V. panifolia*, but present production is only around 500 tons per year (FAO 2013).

The small amount of vanilla beans, powder (ground pods), extract, and sugar (mix of sugar and vanilla extract) on the market are mostly used for food flavoring. Synthetic vanillin, methylvanillin, and ethylvanillin are used by the food, perfume, and pharmaceutical industries. Vanillin is also used as an intermediate in the production of herbicides, home products, and antifoaming agents (Li and Rosazza 2000). Recently, the food industry has been evaluating vanillin derived from microbial biotechnology in order to have the label claim "natural vanilla flavoring" without using actual vanilla beans. So-called "biovanillin" could be derived from microbial fermentation using precursors like ferulic acid, vanillic acid, eugenol, or isoeugenol (Zamzuri and Abd-Aziz 2012). The process may be more environmentally friendly than other synthetic processes, especially if agricultural waste is used as the substrate for the microbes. However, to be economically feasible, it would have to compete with synthetic vanillin at $12/kg or vanilla pods at $120/kg (Zamzuri and Abd-Aziz 2013).

References

De La Cruz Medina, J., Rodriguez Jiménes, G.C., and García, H.S. (2009) *Vanilla–Post Harvest Operations*. Food and Agriculture Organization of the United Nations, Geneva.
Ecott, T. (2004) *Vanilla: Travels in Search of the Ice Cream Orchid*. Grove Press, New York.

Food and Agriculture Organization of the United Nations (2013) FAOSTAT database, http://faostat3.fao.org (accessed September 9, 2015).

John, T.V., and Jamin, E. (2004) Chemical investigation and authenticity of Indian vanilla beans. *Journal of Agricultural and Food Chemistry*, **52**, 7644–7650.

Li, T., and Rosazza, J.P.N. (2000) Biocatalytic synthesis of vanillin. *Applied and Environmental Microbiology*, **66**, 684–687.

Odoux, E., Escoute, J., Verdeil, J.L., and Brillouet, J.M. (2003) Localization of b-D glucosidase activity and glucovanillin in vanilla bean (*Vanilla planifolia* Andrews). *Annals of Botany*, **92**, 437–444.

Odoux, E., Escoute, J., and Verdeil, J.L. (2006) The relation between glucovanillin, b-D-glucosidase activity and cellular compartmentation during the senescence, freezing and traditional curing of vanilla beans. *Annals of Applied Biology*, **149**, 43–52.

Priefert, H., Rabenhorst, J., and Steinbuchel, A. (2001) Minireview: biotechnological production of vanillin. *Applied Microbiology and Biotechnology*, **56**, 296–314.

Remaud, G.S., Martin, Y.L., Martin, G.G., and Martin, G.J. (1997) Detection of sophisticated adulterations of natural vanilla flavors and extracts: application of the SNIF-NMR to method vanillin and p-hydroxybenzaldehyde. *Journal of Agricultural and Food Chemistry*, **45**, 859–866.

Ruiz-Terán, F., Perez-Amador, I., and López-Munguia, A. (2001) Enzymatic extraction and transformation of glucovanillin to vanillin from vanilla green pods. *Journal of Agricultural and Food Chemistry*, **49**, 5207–5209.

Takahashi, M., Inai, Y., Miyazawa, N., Kurobayashi, Y., and Fujita, A. (2013) Key odorants in cured Madagascar vanilla beans (*Vanilla planifioria*) of differing bean quality. *Bioscience, Biotechnology, and Biochemistry*, **77** (3), 606–611.

Zamzuri, N.A., and Abd-Aziz, S. (2013) Biovanillin from agro wastes as an alternative food flavour. *Journal of the Science of Food and Agriculture*, **93** (3), 429–438.

Traditional Treatments for HIV in South Africa

(B. M. Schmidt)

Ethnobotany

In Sub-Saharan Africa, nearly 1 in 20 adults are living with HIV, accounting for about 70% of global HIV infections (WHO 2015). Anti-retroviral treatment (ART) is the standard of care, but coverage is only 37% in Sub-Saharan Africa (UNAIDS 2014). Throughout Africa, medicinal plants are used as treatments for HIV/AIDS and for HIV-related complications including pain, nausea, and opportunistic infections like tuberculosis (TB), diarrhea, and oral candidiasis (Gail *et al.* 2015; Peltzer *et al.* 2011). Some of the most commonly used herbs are listed in Table 4.5. Other common treatments include proprietary herbal mixtures such as Canova® (homeopathic product containing *Thuja occidentalis*, *Bryonia alba*, *Aconitum napellus*, pit viper venom, and arsenic trioxide; Smit *et al.* 2009), *imbiza yomzimba omubi* herb mixture, *izifozonke* (general tonic), *ingwe* tonics, and Stametta™ (aloe-based herbal tonic) (Ndhlala *et al.* 2009; Peltzer *et al.* 2011). One study found that people from a province in South Africa took herbs when pharmaceutical ART was unavailable, but the use of herbal treatments declined once they had access to anti-retroviral drugs (Peltzer *et al.* 2011). Continuing use of herbal treatments was correlated with low ART compliance (Peltzer *et al.* 2011).

Sometimes patients would switch between ART and traditional medicine, hoping to get the best of both worlds.

Despite the beneficial effects of some medicinal herbs, three main issues need to be addressed before recommending herbal medicines for HIV: harmful interactions between herbal medicines and ART, lack of clinical evidence proving the efficacy of medicinal herbs, and ethical issues surrounding the claims made by traditional healers and herbal products. Interaction between herbs and ART drugs could lead to adverse events or an ineffective ART dose. The herbs could also have toxic effects, especially when taken for long periods of time. One example of an unethical practice is when a product or healer claims herbal medicines can prevent HIV infection, instructing users to take the herb before and after sexual activity to "cleanse" the body of HIV. These unfounded claims could lead to increased rates of HIV infection. Furthermore, some people may believe false hopes of a cure offered by herbal medicines, rejecting ART. On the other hand, if any of these herbs truly have a clinical benefit for HIV or HIV-related complications like TB, they could prove to be a valuable source for new drugs or adjuvants to traditional ART.

Phytochemistry and Bioactivity

By Western medical standards, there are hardly any studies to support the use of traditional South African herbal medicines to either treat or prevent HIV. Most are preliminary in vitro or, at best, animal studies (Table 4.5).

There are two primary targets for HIV drugs: direct antiviral activity against HIV and stimulating the immune system to help the body combat HIV. Only one study could be found where a South African herb had a direct antiviral effect on HIV. Kamng'ona *et al.* (2011) found that a polyphenol-rich extract from *Myrothamnus flabellifolia* (Figure 4.19) leaves inhibited HIV-1 reverse transcriptases. After fractionation and purification, the authors discovered the most abundant polyphenol, 3,4,5-tri-*O*-galloylquinic acid, was the active compound. So, it is possible that *M. flabellifolia* taken as an ethnobotanical remedy for HIV may have some benefit. However, the authors stress that several limitations need to be overcome before *M. flabellifolia* or 3,4,5-tri-*O*-galloylquinic acid could be recommended as an antiviral treatment. These limitations include a high effective dose, binding to nontarget proteins like serum albumin, and limited transport within the body (Kamng'ona *et al.* 2011).

Hypoxis hemerocallidea is a popular herbal immunostimulant recommended as a treatment for HIV. Its active component is hypoxoside (Figure 4.20), a glycoside that has been studied for its anticancer effects. Phase I clinical trials for cancer have shown inconclusive results (Smit *et al.* 1995) or toxic effects due to bone marrow suppression (Mills *et al.* 2005). Furthermore, no studies could be found where *H. hemerocallidea* or hypoxoside has an immunostimulatory or antiviral effect. Considering the broad use of *H. hemerocallidea* as a herbal immunostimulant and the lack of information available in the scientific literature, studies are essential to confirm its activity.

Sutherlandia frutescens (Figure 4.21) has been studied more extensively for HIV than any of the herbs in Table 4.5. On one hand, it has demonstrated immunostimulatory activity in several studies (Koffuor *et al.* 2014; Lei *et al.* 2015; Zhang *et al.* 2014). Some suggest polysaccharides are the active components (Lei *et al.* 2015; Zhang *et al.* 2014), while others suggest it is the synergistic effect of several phytochemicals (Koffuor *et al.* 2014). Nevertheless, Africa and Smith (2015) reported *S. frutescens* may increase HIV-associated

Table 4.5 Herbs commonly used in South Africa to treat HIV and/or its complications with scientific research to support the ethnobotanical use.

Plant name	Common name(s)[*]	Traditional use[*]	HIV-related research[†]
Hypoxis hemerocallidea Fisch., C.A.Mey. & Avé-Lall. and *H. colchicifolia*	African potato, Inkomfe	Grated dry roots used to treat HIV symptoms and TB	No studies showing HIV therapeutic activity
Asparagus densiflorus (Kunth) Jessop	Balloon pea, uNwele	Leaves used to treat HIV	No studies showing HIV therapeutic activity
Sutherlandia frutescens (L.) R.Br.	Kankerbossie, uNwele	Leaves used to treat HIV and TB co-infection.	Increased rate of TB in clinical trial (Wilson *et al.* 2015), but in vitro (Lei *et al.* 2015; Zhang *et al.* 2014) and animal studies (Kofftuor *et al.* 2014) show immunostimulatory effects.
Merwilla plumbea (Lindl.)	Blouberglelie, Inguduza	Fresh or dry bulbs warmed and applied to the skin as a poultice for HIV-related shingles	In vitro antimicrobial activity when mixed with two other African herbs (Ncube *et al.* 2012), but no HIV studies
Myrothamnus flabellifolia Welw.	Resurrection plant, uvukwabafile	Leaves used for pain relief, immune booster, relieve side effects from ART, improve overall well-being	Polyphenol from leaves inhibits HIV reverse transcriptases in vitro (Kamng'ona *et al.* 2011)

[*] From Peltzer *et al.* 2011; Gail *et al.* 2015.
[†] From PubMed and Agricola search using the plant name and "HIV" as keywords.

Figure 4.19 *Myrothamnus flabellifolia*, the resurrection plant. © 2011 Marco Schmidt.

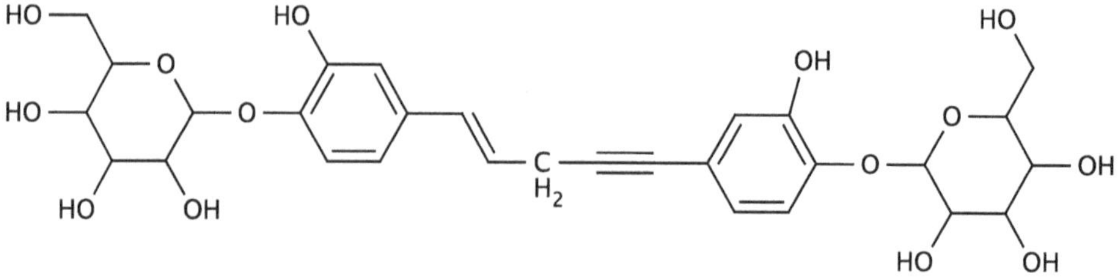

Figure 4.20 Structure of the norlignan diglucoside hypoxoside from *Hypoxis hemerocallidea*.

neuroinflammation by stimulating production of monocyte chemoattractant protein-1 and CD14+ monocyte infiltration across the blood brain barrier. In addition, a clinical trial found higher rates of TB infection in the *S. frutescens* treatment group compared to placebo or ART (Wilson *et al.* 2015). So it is possible *S. frutescens* may be doing more harm than good.

Several studies suggest that some of the commonly used herbs for HIV interact with antiretroviral drugs. For example, when *S. frutescens* was administered with the ART drug atazanavir, it reduced the bioavailability of atazanavir to potentially subtherapeutic levels (Müller *et al.* 2013). Aqueous and methanolic (triterpene glycoside) fractions of *S. frutescens* affected metabolism of atazanavir differently in vitro, making the overall bioavailability of atazanavir unpredictable (Müller *et al.* 2012). In recent studies, *H. hemerocallidea* and *S. frutescens* extracts interfered with in vitro ART drug metabolism by inhibiting several cytochrome P450 enzymes and transport proteins (Fasinu *et al.* 2013a, 2013b) and by slowing the transport of nevirapine across human intestinal epithelial cells (Brown *et al.* 2008). In the Brown *et al.* study (2008), the purported active component of *S. frutescens*,

Figure 4.21 *Sutherlandia frutescens.* © 2012 Christer T. Johansson.

l-canavanine, also slowed nevirapine transport, but the *S. frutescens* crude extract had no effect. In contrast to many of the aforementioned studies, *H. hemerocallidea* had no effect on the plasma concentrations of the ART drug efavirenz (Mogatle *et al.* 2008) or lopinavir/ritonavir (Gwaza *et al.* 2013) in clinical trials. Clearly more studies are required to determine which herbs interfere with ART and at what dose. After safety has been established, clinical trials for herbal efficacy should be a priority.

Modern Use

In 2013, the World Health Organization (WHO) recommended taking two nucleoside reverse-transcriptase inhibitors (NRTIs) and a non-nucleoside reverse-transcriptase inhibitor (NNRTI) as an effective ART. A commonly prescribed "triple therapy" drug combination is tenofovir, lamivudine (or emtricitabine), and efavirenz, but there are several others. In Sub-Saharan Africa, government-sponsored ART programs discourage the use of traditional medicines since they could reduce the efficacy of antiretroviral drugs or cause toxicity (Chinsembu and Hedimbi 2010). Two reasons people may turn to traditional medicine after HIV diagnosis are the limited availability and high cost of ART. Even if people are offered "free" ART at their local hospital or clinic, they still incur costs to get to the clinic, clinic fees to treat complications, long waits, lack of bed space, lack of healthcare workers, and so on. According to one study, once HIV+ patients get access to ART, they typically stop using herbal medicines, or use them less frequently (Peltzer *et al.* 2011). Therefore, it seems South Africans tend to rely on pharmaceutical ART if they have access to the drugs. Another reason people may turn to traditional medicine is to manage the side effects of ART and HIV-related opportunistic infections. So, while a number of the frequently used herbs in Table 4.5 are used for both HIV and related complications, there are dozens of other herbs just used to treat complications not covered in this case study (Chinsembu and Hedimbi 2010; Kisangau *et al.* 2007).

Recent surveys indicate that nearly 70% of South African households turn to public hospitals and clinics for their healthcare needs as opposed to traditional healers (Statistics South Africa 2015). This is not surprising, since South Africa is one of the wealthiest countries in Africa, with modern medical facilities. Other African countries that have fewer healthcare resources and a more rural population may use traditional medicine more frequently. In rural Namibia, Tanzania, and most other rural areas in Africa with high rates of HIV, biomedical services are largely unavailable, so traditional medicine may be the only choice (Chinsembu 2009; Kisangau *et al.* 2007). In these areas, a wealth of ethnobotanical knowledge has been developed out of necessity to help people infected with HIV. Commonly, herbs that had been used as general tonics were repurposed for HIV to increase vitality. Traditional healers likely discovered some herbs that are particularly valuable for managing HIV that remain unknown or unstudied by Western medicine. A few ethnobotanical inventories of herbs used for HIV and related complications in Africa have already been completed (Chinsembu and Hedimbi 2010; Kisangau *et al.* 2007; Semenya *et al.* 2013), but more work is required to preserve this precious knowledge. Moreover, new treatments may be discovered for HIV, TB, or any other complication arising from HIV-related immunodeficiency if we spend time studying these traditional remedies.

References

Africa, L.D., and Smith, C. (2015) *Sutherlandia frutescens* may exacerbate HIV-associated neuroinflammation. *Journal of Negative Results in Biomedicine*, **14**, 14.

Brown, L., Heyneke, O., Brown, D., van Wyk, J.P., and Hamman, J.H. (2008) Impact of traditional medicinal plant extracts on antiretroviral drug absorption. *Journal of Ethnopharmacology*, **119** (3), 588–592.

Chinsembu, K.C. (2009) Model and experiences of initiating collaboration with traditional healers in validation of ethnomedicines for HIV/AIDS in Namibia. *Journal of Ethnobiology and Ethnomedicine*, **5**, 30.

Chinsembu, K.C., and Hedimbi, M. (2010) An ethnobotanical survey of plants used to manage HIV/AIDS opportunistic infections in Katima Mulilo, Caprivi region, Namibia. *Journal of Ethnobiology and Ethnomedicine*, **6**, 25.

Fasinu, P.S., Gutmann, H., Schiller, H., Bouic, P.J., and Rosenkranz, B. (2013a) The potential of *Hypoxis hemerocallidea* for herb-drug interaction. *Pharmaceutical Biology*, **51** (12), 1499–1507.

Fasinu, P.S., Gutmann, H., Schiller, H., James, A.D., Bouic, P.J., and Rosenkranz, B. (2013b) The potential of *Sutherlandia frutescens* for herb-drug interaction. *Drug Metabolism and Disposition.* **41** (2), 488–497.

Gail, H., Tarryn, B., Oluwaseyi, A., Denver, D., Oluchi, M., Charlotte, V.K., Joop de, J., and Diana, G. (2015) An ethnobotanical survey of medicinal plants used by traditional health practitioners to manage HIV and its related opportunistic infections in Mpoza, Eastern Cape Province, South Africa. *Journal of Ethnopharmacology*, **171**, 109–115.

Gwaza, L., Aweeka, F., Greenblatt, R., Lizak, P., Huang, L., and Guglielmo, B.J. (2013) Co-administration of a commonly used Zimbabwean herbal treatment (African potato) does not alter the pharmacokinetics of lopinavir/ritonavir. *International Journal of Infectious Diseases*, **17** (10), e857–861.

Kamng'ona, A., Moore, J.P., Lindsey, G., and Brandt, W. (2011) Inhibition of HIV-1 and M-MLV reverse transcriptases by a major polyphenol (3,4,5 tri-O-galloylquinic acid) present in the leaves of the South African resurrection plant, *Myrothamnus flabellifolia*. *Journal of Enzyme Inhibition and Medicinal Chemistry*, **26** (6), 843–853.

Kisangau, D.P., Lyaruu, H.V.M., Hosea, K.M., and Joseph, C.C. (2007) Use of traditional medicines in the management of HIV/AIDS opportunistic infections in Tanzania: A case in the Bukoba rural district. *Journal of Ethnobiology and Ethnomedicine*, **3**, 29.

Koffuor, G.A., Dickson, R., Gbedema, S.Y., Ekuadzi, E., Dapaah, G., and Otoo, LF. (2014) The immunostimulatory and antimicrobial property of two herbal decoctions used in the management of HIV/AIDS in Ghana. *African Journal of Traditional, Complementary, and Alternative Medicine*, **11** (3), 166–172.

Lei, W., Browning, J.D. Jr., Eichen, P.A., Lu, C.H., Mossine, V.V., Rottinghaus, G.E., Folk W.R., Sun, G.Y., Lubahn, D.B., and Fritsche, K.L. (2015) Immuno-stimulatory activity of a polysaccharide-enriched fraction of *Sutherlandia frutescens* occurs by the toll-like receptor-4 signaling pathway. *Journal of Ethnopharmacology*, **172**, 247–253.

Mills, E., Cooper, C., Seely, D., and Kanfer, I. (2005) African herbal medicines in the treatment of HIV: *Hypoxis* and *Sutherlandia*. An overview of evidence and pharmacology. *Nutrition Journal*, **4**, 19.

Mogatle, S., Skinner M, Mills E, and Kanfer I. (2008) Effect of African potato (*Hypoxis hemerocallidea*) on the pharmacokinetics of efavirenz. *South African Medical Journal*, **98** (12), 945–949.

Müller, A.C., Patnala, S., Kis, O., Bendayan, R., and Kanfer, I. (2012) Interactions between phytochemical components of *Sutherlandia frutescens* and the antiretroviral, atazanavir in vitro: implications for absorption and metabolism. *Journal of Pharmacy and Pharmaceutical Science*, **15** (2), 221–233.

Müller, A.C., Skinner, M.F., and Kanfer, I. (2013) Effect of the African traditional medicine, *Sutherlandia frutescens*, on the bioavailability of the antiretroviral protease inhibitor, atazanavir. *Evidence Based Complementary and Alternative Medicine*, **2013**, 324618.

Ncube, B., Finnie, J.F., and Van Staden, J. (2012) In vitro antimicrobial synergism within plant extract combinations from three South African medicinal bulbs. *Journal of Ethnopharmacology*, **139** (1), 81–89.

Ndhlala, A.R., Stafford, G.I., Finnie, J.F., and Van Staden, J. (2009) In vitro pharmacological effects of manufactured herbal concoctions used in KwaZulu-Natal South Africa. *Journal of Ethnopharmacology*, **122** (1), 117–122.

Peltzer, K., Preez, N.F., Ramlagan, S., Fomundam, H., Anderson, J., and Chanetsa, L. (2011) Antiretrovirals and the use of traditional, complementary and alternative medicine by HIV patients in Kwazulu-Natal, South Africa: a longitudinal study. *African Journal of Traditional, Complementary, and Alternative Medicine*, **8** (4), 337–345.

Semenya, S S., Potgieter, M J., and Erasmus, L.J.C. (2013) Indigenous plant species used by Bapedi healers to treat sexually transmitted infections: Their distribution, harvesting, conservation and threats. *South African Journal of Botany*, **87**, 66–75.

Smit, E., Oberholzer, H.M., and Pretorius, E. (2009) A review of immunomodulators with reference to Canova®. *Homeopathy*, **98** (3), 169–176.

Statistics South Africa (2015) *General Household Survey, July 2014*. Statistical Release P0318. Statistics South Africa, Pretoria.

The Joint United Nations Programme on HIV/AIDS (UNAIDS) (2014) UNAIDS Fact Sheet 2014, http://www.unaids.org/en/resources/campaigns/2014/2014gapreport/factsheet (accessed 17 September 2015).

Wilson, D., Goggin, K., Williams, K., Gerkovich, M.M., Gqaleni, N., Syce, J., Bartman, P., Johnson, Q., and Folk, W.R. (2015) Consumption of *Sutherlandia frutescens* by HIV-seropositive South African adults: an adaptive double-blind randomized placebo controlled trial. *Public Library of Science One*, **10** (7), e0128522.

World Health Organization (2015) Global Health Observatory (GHO) data. http://www.who.int/gho/hiv/en/(accessed 17 September 2015).

Zhang, B., Leung, W.K., Zou, Y., Mabusela, W., Johnson, Q., Michaelsen, T.E., and Paulsen, B.S. (2014) Immunomodulating polysaccharides from *Lessertia frutescens* leaves: Isolation, characterization and structure activity relationship. *Journal of Ethnopharmacology*, **152** (2), 340–348.

Duckweed as a More Sustainable First-Generation Bioethanol Feedstock

(P. Chu)

Ethnobotanical Use of Biofuels

Since ancient times, people have used biofuels for light, heat, and cooking. Solid fuels such as wood and charcoal were some of the earliest biofuels. Vegetable oils have been used as liquid fuel for several thousand years. The oil palm (*Elaeis guineensis* Jacq.) originating in Central or Western Africa, was used for cooking and heating and became a valuable trade commodity with Europe during the 1800s. Palm oil is currently the most productive and cheapest vegetable oil globally (Vijay et al. 2016). However, its ubiquitous use as a cooking oil and more recent application as a biodiesel source encourages further expansion of oil palm production, threatening tropical forests.

Growing African populations and economies drive rising liquid fuel production and consumption (U.S. Energy Information Administration 2016). Two of the most common types of liquid transportation biofuels are bioethanol and biodiesel. This case study will focus on bioethanol. Ethanol is regarded as a high-performance renewable fuel that boosts engine performance and lowers greenhouse gas emissions. Starting in the 1980s, bioethanol facilities were built in Zimbabwe, Malawi, and Kenya (Amigun *et al.* 2011). Currently, Angola, Kenya, Malawi, Mozambique, South Africa and Zimbabwe have E10 (10% ethanol blend with gasoline) mandates, while Ethiopia, Mauritius, and Sudan mandate E5 (Global Renewable Fuels Alliance 2017). Sugarcane, the main bioethanol feedstock, is often discussed in the context of the water-food-energy debate, because of its potential to have significant negative environmental impacts on land use change, water quality, and soil quality (Hess *et al.* 2016).

Efforts to find a feedstock that doesn't compete with food crops is of great importance in Africa and beyond. A promising alternative feedstock is duckweed, which has been garnering more attention as a renewable and sustainable biofuel feedstock in recent years. Duckweed is a miniature, aquatic basal monocot that floats on stationary to slow-moving freshwater (Figure 4.22). Waterfowl disperse the plants, promoting its

Figure 4.22 A species of duckweed, *Spirodela polyrhiza*, floating on liquid media in a petri dish. Each cluster comprises multiple plants.

widespread distribution. Located on every continent except Antarctica, duckweeds have adapted to a wide range of environments. Belonging to the family Lemnaceae, there are five genera (*Lemna, Landoltia, Spirodela, Wolffiella,* and *Wolffia*) and 37 species of duckweed and countless ecotypes, or natural strains, which provide a diverse gene pool and a plethora of biochemical properties that can be exploited for commercial use. In addition, its simple body architecture and high surface-area-to-volume ratio help duckweed achieve an extremely fast doubling time of ~24 hours to three days (Landolt and Kandeler 1987).

Chemical Constituents and Bioactivities

Starch is a storage polysaccharide used by plants to stockpile excess glucose. Starch comes in two types: linear and branched. The former, called amylose, is an unbranched chain of glucose units covalently linked by α-(1,4) glycosidic bonds, whereas amylopectin includes side chains that branch at α-(1,6) linkages. Once extracted from plant tissue, starch must be depolymerized into individual sugar units in order to be amenable to downstream fermentation processes.

During the biological conversion process, an enzymatic cocktail is required to efficiently hydrolyze polysaccharides into glucose. First, starch is liquefied by α-amylase, which attacks interior bonds, yielding a mixture of oligosaccharides and dextrins. Then glucoamylase, an exo-acting amylase, hydrolyzes the terminal α-(1,4) bonds from the nonreducing ends of starch to yield glucose units. The enzyme also has activity toward cleaving α-(1,6) linkages. Debranching enzymes such as pullulanase can be added to boost glucose yields.

Studies have examined ethanol production from duckweed grown on wastewater sources. A study comparing ethanol production of a local *Lemna aequinoctialis* Welw. ecotype grown on nutrient-rich liquid Schenk & Hildebrandt (SH) plant growth media versus sewage water (SW) from a wastewater treatment facility in Qingdao, China, was conducted (Yu *et al.* 2014). First, *Lemna* was grown on either SH or SW for 30 days. By day 18, nitrogen and phosphorous levels in the media were 80% removed in SH and 90% for SW. After this point, starch content in the *Lemna* tissue steadily increased, peaking at ~40% on SH and ~36% on SW 24 days after cultivation. The amylopectin content slightly increased after cultivation on either SH and SW, which might result in easier hydrolysis since lower amylose content has previously been associated with greater starch digestibility (Okuda *et al.* 2005; Wu *et al.* 2006; Yu *et al.* 2014). After a one-step hydrolysis with α-amylase, α-amyloglucosidase, and pullanase, ~94% reducing sugars were recovered from duckweed grown on SH and on SW. The majority of the reducing sugars (94–96%) were glucose. Glucose is then fermented into ethanol and carbon dioxide by yeast.

Modern-Day Use

Integrating duckweed into a wastewater system to support its biomass growth while remediating wastewater would create a more sustainable bioethanol feedstock (Cheng and Stomp 2009) (Figure 4.23). Duckweed is therefore an ideal candidate: rapid doubling

Figure 4.23 Commercial-scale duckweed wastewater treatment project between MamaGrande, a social biotechnology company, and Aguas Del Norte. Salta, Argentina. © 2015 Ryan Gutierrez.

rate, easy harvestability, renewability, sustainability, high starch accumulation, nonfood crop, and nearly worldwide distribution. In Uganda, a pilot project to produce duck-weed for animal feed produced 800–1250 kg biomass/acre/day (The State House of Uganda 2014).

Duckweed cultivation does not require arable land or freshwater. Most notably, duckweed can be grown on abundant and readily accessible municipal or agricultural wastewater, providing an inexpensive resource for biofuels, while removing excess nitrogen and phosphorous that could pollute waterways (Cheng and Stomp 2009). That remediated water can then be recycled. A pilot-scale study growing *Spirodela polyrhiza* on dilute swine wastewater produced 50% more starch per hectare than that achievable with corn, the main bioethanol feedstock in the US (Xu *et al.* 2011). Thus, duckweed can be a viable, sustainable ethanol production platform for our cur-rent needs. Although established industrial processes may be exploited for converting duckweed-derived starch into ethanol, additional technological innovations could optimize a novel duckweed biofuel platform to quickly reach a point of economic viability.

References

Amigun, B., Musango, J.K., and Stafford, W. (2011). Biofuels and sustainability in Africa. *Renewable and Sustainable Energy Reviews*, **15** (2), 1360–1372.

Cheng, J.J., and Stomp, A.-M. (2009) Growing duckweed to recover nutrients from wastewaters and for production of fuel ethanol and animal feed. *Clean*, **37** (1), 17–26.

Global Renewable Fuels Alliance. Global Biofuel Mandates, http://globalrfa.org/biofuels-map/. (accessed 30 April 2017).

Hess, T.M., Sumberg, J., Biggs, T. *et al.* (2016) A sweet deal? Sugarcane, water and agricultural transformation in Sub-Saharan Africa. *Global Environmental Change*, **39**, 181–194.

Landolt, E., and Kandeler, R. (1987) *The Family of Lemnaceae, a Monographic Study*. Geobotanischen Instutites der ETH, Stiftung Rubel, Zurich.

Okuda, M., Aramaki, I., Koseki, T., Satoh, H., and Hashizume, K. (2005) Structural characteristics, properties, and in vitro digestibility of rice. *Cereal Chemistry*, **82**, 361–368.

The State House of Uganda (2014) President hails project to turn Lake Weed into animal feed. http://www.statehouse.go.ug/media/news/2014/06/26/president-hails-project-turn-lake-weed-animal-feed (accessed 01 April 2016).

U.S. Energy Information Administration. Petroleum and other liquid fuels, in *International Energy Outlook 2016*, https://www.eia.gov/outlooks/ieo/pdf/0484(2016).pdf (accessed 30 April, 2017).

Vijay, V., Pimm, S.L., Jenkins, C.N., and Smith, S.J. (2016) The Impacts of oil palm on recent deforestation and biodiversity loss. *PLoS ONE*, **11** (7), e0159668.

Wu, X., Zhao, R., Wang, D. *et al.* (2006) Effects of amylose, corn protein, and corn fiber contents on production of ethanol from starch-rich media. *Cereal Chemistry*, **83**, 569–575.

Xu, J., Cui, W., Cheng. J.J., and Stomp, A.M. (2011). Production of high-starch duckweed and its conversion to bioethanol. *Biosystems Engineering*, **110**, 67–72.

Yu, C., Sun, C., Yu, L. *et al.* (2014) Comparative analysis of duckweed cultivation with sewage water and SH media for production of fuel ethanol. *PLoS ONE*, **9** (12), e115023.

5

The Americas

J. Kellogg, B. M. Schmidt, B. L. Graf, C. Jofré-Jimenez, H. Aguayo-Cumplido, C. Calfío, L. E. Rojo, K. Cubillos-Roble, A. Troncoso-Fonseca, and J. Delatorre-Herrera

Introduction

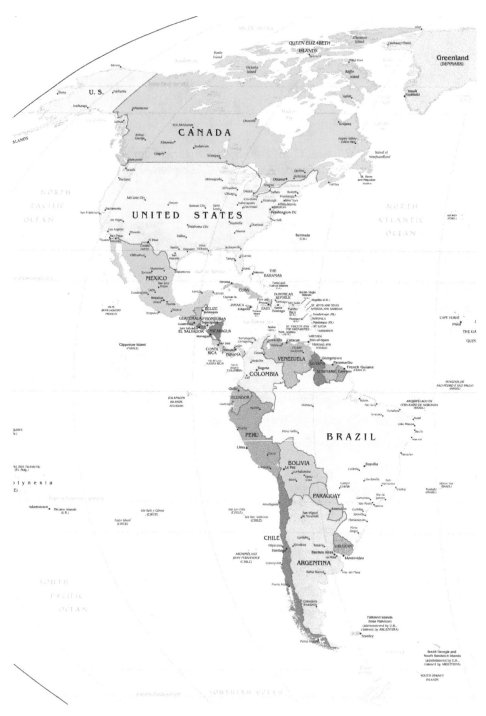

Figure 5.1 Political map of the Americas (courtesy of United States of America Central Intelligence Agency).

Europeans called America the "New World" because it was unknown to them prior to the sixteenth century. Today the term is still used, especially when referring to the origin of plants. America is named after Italian explorer Amerigo Vespucci, who visited South America from 1497 to 1502. He speculated that South America was not the East Indies, but an entirely new continent unknown to Europe. Christopher Columbus used the term "Indian" to describe Native Americans because he thought he had reached the Indian Ocean. The islands where Columbus first landed are still called the West Indies. Most likely the first humans came to America from Asia during the last Ice Age, either walking across the Bering Strait land bridge or taking boats across the shallow waters. They eventually migrated south and east, forming tribal groups and giant empires over the course of several millennia.

Ranging from the Arctic Circle across the equator to the southernmost tip of Argentina, the Americas have a wide range of topography, climates, and ecosystems (Figure 5.1). Likewise, there are a variety of languages, religions, and cultures. In general, native tribes lived sustainable, subsistence lifestyles where ethnobotany was critical to survival. It was an oral tradition, passed down through the generations. This chapter will focus on Native American ethnobotany, with a smaller amount of information on the European Diaspora that dominates the Americas today. The Americas can be separated into the following regions: North, Central, and South America, and the Caribbean islands.

North America

Before the arrival of Europeans during the sixteenth to nineteenth centuries, North America was home to an estimated 18 million indigenous people from hundreds of different tribes. North America can be divided into separate cultural areas with different lifestyles. Tribes in areas where resources were scarce, such as the Great Basin, were nomadic, but most other tribes lived in settled villages. Some were farmers (e.g., Algonquian and Muskogean language tribes) but a great number were hunter-gatherers, collecting wild plants for food and medicine. Great Plains inhabitants were traditionally hunters and farmers in settled villages. But after the arrival of the Europeans disrupted their way of life, they became nomadic. The Navajo and Apache of the Southwest were somewhat nomadic, hunting, gathering, and raiding villages for crops.

Many farming tribes were known for cultivating the "three sisters" companion crops: squash (*Cucurbita* spp.), maize (*Zea mays* L.), and beans (several genera in Fabaceae). Mounds of maize were planted first. After the plants grew approximately six inches in height, the other two crops were planted at the base of the cornstalks. The corn provided a trellis for the climbing beans, the beans added nitrogen to the soil, and the squash vines reduced weed competition. The three sisters were a staple throughout North America. Other important crops are listed in Table 5.1.

Medicinal herbs were collected from the wild and considered sacred by many tribes. They were often prescribed in conjunction with prayers, chanting, or ceremonies to speak with the spiritual realm. Southwest tribes used mescal beans from *Dermatophyllum* spp. as a hallucinogen. The beans do not contain mescaline, but they do contain hallucinogenic quinolizidine alkaloids. Four plants were commonly used during smudging (cleansing

Table 5.1 Crops originating in North America.

Fruits and vegetables		Grains and legumes	
Blueberry	*Vaccinium* spp., Ericaceae	Sunflower	*Helianthus annuus* L., Asteraceae
Cranberry	*Vaccinium macrocarpon* Aiton, Ericaceae	Wild rice	*Zizania* spp., Poaceae
Huckleberry	*Vaccinium* and *Gaylussacia* spp., Ericaceae	**Sweeteners and flavorings**	
		Maple syrup	*Acer saccharum* Marshall., Sapindaceae
Jerusalem artichoke	*Helianthus tuberosus* L., Asteraceae		
		Sassafras	*Sassafras albidum* (Nutt.) Nees, Lauraceae
Pawpaw	*Asimina triloba* (L.) Dunal, Annonaceae	**Stimulants**	
Prickly pear	*Opuntia* spp., Cactaceae	Tobacco	*Nicotiana tabacum* L., Solanaceae
Raspberry	*Rubus* spp., Rosaceae		

smoke) ceremonies: tobacco (*Nicotiana tabacum* L.), sweet grass [*Hierochloe odorata* (L.) P.Beauv.], sage (*Salvia apiana* Jeps. or other North American *Salvia* spp.), and cedar (*Juniperus virginiana* L., *Thuja plicata* Donn ex D.Don, or *T. occidentalis* L.). Some tribes would use red willow bark (*Cornus amomum* Mill.), compass plant (*Silphium laciniatum* L.), and/or osha root (*Ligusticum porteri* J.M.Coult. & Rose). In addition to smudging, tobacco was used for wound dressings, toothache relief (chewing tobacco), and as a general painkiller. It was only smoked occasionally for special medical and ceremonial purposes. When tobacco was taken back to Europe by early explorers, it was promoted as a cure-all. Early American settlers grew tobacco as a cash crop. It was not until the early 1800s that scientists discovered nicotine along with the addictive and detrimental effects of habitual tobacco consumption. It is still widely cultivated in North America today (Figure 5.2).

Bark of the Pacific yew (*Taxus brevifolia* Nutt.) contains the diterpene taxol, an effective anticancer agent. The berries were used in traditional medicine as birth control; other parts were used for hair removal and a range of ailments. The needles were reportedly smoked, causing dizziness (Bolsinger and Jaramillo, 1990). Native Americans used mayapple (*Podophyllum peltatum* L.) root as an emetic and antihelminthic agent. The lignan podophyllotoxin is the active component, sold as a topical pharmaceutical for warts. The Omaha tribe used pleurisy root (*Asclepias tuberosa* L.) as an expectorant for bronchitis. It contains pregnane glycosides that may be responsible for its medicinal properties. Oswego tea (*Monarda didyma* L.) was another botanical cold remedy. It was used as a tea replacement after the Boston Tea Party of 1773. Purple coneflower [*Echinacea purpurea* (L.) Moench.] was used for a wide range of ailments including insect and snake bites, burns and wounds, stomach ailments, and as a general "detoxicant."

Several North American nuts provided high-fat dietary sustenance such as hickory (*Carya* spp.), pecans [*Carya illinoinensis* (Wangenh.) K.Koch], piñon nuts (particularly from Great Basin *Pinus* spp.), American chestnut [*Castanea dentata* (Marshall) Borkh.], black walnut (*Juglans nigra* L.), and acorns (*Quercus* spp.). Maple syrup is derived from the sap of *Acer saccharum* Marshall (Figure 5.3). Native Americans had knowledge of

Figure 5.2 North American farm growing corn (*Zea mays*) and tobacco (*Nicotiana tabacum*).

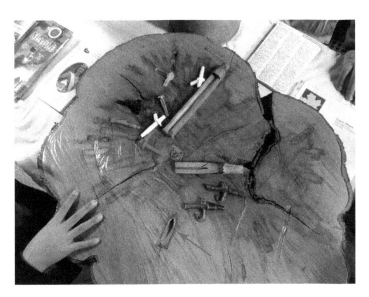

Figure 5.3 Cross section of a sugar maple (*Acer saccharum*) with spiles or tapping spouts. The red arrow indicates one of the many places where the tree had been tapped.

maple syrup before the arrival of the Europeans, but it is unclear if they boiled down the sap to produce maple sugar. French colonists and missionaries may have introduced these techniques. By the late 1700s, maple sugar was sold commercially and promoted by Quakers and abolitionists to protest the use of slaves in white cane sugar production. Today the process of producing maple syrup has advanced from tin cans and metal spouts to vacuum pumps and tubing systems. Yet many small farmers still use the traditional metal spout and can method.

Wild berries were important to the health of Native Americans. In 1615, Samuel de Champlain observed Algonquin women drying a type of blueberry in the sun. They used these berries to make bread with cornmeal and beans, "manna in winter" when other food was scarce (Hummer 2013). They used cranberries (*Vaccinium macrocarpon* Aiton) (Figure 5.4), blueberries (*V. corymbosum* L., *V. angustifolium* Aiton) (Figure 5.5), and other *Vaccinium* spp. as food, dye, and medicine. In his writings, American botanist John Bartram described an Iroquois woman and her children drying huckleberries over a smoky fire. Henry David Thoreau stated that whortleberries were a major food source for Native Americans, sustaining them through thousands of years. Some tribes used *Vaccinium* berries as ceremonial food and medicinally as a health-promoting tea, for gastrointestinal and gynecological problems, and for children's health. The Newberry Crater archeological site located in central Oregon has produced a wealth of information about Paleolithic Native Americans. Pollen and phytolith analysis at the site revealed the presence of chokecherry [*Prunus virginiana* var. demissa (Nutt. ex Torr. & A.Gray) Torr.], salmonberry (*Rubus spectabilis* Pursh), thimbleberry (*Rubus parviflorus* Nutt), raspberry [*Rubus idaeus* subsp. *strigosus* (Michx.) Focke], blackberry (*Rubus* spp.), huckleberry, (*Vaccinium* spp.), and possibly currants (*Ribes* spp.) (Hummer 2013).

Native Americans used sassafras [*Sassafras albidum* (Nutt.) Nees] as flavoring and medicine. The roots are the basis of root beer flavor and are an important seasoning in Cajun and Creole gumbos. Some sources say the word *gumbo* is derived from the

Figure 5.4 Harvested cranberries (*Vaccinium macrocarpon*) floating in a bog in New England, the United States © 2015 Mary Ann Lila.

Figure 5.5 Wild blueberries (*V. angustifolium*) are typically smaller and darker than highbush (*V. corymbosum*) or rabbiteye (*V. virgatum*) blueberries. © 2006 Mary Ann Lila.

Choctaw (tribe native to Louisiana) name for sassafras, *kombo* (Buisseret and Reinhardt 2000). But early French texts such as Baudry des Lozières's *Second Voyage À La Louisiane* (1803) mentioned gumbo as the West African Bantu word for okra (*ki ngombo* or *quingombó*). Both groups were used as slave labor in the south, so it is possible the two groups shared knowledge and developed their own versions of gumbo together. The Meskwaki tribe reportedly used Indian turnip [*Arisaema triphyllum* (L.) Schott], also called Iroquois breadroot or Jack-in-the-pulpit, to poison their enemies. The corm contains large amounts of calcium oxalate that produces a burning, needle-like sensation when ingested. Dried and aged corms evidently contain less calcium oxalate, as they were ground and baked in bread. The dried corms were also taken as a treatment for coughs and colds and applied topically as a poultice for sores, inflamed joints, and snakebites.

Central America and the Caribbean

Mesoamerica is the cultural region extending from central Mexico south to Costa Rica. It included several advanced pre-Columbian societies such as the Olmec, Teotihuacan, Maya, and Aztecs (Nahuatl language groups). By 7000 BCE, Mesoamericans were domesticating maize, beans, squash, chili peppers (*Capsicum annuum* L.) and other crops (Table 5.2), changing societies from hunter-gatherer to simple farming villages.

Archaic period archeological sites throughout southern Mexico contain evidence of domesticated maize, squash, and beans. Pollen and cob remains from teosinte, the ancient progenitor to maize, date back to c. 9000 BCE. Fully domesticated maize appeared around 4000–3600 BCE, according to carbon dating (Pipierno and Flannery 2001; Kraft *et al.* 2014). Squash was domesticated c. 4000–4400 BCE. The common bean (*Phaseolus vulgaris* L.) was domesticated in southern Mexico c. 5000–3500 BCE and independently in the Andes of South America. There may have been some

Table 5.2 Crops originating in Central America or the Caribbean.

Fruits and vegetables		Grains and legumes	
Bottle gourd	*Lagenaria siceraria* (Molina) Standl., Cucurbitaceae	Beans	Fabaceae species
Chili peppers	*Capsicum* spp., Solanaceae	Cotton	*Gossypium hirsutum* L., Malvaceae
Culantro	*Eryngium foetidum* L., Apiaceae	Maize	*Zea mays* L., Poaceae
Dragonfruit	*Hylocereus* spp., Cactaceae	**Medicines, resins, dyes, and flavorings**	
Guava	*Psidium guajava* L., Myrtaceae		
Jicama	*Pachyrhizus erosus* (L.) Urb., Fabaceae	Chicle	*Manilkara chicle* (Pittier) Gilly Sapotaceae
Passionfruit	*Passiflora edulis* Sims, Passifloraceae	Copal	*Bursera* spp., Burseraceae
Papaya	*Carica papaya* L. Caricaceae	Fustic	*Maclura tinctoria* (L.) D.Don ex Steud., Moraceae
Squash	*Cucurbita* spp., Cucurbitaceae	Peyote	*Lophophora williamsii* (Lem. ex Salm-Dyck) J.M. Coult, Cactaceae
Star apple	*Chrysophyllum cainito* L., Sapotaceae	Vanilla	*Vanilla planifolia* Jacks. ex Andrews, Orchidaceae
Tomatillos	*Physalis philadelphica* Lam., Solanaceae		

Figure 5.6 Immature avocado (*Persea americana*) fruit.

hybridization with *P. coccineus*, resulting in larger bean sizes in Mesoamerican varieties. Avocado (*Persea americana* Mill; Figure 5.6) remains were found in a cave in Coxcatlán, Puebla, Mexico, dating to 10,000 BCE, but domestication likely occurred closer to 5000 BCE. Some believe California was its northern range; however, the US avocado industry did not get off the ground until the 1900s. Up until then, the fruit was known by its Spanish name, *ahuacate*, derived from the Nahuatl word *ahuacatl*, meaning "testicle." A California growers association formerly changed the name to avocado. The popular 'Hass' cultivar is named after Rudolph Hass, who purchased the industry-changing seedling from a California farmer in 1926.

Chili peppers (Figure 5.7) were domesticated in northeastern Mexico and/or east central Mexico c. 4000 BCE (Kraft *et al.* 2014). *C. annuum* is one of five domesticated species of *Capsicum* native

Figure 5.7 *Capsicum annuum* varieties include hot chili peppers, mild bell peppers, and a range of peppers in between.

to the Americas, which include *C. baccatum* L., *C. chinense* Jacq., *C. fructescens* L., and *C. pubescens* Ruiz & Pav. In addition to food and flavoring, *Capsicum* peppers were used medicinally to treat asthma, coughs, convulsions, dysentery, liver disease, toothache, and skin diseases. Tomatillos (*Physalis philadelphica* Lam.) were domesticated in southern Mexico by the Mexica people c. 800 BCE and remain a staple in Mexican cuisine. Cilantro (*Coriandrum sativum* L.) is a herb frequently found in modern Mexican cuisine, but it is not native to Central America. Culantro (*Eryngium foetidum* L.; Figure 5.8) is native to Central and South America. It has a similar flavor to cilantro and a wide range of ethnomedical applications. Tomatoes (*Solanum lycopersicum* L.) are another staple in Mexican cuisine that is probably not native to the region. While their time and place of domestication is still uncertain, it most likely occurred somewhere in South America.

Figure 5.8 Culantro or recao (*Eryngium foetidum*) is a biennial herb with a flavor similar to cilantro (*Coriandrum sativum*).

There are several tropical fruits that originated in southern Mexico including guava (*Psidium guajava* L.) (Figure 5.9), papaya (*Carica papaya* L.) (Figure 5.10), and yellow passionfruit (*Passiflora edulis* Sims). Agave spp. were harvested for sweet sap and fiber (see Agave case study in the section titled "Agave: More Than Just Tequila" in Chapter 5). A species of vanilla orchid, *Vanilla planifolia* Jacks. ex Andrews, is native to Veracruz, Mexico. It is believed that the Totonac people were the first to cultivate vanilla (see vanilla case study in the section titled "Vanilla: Madagascar's Orchid Economy" in Chapter 4).

Figure 5.9 Guava (*Psidium guajava*).

Figure 5.10 Papaya (*Carica papaya*)

Chicle, a sweet gum from *Manilkara chicle* (Pittier) Gilly, originated in Mesoamerica. Maya, Nahuatl, and undoubtedly other indigenous groups chewed chicle as a breath freshener and to quench thirst and hunger. American Thomas Adams created the modern chewing gum "Chicklets" from chicle after a brainstorming session with Mexican president Antonio Lopez de Santa Anna to use chicle as a type of vulcanized rubber. William Wrigley formed the William Wrigley Jr. Company, creating a massive demand for chicle in North America. By the 1930s, most of Mexico's chicle trees had been destroyed by unsustainable harvesting practices. Chewing gum companies switched to synthetic bases, destroying the chicle economy in Central America, but saving the remaining trees. American scientist Russell Marker used Mexican yams (*Dioscorea mexicana* Scheidw.) to synthesize progesterone from diosgenin in 1942, leading to the first birth control pill as well as synthetic testosterone, various estrogens, and cortisone.

Mesoamericans used tree resin primarily from *Bursera bipinnata* (Moc. & Sessé ex DC.) Engl. and other Burseraceae species called copal for incense, chewing, glue, paint binder, preservative, and medicine. Zinacantecos Mayans used resin from *Bursera excelsa* (Kunth) Engl., *B. tomentosa* (Jacq.) Triana & Planch., and *B. bipinnata* to fill tooth cavities and repair loose teeth. Huastec Mayans used *Bursera simaruba* (L.) Sarg. to treat burns, headache, nosebleed, fever, and stomachache and *Protium copal* (Schltdl. & Cham.) Engl. for fright, dizziness, and stomachache. The Lacandón Maya offered copal incense made from pitch pine (*Pinus pseudostrobus* Lindl.) to their gods. Chortí Mayans fashioned a maize cob out of *Bursera* copal to be placed in the granary to protect their harvest from harmful spirits. Several Mesoamerican communities smoked their maize seeds with copal smoke before planting as a form of spiritual protection. Mayans used smoke from *Bursera* copal to induce a trance-like hypnotic state in human sacrifices and shamans, although there are no known hypnotic or psychoactive compounds present.

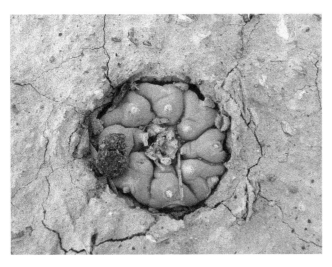

Figure 5.11 Peyote cactus (*Lophophora williamsii*). © 2011 Amante Darmanin.

The peyote cactus [*Lophophora williamsii* (Lem. ex Salm-Dyck) J.M. Coult.] is native to the Chihuahua Desert in northern Mexico (Figure 5.11). Mescaline is the primary psychoactive alkaloid. It binds to serotonin receptors, affecting the frontal cortex, causing hallucinations. Native Americans used peyote in ceremonies, rituals, and for medicine. Likewise, morning glory seeds from several species in Convolvulaceae were used by Oaxaca Mexican tribes for divination, prophesy, and to treat illnesses. They contain ergoline alkaloids including the particularly hallucinogenic ergine, also known as lysergic acid (LSA). Mazatec shamans used *Salvia divinorum* Epling & Játiva, as its name implies, for divination. The diterpene salvinorin A is the active component. It is a potent κ opioid receptor agonist with no 5-HT2A serotonin receptor activity, the principal molecular target of classical hallucinogens (Roth *et al.* 2002).

Many of the crops and medicinal plants grown in the Caribbean were brought from Africa via the slave trade or by European explorers expanding their agricultural production (e.g., sugarcane and indigo). The Europeans decimated much of the forests for lumber and firewood, taking land away from native Caribs and Arawaks. Several native trees were used for timber including Caribbean mahogany [*Swietenia mahagoni* (L.) Jacq.], West Indian ebony [*Brya ebenus* (L.) DC.], walnut (*Juglans jamaicensis* C.DC.), and blue mahoe (*Hibiscus elatus* Sw.). The star apple (*Chrysophyllum cainito* L.) and Caribbean jujube soana (*Ziziphus rignonii* Delponte) are the only two endemic fruits valued for fresh consumption. Other native/naturalized fruit species from Central or South America include allspice [*Pimenta dioica* (L.) Merr.] (Figure 5.12), cashew (*Anacardium occidentale* L.) (Figure 5.13), annatto (*Bixa orellana* L.) (Figure 5.14), mammey (*Mammea americana* L.), soursop (*Annona muricata* L.), papaya, guava, and Jamaican plum (*Spondias purpurea* L.). Medicinal plants tend to be weedy, common species like *Dysphania ambrosioides* (L.) Mosyakin & Clemants and *Aristolochia trilobata* L. Several naturalized species such as the Pará rubber tree [*Hevea brasiliensis*, (Willd. ex A.Juss.) Müll.Arg.] from the Amazon, mango (*Mangifera indica* L.), and breadfruit [*Artocarpus altilis* (Parkinson ex F.A.Zorn) Fosberg] have historically been important crops.

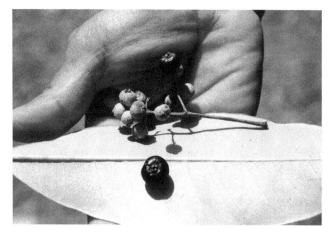

Figure 5.12 **Allspice** (*Pimenta dioica*) leaf and fruit.

Figure 5.13 **Cashew** (*Anacardium occidentale*) has a pseudo-fruit "apple" born on top of the true cashew fruit, a kidney-shaped nut enclosed in a double shell containing irritating phenolic resin. © 2015 Maya Pujiati.

Figure 5.14 **Annatto** (*Bixa orellana*) fruit. Food coloring is obtained from a carotenoid-rich resinous coating on the seeds.

South America

Although there were hundreds of pre-Columbian native tribes living in South America, the Inca Empire was the largest and most sophisticated civilization in the New World. Centered in Peru's Cusco valley, the Inca Empire arose sometime between 1200 and 1300 CE and lasted until the Spanish eventually conquered them in 1572. In part, their rise to power is attributed to warming climate and terraced agriculture, leading to a surplus in maize. The Inca conquered the kingdoms in the Titicaca Basin, gaining a surplus of llamas and alpacas as draft animals, meat, leather, and fiber for clothing. More than 70 native crops were grown in their high-altitude agriculture including cotton, peanuts, peppers, potatoes, and yuca (cassava, *Manihot esculenta* Crantz). Cassava was domesticated c. 8000 BCE in western Brazil. There is evidence of cassava cultivation at Inca and Mesoamerican Mayan sites. The starchy roots can be cooked and consumed as a vegetable. When the root is dried and powdered, it is known as tapioca or manioc flour. Cassava starch can be used for baking, as a thickener, or sizing for fabrics. Potatoes (*Solanum tuberosum* L.) originated in the Andes and were domesticated between 8000 and 5000 BCE. The Incas cultivated over 200 varieties of potatoes. They produced a freeze-dried potato product called *chuño* that could keep for many months without spoiling. Table 5.3 provides a summary of crops that originated in South America.

Many plants were used for religious rituals or warfare by South American tribes. The Incas poured *Chicha de jora*, a fermented maize beverage, in the rivers as an offering to

Table 5.3 Crops originating in South America.

Fruits and vegetables		Grains, legumes, and fiber	
Brazil nut	*Bertholletia excelsa* Bonpl., Lecythidaceae	Beans	*Phaseolus* and other Fabaceae species
Cashew	*Anacardium occidentale* L., Anacardiaceae	Cotton	*Gossypium barbadense* L., Malvaceae
Cassava	*Manihot esculenta* Crantz, Euphorbiaceae	Peanut	*Arachis hypogaea* L., Fabaceae
Cherimoya	*Annona cherimola* Mill., Annonaceae	Quinoa	*Chenopodium quinoa* Willd., Amaranthaceae
Lúcuma	*Pouteria lucuma* (Ruiz & Pav.) Kuntze, Sapotaceae	Tonka beans	*Dipteryx odorata* (Aubl.) Willd., Fabaceae
Pineapple	*Ananas comosus* (L.) Merr., Bromeliaceae	**Stimulants and medicines**	
Potato	*Solanum tuberosum* L.	Coca	*Erythroxylum coca* Lam., Erythroxylaceae
Squash	*Cucurbita* spp., Cucurbitaceae	Cocoa	*Theobroma cacao* L., Malvaceae
Sweet potato	*Ipomoea batatas* (L.) Lam., Convolvulaceae	Guarana	*Paullinia cupana* Kunth, Sapindaceae
Tomato	*Lycopersicon esculentum* Mill., Solanaceae	Ipecac	*Carapichea ipecacuanha* (Brot.) L.Andersson, Rubiaceae
Dyes and latex		Jaborandi	*Pilocarpus* spp., Rutaceae
Annatto	*Bixa orellana* L., Bixaceae	Quinine	*Cinchona* spp., Rubiaceae
Rubber	*Hevea brasiliensis* (Willd. ex A.Juss.) Müll.Arg., Euphorbiaceae	Yerba mate	*Ilex paraguariensis* A.St.-Hil., Aquifoliaceae

the gods and consumed it during religious ceremonies. In sacred fields, crops were grown for the sole purpose of offering to the gods. Powdered bark resin from *Virola* spp. (Myristicaceae) trees were used to induce hallucinations in shamans. 5-MeO-DMT (5-methoxy-*N,N*-dimethyltryptamine) and DMT (*N,N*-dimethyltryptamine) are the psychoactive agents. Ayahuasca (Figure 5.15) is a South American concoction typically containing yagé vine [*Banisteriopsis caapi* (Spruce ex Griseb.) Morton], chaliponga [*Diplopterys cabrerana* (Cuatrec.) B.Gates] leaves, and/or chacruna (*Psychotria viridis* Ruiz & Pav.) leaves. The main active constituents of Ayahuasca are β-carboline harmala alkaloids and monoamine oxidase inhibitors from *B. caapi* and DMT from *D. cabrerana* and *P. viridis*. Sometimes *Brugmansia* sp. was added, contributing atropine, scopolamine, and hyoscyamine to the hallucinogenic mix. Curare is a generic term for isoquinoline or indole alkaloid arrow poisons. It can be prepared from *Strychnos* spp., *Chondrodendron tomentosum* Ruiz & Pav., and more than a dozen other species found in South America. d-Tubocurarine is the primary toxic compound. It is a muscle relaxant that was briefly used in anesthesia during the early twentieth century. San Pedro cactus [*Echinopsis pachanoi* (Britton & Rose) Friedrich & G.D.Rowley], and the Peruvian torch [*Echinopsis peruviana* (Britton & Rose) Friedrich & G.D.Rowley] contain mescaline and were used as hallucinogens throughout South America. Fruits from the sorcerers' tree [*Latua pubiflora* (Griseb.) Baill.] were used by Mapuche medicine men in Chile to induce delirium and hallucinations for medicinal purposes; hyoscyamine and scopolamine are responsible for the observed effects. In the Amazon, *Brunfelsia* spp. had a wide range of medicinal and hallucinogenic uses. *Brunfelsia* spp. contain the alkaloids manaceine, manacine, scopoletin, and the coumarin aesculetin.

The Amazon rainforest contains an abundance of biodiversity including important medicinal plants, some of which gained global importance. Cocaine from coca leaves (*Erythroxylum coca* Lam.), emetine from ipecacuanha rhizomes [*Carapichea ipecacuanha* (Brot.) L.Andersson.], pilocarpine from jaborandi (*Pilocarpus* spp.), and quinine

Figure 5.15 A pot of Ayahuasca. © 2015 Paul Hessell.

from *Cinchona* spp. bark are four well-known alkaloids with medicinal properties. Cocaine is a highly addictive recreational drug, but it also has been used as an anesthetic, analgesic, and to stop bleeding. Chewing coca leaves or drinking coca tea was and still is popular among native South American tribes for medicinal and religious purposes. Ipecacuanha has a long history of use as an emetic in South America. Up until the later part of the twentieth century, syrup of ipecac was prescribed throughout the world as a purgative for accidental poisoning. From the eighteenth to the twentieth centuries, it was used as an emetic and expectorant and mixed with opium to produce "Dover's powder," a common cold remedy (Figure 5.16). Pilocarpine from the bark of jaborandi trees is used in ophthalmology to relieve glaucoma pressure. "Jaborandi" is a Tupi word meaning "causes slobbering," describing its traditional and modern-day use as a treatment for dry mouth. Quinine is a well-known treatment for malaria that has made a colossal impact on the world of medicine and society as a whole. Native South Americans likely used cinchona or quina-quina bark long before the arrival of Europeans. Yet cinchona bark is sometimes called "Jesuit's bark" in reference to the Jesuit missionaries who used it to treat malaria during their missions to South America during the 1600s.

The Amazon rainforest is also the native habitat for the Pará rubber tree (*Hevea brasiliensis*). While native people used the natural latex for balls, shoes, and so on, for centuries, the vulcanization process developed in 1839 greatly increased its versatility, especially for industrial applications. The British Empire smuggled out seeds and eventually succeeded in growing rubber trees at their colonies in modern-day Singapore, Malaysia, Java, and Sri Lanka.

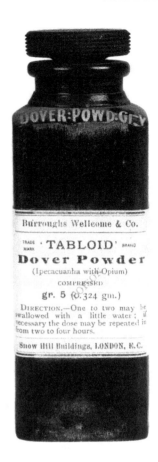

Figure 5.16 A bottle of Dover's Powder cold remedy: opium (*Papaver somniferum*) with ipecacuanha (*Carapichea ipecacuanha*). © 2014 Wellcome Library, London.

Yoco (*Paullinia yoco* R.E. Schult. & Killip), guarana (*Paullinia cupana* Kunth), yerba mate (*Ilex paraguariensis* A.St.-Hil.), and cocoa (*Theobroma cacao* L.) are four South American plants known for their stimulating properties. Xanthine alkaloids caffeine, theobromine, and theophylline are the active components. The Pre-Columbian Tupi and Guarani people of Paraguay and Brazil used guarana seeds and yerba mate leaves to make stimulating teas. Guarana seeds (Figure 5.17) are dried and pounded into a fine powder that can be added to teas, bread, and other foods. Yerba mate was primarily used as a stimulant and diuretic, but also for depression, pain, as an appetite suppressant, and topically as a poultice for ulcers. Yerba wildcrafters called *yerbateros* or *tarrafeiros* collect wild leaves that are said have superior flavor to plantation-grown yerba mate. The world produced over 4.5 million tons of cocoa beans in 2013 (FAO 2013). And while the top producers are not South American countries, the crop

Figure 5.17 Guarana (*Paullinia cupana*) fruit have black seeds covered by white aril. © 2012 Geoff Gallice.

originated in northern South America and Central America. The Latin name *Theobroma* means "food of the gods," and the species name *cacao* is derived from the Nahuatl word *xocolatl*, meaning "bitter water."

References and Additional Reading

Balick, M.J., and Arvigo, R. (2015) *Messages from the Gods: A Guide to the Useful Plants of Belize*. Oxford University Press, Oxford.

Baudry Des Lozières, L.-N. (1803) *Second Voyage à la Louisiane*. Chez Charles, Paris.

Bolsinger, C.L., and Jaramillo, A.E. (1990) *Taxus brevifolia* Nutt. Pacific yew, in *Silvics of North America Volume 1, Conifers*. (eds R.M. Burns and B.H. Honkala), Agriculture Handbook 654. United States Department of Agriculture Forest Service, Washington, DC.

Buisseret, D., and Reinhardt, S.G. (2000) *Creolization in the Americas*. Texas A&M University Press, College Station.

Connolly, T. (1999) *Newberry Crater: Anthropological Papers Number 121*. University of Utah Press, Salt Lake City.

Food and Agriculture Organization of the United Nations (2013). FAOSTAT database collections. Food and Agriculture Organization of the United Nations, Rome, http://faostat.fao.org (accessed 14 September, 2015).

Gurib-Fakim, A. (2006) Medicinal plants: Traditions of yesterday and drugs of tomorrow. *Molecular Aspects of Medicine*, **27** (1), 1–93.

Hummer, K.E. (2013) Manna in winter: Indigenous Americans, huckleberries, and blueberries. *HortScience*, **48** (4), 413–417.

Kraft, K.H., Brown, C.H., Nabhan, G.P. *et al.* (2014) Multiple lines of evidence for the origin of domesticated chili pepper, *Capsicum annuum*, in Mexico. *Proceedings of the National Academy of Sciences*, **111** (17), 6165–6170.

Piperno, D.R., and Flannery, K.V. (2001) The earliest archaeological maize (*Zea mays* L.) from highland Mexico: New accelerator mass spectrometry dates and their implications. *Proceedings of the National Academy of Sciences*, **98**, 2101–2103.

Pope, K.O., Pohl, M.E.D., Jones, J.G. *et al.* (2001) Origin and environmental setting of ancient agriculture in the lowlands of Mesoamerica. *Science*, **292** (5520), 1370–1373.

Roth, B.L., Baner, K., Westkaemper, R. *et al.* (2002) Salvinorin A: A potent naturally occurring nonnitrogenous κ opioid selective agonist. *Proceedings of the National Academy of Sciences*, **99** (18), 11934–11939.

Schultes, R.E. (1969) Hallucinogens of plant origin. *Science*, **163** (3864), 245–254.

Smith, B.D. (1989) Origins of agriculture in Eastern North America. *Science*, **246** (4937), 1566–1571.

Stross, B. (1994) Mesoamerican copal resins. In *U Mut Maya lV: Reports and Readings Inspired by the Advanced Seminars led by Linda Schele at the University of Texas at Austin* (ed. Waynerka, P.). University of Texas, Austin, pp. 177–186.

Phlorotannins in Seaweed

(J. Kellogg)

Ethnobotany and ethnopharmacology

Marine seaweeds have been traditionally harvested as a source of food, livestock fodder, and pharmaceuticals by coastal communities across the globe, especially those situated around the Pacific Rim (Widjaja-Adhi Airanthi *et al.* 2011). From arctic Alaska to the temperate Pacific Northwest, seaweeds have played a large role in the traditional cultures of multiple Native American/Alaska Native (NA/AN) and First Nation communities, including the Haida, Inuit, Tlingit, Tsimshian, Eyak, and Alutiiq peoples, as a ubiquitous source of macro- and micronutrients (Garza 2005; Turner and Bell 1973). Indigenous populations have harvested and consumed red, brown, and green seaweeds for generations, and these marine resources form an important part of the communities' traditional ecological knowledge.

Communities in Alaska's Bristol Bay region consumed *Fucus* spp. (Figure 5.18) and *Macrocystis* spp. covered with herring eggs as a springtime delicacy. First Nations in British Columbia prepare soups with species of the red alga *Porphyra* and clams, salmon

Figure 5.18 Fresh *Fucus distichus* ("bladder wrack" or "popweed") harvested near Whittier, Alaska, the United States.

eggs, or fish. The thalli (filaments of undifferentiated vegetative tissue) of red and brown algae, including *Porphyra* spp. and *Laminaria* spp., are dried and toasted to eat as a snack or sprinkled over other foods (Turner 2003). Seaweeds are also gathered and fermented to improve their shelf life for long-term use, then reconstituted as a soup flavored with oulachen oil (Turner and Bell 1973). To this day, nearly 60% of Inuit households among the Canadian Arctic's Belcher Islands regularly consume *Rhodymenia* spp. and *Laminaria* spp. (Wein *et al.* 1996).

In addition to their consumption as a part of traditional diets, brown and red seaweeds have been utilized as part of many communities' ethnomedical pharmacopeia. For example, hot baths of *Fucus* spp. were employed with yellow cedar boughs [*Chamaecyparis nootkatensis* (D.Don) Spach] in order to create a therapeutic steam to help cure bronchial infections and rheumatism (Boas 1966). In addition, *Fucus* were chopped and heated in conjunction with dried tobacco (*Nicotiana* spp.), alder bark (*Alnus rubra* Bong.), and twinberry cuttings [*Lonicera involucrate* (Richardson) Banks ex Spreng.] and applied as a poultice compress to relieve inflammation, aches, and pains (Turner and Bell 1973). Similarly, children with sores or itchy scabs were rubbed with *Fucus* thalli followed by catfish oil and burnt red ochre, or with strips of the brown seaweed *Nereocystis luetkeana* (K.Mertens) Postels & Ruprecht (Boas 1966). *N. luetkeana* was also applied externally to the stomach of a pregnant woman to ease childbirth, so that the child would become as slippery as the seaweed. The green seaweed *Ulva lactuca* L. was mixed with twinberry bark and applied to a woman's breasts after delivery to relieve soreness and inflammation (Turner and Bell 1973).

Chemistry and Bioactivity

Seaweeds, especially brown algae, have been shown to contain high levels of polyphenols, which can account for as much as 20% by dry weight of the seaweed (Ragan and Glombitza 1986). In brown seaweeds, the predominant polyphenols are a family of tannin-like structures known as phlorotannins, with nearly 150 unique phlorotannin structures identified that range from 126 Da to over 500 kDa (Martínez and Castañeda 2013). Phlorotannins are oligomeric constructs based upon the monomer phloroglucinol (1,3,5-trihydroxybenzene), which is biosynthesized via the polyketide pathway (Meslet-Cladière *et al.* 2013). The polymers are primarily stored in the thalli, and their composition and quantity exhibit internal, geographic, and temporal variability. The monomeric units are linked through aryl–aryl bonds and diarylether bonds, forming four phlorotannin subgroups differentiated by their means of linkage: phlorotannins with an ether bridge (fuhalols and phlorethols), with an aryl–aryl linkage (fucols), those with ether and phenyl links (fucophlorethols), and phlorotannins with a dibenzodioxin linkage (eckols and carmalols) (Glombitza and Pauli 2003) (Figure 5.19).

The myriads of phlorotannin structures in brown algae have demonstrated numerous bioactive properties. They have strong antioxidant activities against free-radical-mediated oxidation by scavenging radicals and inhibiting peroxidation (Shibata *et al.* 2008). Like terrestrial polyphenolic compounds, phlorotannins are also ubiquitous enzyme inhibitors that have been shown to modulate the activity of carbohydrate-hydrolyzing

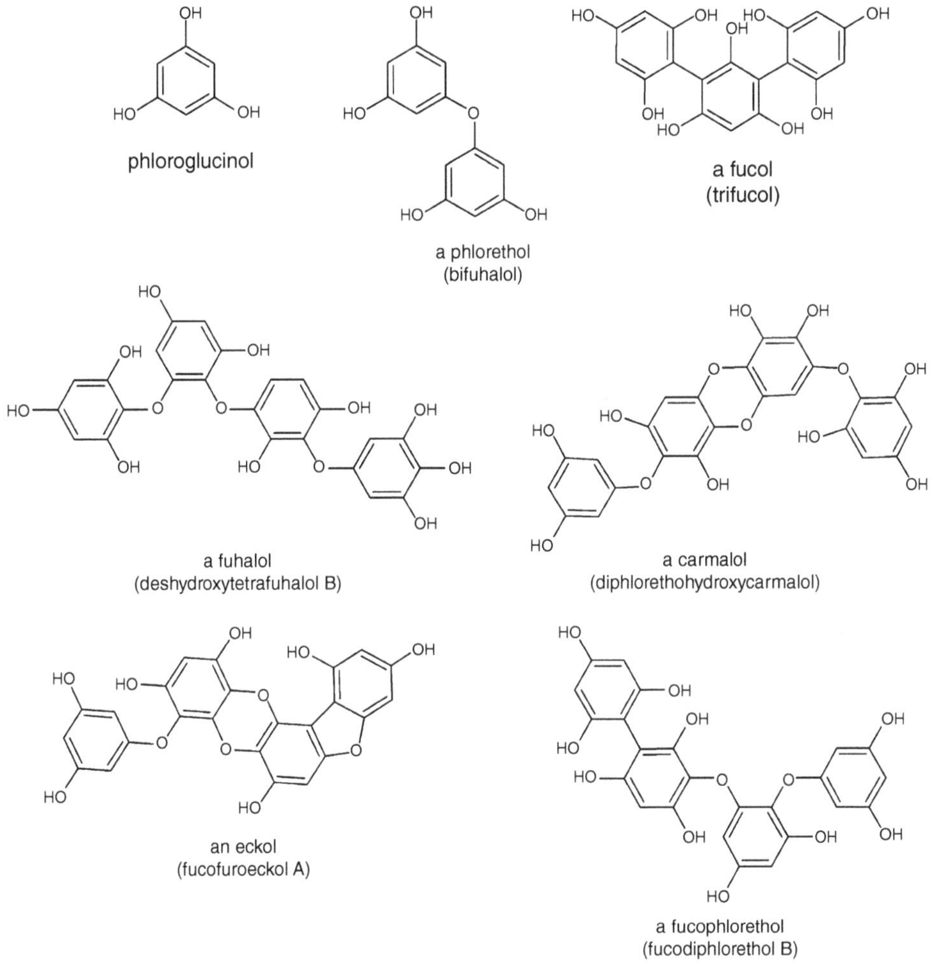

Figure 5.19 Structural classes of phlorotannins, oligomers of phloroglucinol.

enzymes α-glucosidase and α-amylase, thereby decreasing hyperglycemia (Eom *et al.* 2012), inhibiting the angiotensinogen-I-converting enzyme to regulate blood pressure (Jung *et al.* 2006), blocking the digestive enzyme lipase and lowering dyslipidemia (Eom *et al.* 2013), and arresting tyrosinase activity, preventing the synthesis of melanin and subsequent hyperpigmentation (Yoon *et al.* 2009). Furthermore, phlorotannins have exhibited strong anti-inflammatory properties, blocking production of pro-inflammatory cytokines such as prostaglandins and nitric oxide (Kim *et al.* 2009). In addition, phlorotannins have reduced growth of certain cancers, including MCF-7, HeLa, HT1080, A549, and HT-29 cells (Li *et al.* 2011). More recently, evidence has emerged that phlorotannins may reduce allergic reactions by blocking histamine release from basophils (Sugiura *et al.* 2006).

Brown seaweed consumption around the globe remains a dietary source of phlorotannins. However, concentrated phlorotannin preparations have begun to be commercialized as ingredients within various health and beauty products. Commercial extracts of the brown algae prepared from *Ascophyllum nodosum* (L.) Le Jolis and *Fucus vesiculosus* L. are marketed to reduce postprandial serum glucose levels (Roy *et al.* 2011). Phlorotannins have also been incorporated into cosmeceutical formulations for their tyrosinase inhibiting (Thomas and Kim 2013) and anti-photoaging properties (Pallela *et al.* 2010). As such, they have efficacy in preventing the development of wrinkles and skin darkening, making them useful additives in skin-whitening creams and anti-wrinkle formulations (Thomas and Kim 2013). Marine-based phlorotannins hold great potential for continued development as nutraceutical supplements and other therapeutic formulations that impact multiple chronic human conditions.

References

Boas, F. (1966) *Kwakiutl Ethnography*. University of Chicago Press, Chicago, IL.

Eom, S.H., Lee, M.S., Lee, E.W., Kim, Y.M., & Kim, T.H. (2013) Pancreatic lipase inhibitory activity of phlorotannins isolated from *Eisenia bicyclis*. *Phytotherapy Research*, **27**, 148–151.

Eom, S.H., Lee, S.H., Yoon, N.Y. *et al.* (2012) α-Glucosidase- and α-amylase-inhibitory activities of phlorotannins from *Eisenia bicyclis*. *Journal of the Science of Food and Agriculture*, **92**, 2084–2090.

Garza, D. (2005) *Common Edible Seaweeds in the Gulf of Alaska*. Alaska Sea Grant College Program, Fairbanks, AK.

Glombitza, K.W., and Pauli, K. (2003) Fucols and phlorethols from the brown alga *Scytothamnus australis* Hook. et Harv. (Chnoosporaceae). *Botanica Marina*, **46**, 315–320.

Jung, H.A., Hyun, S.K., Kim, H.R., and Choi, J.S. (2006) Angiotensin-converting enzyme I inhibitory activity of phlorotannins from *Ecklonia stolonifera*. *Fisheries Science*, **72**, 1292–1299.

Kim, A.R., Shin, T.S., Lee, M.S. *et al.* (2009) Isolation and identification of phlorotannins from *Ecklonia stolonifera* with antioxidant and anti-inflammatory properties. *Journal of Agricultural and Food Chemistry*, **57**, 3483–3489.

Li, Y., Qian, Z.J., Kim, M.M., and Kim, S.K. (2011). Cytotoxic activities of phlorethol and fucophlorethol derivatives isolated from Laminariaceae Ecklonia cava. *Journal of Food Biochemistry*, **35**, 357–369.

Martínez, J.H.I., and Castañeda, H.G.T. (2013) Preparation and chromatographic analysis of phlorotannins. *Journal of Chromtographic Science*, **51**, 825–838.

Meslet-Cladière, L., Delage, L., and Leroux, C.J.J., (2013) Structure/function analysis of a type III polyketide synthase in the brown alga *Ectocarpus siliculosus* reveals a biochemical pathway in phlorotannin monomer biosynthesis. *The Plant Cell*, **25**, 3089–3103.

Pallela, R., Na-Young, Y., and Kim, S.K. (2010) Anti-photoaging and photoprotective compounds derived from marine organisms. *Marine Drugs*, **8**, 1189–1202.

Ragan, M.A., and Glombitza, K.W. (1986) Phlorotannins, brown algal polyphenols. *Progress in Phycological Research*, **4**, 129–241.

Roy, M.C., Anguenot, R., Fillion, C., Beaulieu, M., Bérubé, J., and Richard, D. (2011) Effect of a commercially-available algal phlorotannins extract on digestive enzymes and carbohydrate absorption in vivo. *Food Research International*, **44**, 3026–3029.

Shibata, T., Ishimaru, K., Kawaguchi, S., Yoshikawa, H., and Hama, Y. (2008) Antioxidant activities of phlorotannins isolated from Japanese Laminariaceae. *Journal of Applied Phycology*, **20**, 705–711.

Sugiura, Y., Matsuda, K., Yamada, Y. *et al.* (2006) Isolation of a new anti-allergic phlorotannin, phlorofucofuroeckol B, from an edible brown alga. *Bioscience Biotechnology and Biochemistry*, **70**, 2807–2811.

Thomas, N.V., and Kim, S.K. (2013) Beneficial effects of marine algal compounds in cosmeceuticals. *Marine Drugs*, **11**, 146–164.

Turner, N.C., and Bell, M.A.M. (1973) The ethnobotany of the Southern Kwakiutl Indians of British Columbia. *Economic Botany*, **27**, 257–310.

Turner, N.J. (2003) The ethnobotany of edible seaweed (*Porphyra abbottae* and related species; Rhodophyta: Bangiales) and its use by First Nations on the Pacific Coast of Canada. *Canadian Journal of Botany*, **81**, 283–293.

Wein, E.E., Freeman, M.M.R., and Makus, J.C. (1996) Use of and preference for traditional foods among the Belcher Island Inuit. *Arctic*, **49**, 256–264.

Widjaja-Adhi Airanthi, M.K., Hosokawa, M., and Miyashita, K. (2011) Comparative antioxidant activity of edible Japanese brown seaweeds. *Journal of Food Science*, **76**, C104–111.

Yoon, N.Y., Eom, T.K., Kim, M.-M., and Kim, S.K. (2009) Inhibitory effect of phlorotannins isolated from Ecklonia cava on mushroom tyrosinase activity and melanin formation in mouse B16F10 melanoma cells. *Journal of Agricultural and Food Chemistry*, **57**, 4124–4129.

Agave: More Than Just Tequila

(B. M. Schmidt)

Ethnobotany

Agave is a genus of succulent plants with a geographic center of origin in Mexico (Good-Avila *et al.* 2006). Populations have spread north to the deserts of the southwestern United States, east into the Caribbean, and south to Central and South America. The genus was formerly placed in the family Agavaceae, but has since been moved to Asparagaceae. Perhaps the most well-known member of the genus is *Agave tequilana* F.A.C.Weber, the blue agave. It is native to Jalisco, Mexico, and is the primary ingredient in the alcoholic beverage tequila. To make tequila, the heart or *piña* of the plant is harvested around the twelfth year of growth by *jimadores*, Mexican agave farmers. Agave juice is extracted from the pulped *piña* (Figure 5.20) then fermented, distilled, and often aged to form tequila. Mescal is a similar distilled alcoholic beverage produced from *Agave parryi* Engelm. and other *Agave* spp.

Figure 5.20 *Jimadores* use a dried calabash gourd [*Lagenaria siceraria* (Molina) Standl.] called an *acocote* to suck *aguamiel* out of the center of an Agave stem. © 2012 Nacho Pintos.

Before the Spaniards arrived, native Mesoamericans did not have knowledge of distillation. For centuries, they consumed fermented beverages from *Agave* (locally called *maguey*) known as *pulque* (Figure 5.21). Murals in Teotihuacan (c. 150 BCE–650 CE) depict maguey plants and scenes of possible pulque consumption (Correa-Ascencio *et al.* 2014). Some speculate pulque served as probiotic medicine and a source of micronutrients in early Mesoamerica (Correa-Ascencio *et al.* 2014). Pulque was certainly an important element of many Mesoamerican rituals; both the Maya and Aztecs had pulque deities (Henderson 2008). To make pulque, a juice called *aguamiel* is collected from the center of *A. americana* L., *A. salmiana* Otto ex Salm-Dyck, *A. sisalana* Perrine, or other *Agave* species and fermented to a milky, viscous liquid with a short shelf life. The short shelf life meant that Mesoamericans often drank to the point of vomiting during ceremonies, as to not let any go to waste. But ethnobotanists also speculate vomiting from pulque was viewed as a cleansing necessary to communicate with the gods (Henderson 2008).

A. sisalana is also the source of sisal, a fiber used for rope, twine, nets, baskets, and clothing. The Aztecs and Mayans extracted the fibers for use as rough garments and used the spines as needles. Sometimes it is referred to as the "needle and thread plant." Sisal fibers were originally exported from the port of Sisal in the Yucatan, but the plant does not grow in that region. Yucatan farmers instead have fiber plantations growing *A. fourcroydes* Lem. for henequen production. Henequen is a similar type of fiber but considered lower quality than sisal. It is primarily used for burlap bags, ropes, and mooring. The most likely place of origin for *A. sisalana* is the neighboring state of Chiapas, where farmers grow it as fencerows and for fiber. Each plant will produce about 220 leaves before bolting at around seven years of age. If leaves are regularly harvested, bolting may be delayed for 15–20 years. After bolting, the plant dies. Under the hot and dry native conditions in Mexico, *A. sisalana* produces around one ton of dried fibers per hectare. In East Africa, especially regions in Kenya and Tanzania with much higher annual rainfall than Mexico, yields of dry fiber can reach two to five tons per hectare.

The namesake of *A. sisalana*, Dr. Henry E. Perrine, was a French doctor from New Brunswick, New Jersey, appointed as US Consul at Campeche, Yucatan in 1827. He made extensive plant collections, many of which can be found at the New York Botanical Gardens today. In 1838, Congress awarded him a six square mile plot of land on lower Biscayne Bay south of Miami, Florida, for the propagation and cultivation of tropical plants. He and his family first settled on Indian Key, where he had sent many of his plant and seed specimens for safe keeping until the end of the Seminole War. Unfortunately, on August 7, 1840, several canoes of Seminoles landed on the shores and began a nighttime siege on the outpost (Perrine 1885). Dr. Perrine was killed, and many of his specimens destroyed. A few survived, including *A. sisalana*, some date palms, and wild limes, known as "key lime" today (Robinson 1937). About 50 years later, sisal became important as binder or reaper twine for US agriculture. Mexico

Figure 5.21 Bottle of pulque in Zacatlán, Puebla, Mexico. © 2015 Alejandro Linares Garcia.

banned export of plants for cultivation and enjoyed a brief but profitable monopoly. But American horticulturists had Perrine's stock of wild *A. sisalana* and were soon able to produce thousands of plants. In 1893, 1000 bulbils were sent from Florida to Germany and onward to Tanzania (Brown 2002). The 62 plants that survived started large plantations in East Africa. Around 1919, the United States Sisal Trust was formed to start commercial production in Florida (Brown 2002). They imported 220,000 tons of *A. sisalana* from Mexico and the Caribbean to grow on huge plantations in Dade County. Production eventually waned owing to the introduction of synthetic fibers after World War II and unsuitable environmental conditions such as hurricanes.

Phytochemistry

Alcoholic beverage production from *Agave* spp. relies on fermentation of the plant sugars, mostly fructose. In contrast to beer and wine, which rely heavily on yeast for fermentation, pulque is primarily produced by bacterial fermentation. Reported species include dextran-producing lactic acid bacteria such as *Leuconostoc* (Correa-Ascencio *et al.* 2014; Escalante *et al.* 2008). Dextran, a branched glucose polysaccharide from 3–2000 kDa, is largely responsible for the viscous texture of pulque. Dextran along with several micro/macronutrients and lactic acid bacteria are thought to be both pre- and probiotic for gut bacteria (Correa-Ascencio *et al.* 2014). The presence of phytase would have helped Mesoamericans better digest maize by increasing the bioavailability of iron, phosphorous, and zinc (Correa-Ascencio *et al.* 2014).

Sisal (and henequen) fibers are removed from mature *Agave* leaves cut off at the base of the stem. The leaves are squeezed between two large rollers to press out the water and turn the soft tissues into a pulp that can be scraped away from fibers. Fibers account for only 4% of the harvested biomass; the remaining pulp is an agricultural waste that has recently been evaluated for use as mulch, animal feed, or biofuel (Machin 2008). After pulp removal, sisal fibers are washed and hung in the sun to dry. The result is a creamy white fiber up to one meter in length. Scanning electron microscope (SEM) micrographs of fiber strand cross sections reveal bundles of about 100 hollow cellulosic sub-fiber cells that run the length of the fiber strand (Fávaro *et al.* 2010, Lacerda *et al.* 2013). These individual fibers are 1.0–1.5 cm in length, 100–300 μm in diameter (Fávaro *et al.* 2010), and primarily composed of cellulose (~70%). A middle lamella composed of hemicellulose, lignin, and pectin link the fiber cells together (Figure 5.22).

While cellulose is only composed of glucose, hemicellulose is a heterogeneous molecule, composed of an assortment of sugar monomers in a random, branched pattern. Hemicellulose retains water (a disadvantage in composites), and even though lignin is the "glue" that attaches the cellulosic bundles, it can be a source of weakness in composites. Therefore, hemicellulose and lignin are typically removed by mercerization with alkali or

(A)

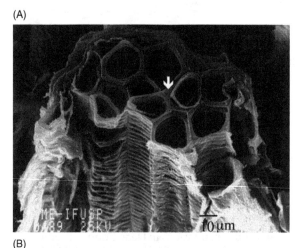

(B)

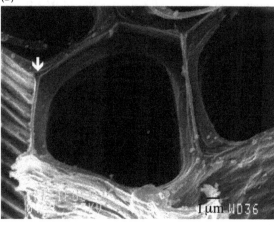

Figure 5.22 SEM micrograph (scale bar = 10 μm) of sisal fiber conductive vessels (A). Micrograph (B) (scale bar = 1 μm) is a detail of (A). The arrows show the middle lamella. Martins *et al.* 2004. Reproduced with permission of John Wiley & Sons, Ltd.

by hydrothermolysis before creating the composite (Brouwer 2000). Mercerization also improves the fiber tensile strength dimensional stability, elasticity, uniformity, and the success of binding cellulose fibers to polymer monomers (Khan *et al.* 2005). The fibers must be dried to 2–3% moisture before mixing with thermoplastic polymers (Brouwer 2000). After the sisal fibers and thermoplastic are mixed and molded, they are often cured with UV radiation (Khan *et al.* 2005).

Compared to glass composites such as fiberglass, sisal composites are only half as strong, but they are also less dense. When accounting for density, their tensile strength is similar. Their impact strength is less than fiberglass, but three times higher than polyester composites and higher than other natural fibers like coir, banana, and pineapple (Joseph *et al.* 1999). New developments in thermoplastic technology and fiber modifications have greatly improved composite mechanical properties (e.g., impact and tensile strength), making sisal composites competitive with fiberglass and other similar materials (Fávaro *et al.* 2010).

Mexican Uses

According to the Distilled Spirits Council of the United States, the United States imported 13.8 million 9 L cases of tequila in 2014. This accounts for roughly 52% of global sales. Pulque is still consumed throughout Mexico (Figure 5.23) and is gaining popularity with Mexico's youth at century-old "pulquerias" in Mexico City.

Currently, sisal fiber is produced in East Africa, Brazil, Haiti, India, and Indonesia. Mexico no longer produces much sisal, but henequen production continues in the Yucatan region. Several German automobile companies, including Audi, BMW, and Mercedes-Benz, use sisal textiles and mats, and sisal fiber composites for body panels and trim. The

Figure 5.23 Pulque vendor in Zacatecas, Mexico. © 2007 Tomas Castelazo.

composites have high strength and impact resistance while reducing the overall weight and cost of the vehicle. The Lotus Eco Elise contains several different natural fibers including wool, hemp, and sisal. In civil engineering, sisal can be used in building materials (panels, floor tiles, etc.) as a low-cost and eco-friendly alternative to materials like fiberglass.

References

Brouwer, W.D. (2000) Natural Fibre Composites in Structural Components: Alternative Applications for Sisal? in *Alternative Applications for sisal and henequen*. Technical paper #14, Common Fund for Commodities and Food and Agriculture Fund for Commodities, Rome, pp. 75–82.

Brown, K. (2002) *Agave sisalana* Perrine. *Wildland Weeds*. Summer, pp. 18–21.

Correa-Ascencio, M., Robertson, I.G., Cabrera-Cortés, O., Cabrera-Castro, R., and Evershed, R.P. (2014) Pulque production from fermented agave sap as a dietary supplement in Prehispanic Mesoamerica. *Proceedings of the National Academy of Sciences of the United States of America*, **111** (39), 14223–14228.

Escalante, A., Giles-Gómez,M., Hernández, G., Córdova-Aguilar, M.S., López-Munguía, A., Gosset, G., and Bolívar, F. (2008) Analysis of bacterial community during the fermentation of pulque, a traditional Mexican alcoholic beverage, using a polyphasic approach. *International Journal of Food Microbiology*, **124** (2), 126–134.

Fávaro, S.L., Ganzerli, T.A., de Carvalho Neto, A.G.V., da Silva, O.R.R.F., and Radovanovic, E. (2010) Chemical, morphological and mechanical analysis of sisal fiber-reinforced recycled high-density polyethylene composites. *eXPRESS Polymer Letters*, **4** (8), 465–473.

Good-Avila, S.V., Souza, V., Gaut, B.S., and Eguiarte, L.E. (2006) Timing and rate of speciation in *Agave* (Agavaceae). *Proceedings of the National Academy of Sciences of the United States of America*, **103** (24), 9124–9129.

Henderson, L. (2008) Blood, water, vomit, and wine. in *Mesoamerican Voices*, 3rd edition (ed. J Palka). Beaker Press, Chicago, pp. 53–76.

Joseph, K., Toledo Filho, R.D., James, B., Thomas, S., and Carvalho, L.H. (1999) A review on sisal fiber reinforced polymer composites. *Revista Brasileira de Engenharia Agrícola e Ambiental*, **3** (3), 367–379.

Khan, M.A., Bhattacharia, S.K., Kabir, M.H., Chowdhury, A.M.S.A., and Rahman, M.M. (2005) Effect of mercerization on surface modification of henequen (*Agave fourcroydes*) fiber by photo-curing with 2-hydroxyethyl methacrylate (HEMA). *Polymer-Plastics Technology & Engineering*, **44** (6), 1079–1093.

Lacerda, T.M., Zambon, M.D., and Frollini, E. (2013) Effect of acid concentration and pulp properties on hydrolysis reactions of mercerized sisal. *Carbohydrate Polymers* **93** (1), 347–356.

Machin, D. (2008) Sisal: Small farmers and plantation workers. In *Proceedings of the Symposium on Natural Fibers*. Common Fund for Commodities and Food and Agriculture Fund for Commodities, Rome, pp. 39–42.

Martins, M.A., Kiyohara, P.K., and Joekes, I. (2004) Scanning electron microscopy study of raw and chemically modified sisal fibers. *Journal of Applied Polymer*, **94** (6), 2333–2340.

Perrine, H.E. (1885) *The True Story of Some Eventful Years in Grandpa's Life*. Press of E.H. Hutchinson, Buffalo, N.Y.

Robinson, T.R. (1937) Henry Perrine, Pioneer Horticulturist of Florida. *Proceedings of the Florida State Horticultural Society*, **50**, 16–24.

Quinoa: A Source of Human Sustenance and Endurance in the High Andes

(B. L. Graf)

Quinoa (*Chenopodium quinoa* Willd., Amaranthaceae) (Figure 5.24), a South American subsistence crop, has been cultivated for over 5000 years in the Andes Mountains up to 4500 m in altitude among the modern-day countries of Argentina, Bolivia, Chile, Colombia, Ecuador, and Peru (Cusack 1984; Dillehay *et al.* 2007; Fuentes *et al.* 2009; Vega-Galvez *et al.* 2010). Quinoa was a sacred staple crop of the ancient Incas (Cusack 1984) and continues to be utilized by the Quechua, Aymara, Tiahuancota, Chibcha, and Mapuche indigenous peoples (Bhargava and Srivastava 2013; Vega-Galvez *et al.* 2010). In many regions of the Andean Altiplano, quinoa is the only crop viable for cultivation under the extreme environmental conditions, characterized by high altitude, low rainfall, arid climate, saline soils, intense temperature fluctuations, and high ultraviolet radiation (Figure 5.25) (FAO 2011; Hellin and Higman 2003; Vega-Galvez *et al.* 2010).

Traditionally, quinoa has been used as a wellness-promoting and endurance-enhancing food, especially directed toward vulnerable populations such as children and the elderly (FAO 2011; Gorelick-Feldman *et al.* 2008; Kokoska and Janovska 2009; Lafont 1998; NRC 1989). Quinoa seeds are typically consumed similarly to rice: prepared in soup, puffed to make breakfast cereal, or ground to flour to produce toasted and baked goods (Bhargava 2006; NRC 1989). "Llipta," the pungent ash of quinoa stems, mixed with the leaves of the coca plant (*Erythroxylum coca* Lam.), has been chewed by Andean farmers to sustain their energy during hard labor (Martindale 1894). Furthermore, the Incan armies were sustained for days as they marched over the Andes Mountains by a mixture of quinoa and fat called "war balls" (Small 2013). Due to these traditional uses, quinoa may be considered an "adaptogen." Adaptogens are defined as agents that reduce stress-induced damage (e.g., anti-fatigue, anti-depressant) and exhibit stimulating effects (e.g., increased working capacity) within a stressful context without depleting energetic resources or perturbing normal bodily functions (Panossian and Wikman 2009, 2010; Brekhman and Dardymov 1968).

Figure 5.24 A red variety of *Chenopodium quinoa* Willd. (quinoa). © 2014 David Wu.

Figure 5.25 A quinoa field in the Aymara village of Ancovinto, Tarapacá, northern Chile, a village situated 3681 m above sea level that receives less than 200 mm of rainfall annually. January 2013.

Chemistry and Bioactivities

Quinoa has been shown to contain two major classes of biologically active compounds that likely play a role in its adaptogenic properties: phenolics and phytoecdysteroids. Phenolics, mainly flavonoid glycosides (Figure 5.26A) and phenolic acids (Figure 5.26B), are biosynthesized through the shikimate pathway, yielding total levels as high as 3996 μg/g seed (Gomez-Caravaca *et al.* 2011). Phytoecdysteroids, among which 20-hydroxyecdysone (20HE) (Figure 5.26C) is most abundant of 13 different analogs (Kumpun *et al.* 2011), are biosynthesized through the melavonate pathway with total levels as high as 570 μg/g seed (Adler and Grebenok 1995; Graf *et al.* 2016).

Phenolics and phytoecdysteroids are thought to be the major bioactive constituents of the traditional "adaptogenic" herbs such as *Rhaponticum carthamoides* (maral rool or Russian *Leuzea*) (Brekhman and Dardymov 1968; Panossian and Wikman 2009, 2010) and *Ajuga turkestanica* (sanabor) (Arthur *et al.* 2014). Flavonoid, phenolic acid, and phytoecdysteroid-containing extracts of *R. carthamoides* have exerted significant positive effects on running and swimming endurance, working capacity, rehabilitation from intense physical activity, and mental capacity (learning and memory) in animal and human clinical trials (Kokoska and Janovska 2009). Phytoecdysteroid-enriched extracts of *A. turkestanica* have been shown to stimulate protein synthesis in muscle cells (Gorelick-Feldman *et al.* 2008) and promote muscle integrity in aged mice (Arthur *et al.* 2014).

Individual phenolics and phytoecdysteroids have demonstrated a range of biological activities in mammals that may contribute to the endurance- and wellness-promoting effect of quinoa consumption (Dinan 2009; Dinan and Lafont 2006; Klein 2004). Phenolics are particularly renowned for their anti-inflammatory and immunomodulatory

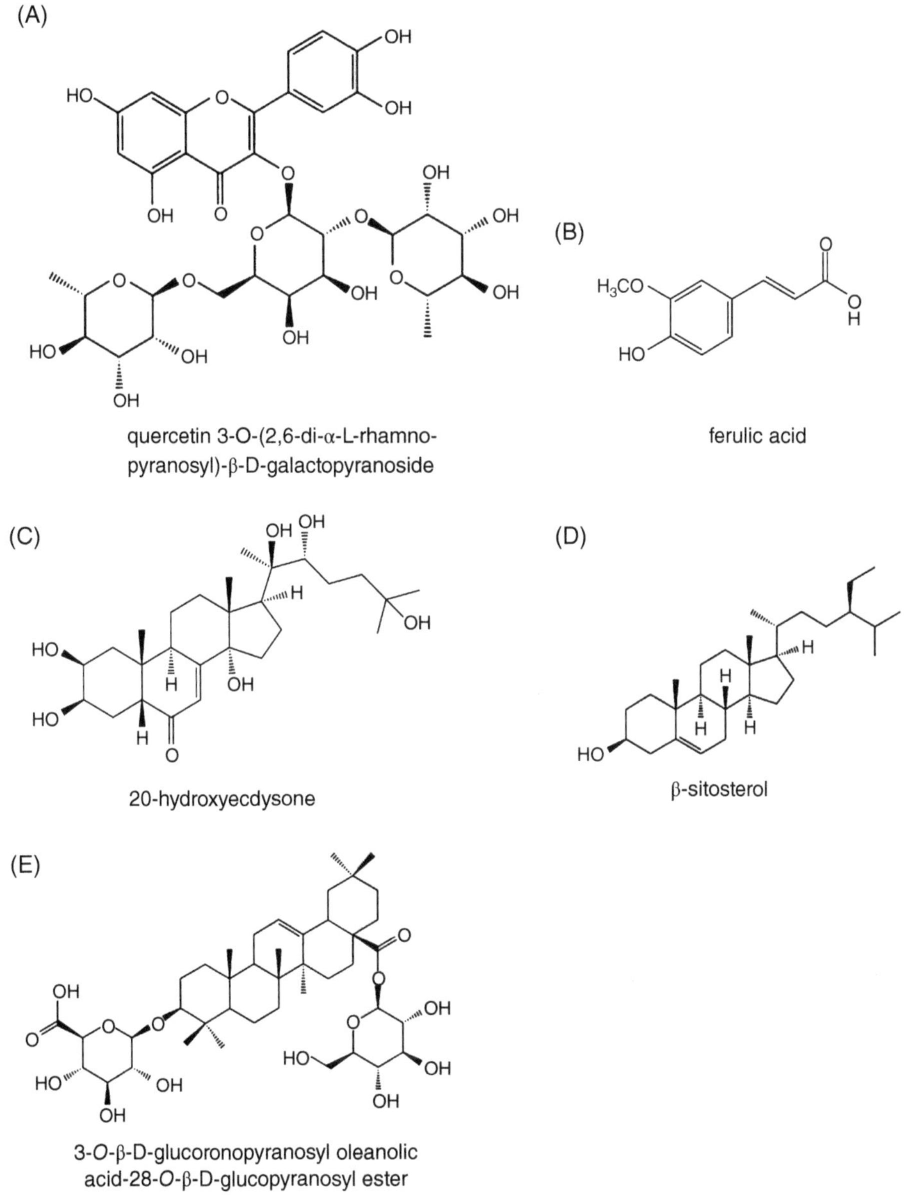

(A)

(B)

quercetin 3-O-(2,6-di-α-L-rhamno-
pyranosyl)-β-D-galactopyranoside

ferulic acid

(C)

(D)

20-hydroxyecdysone

β-sitosterol

(E)

3-O-β-D-glucoronopyranosyl oleanolic
acid-28-O-β-D-glucopyranosyl ester

Figure 5.26 Examples of biologically active compounds in quinoa from five classes of molecules:
(A) flavonoid glycoside, (B) phenolic acid, (C) phytoecdysteroid, (D) phytosterol, and (E) saponin.

capacities. Quercetin, the most widely studied phenolic, reduced expression of inflammatory genes in human macrophages and adipocytes, protected β-cells from oxidative
damage, reduced lipid peroxidation, and prevented myocardial infarction in rats, lowered circulatory inflammatory risk factors in mice, and lowered blood pressure in
hypertensive humans (Russo et al. 2012). Meanwhile, isolated phytoecdysteroids have
been shown to reduce stress-induced reactive oxygen species production in human cells

(Graf *et al.* 2014), accelerate wound healing (Syrov and Khushbaktova 1996), promote growth, and enhance working capacity in animals (Cheng *et al.* 2013; Gorelick-Feldman *et al.* 2008; Syrov *et al.* 2008). Both phenolics and phytoecdysteroids have also demonstrated anti-diabetic, hepatoprotective, neuroprotective, anti-cancer, and antioxidant properties in vitro and in vivo (Dinan 2009; Dinan and Lafont 2006; Klein 2004). Other phytochemical constituents of quinoa seeds, including phytosterols (Figure 5.26D) and saponins (Figure 5.26E), have been hypothesized to potentiate the biological activities of phenolics and phytoecdsyteroids by promoting their cellular bioavailability (Klein 2004), though further research is needed to determine the interactions and possible synergistic effects of these molecules.

Modern and Prospective Uses

Over the past few decades, quinoa production and consumption has expanded from its traditional cultivation regions in the Andes to Africa, Asia, Europe, and North America, thereby offering nutritional benefits worldwide (Bhargava *et al.* 2006). Global research and development of quinoa was promoted in 2013 by the United Nations' program entitled the "International Year of Quinoa" (FAO 2011). Novel means of incorporating the seeds in the diet have been developed to increase the market appeal and convenience of quinoa consumption, such as quinoa-containing fruit compotes and energy bars (Figure 5.27). Furthermore, numerous methods to concentrate and/or deliver bioactive components from quinoa have been invented and commercialized for therapeutic uses (Bhargava *et al.* 2006; Graf *et al.* 2014b). For example, phytoecdysteroid-enriched extracts and purified saponins have been implicated in weight loss and drug absorption

Figure 5.27 Ecuadorian fruit compote for children. © 2013 Cancillería del Ecuador.

applications, respectively (Estrada *et al.* 1997; Foucault *et al.* 2011; Msika 2012; Scanlin and Stone 2008; Veillet and Lafont 2012). Recent clinical research has shown potential for quinoa to play a role in the prevention of metabolic disease (De Carvalho *et al.* 2014) and the promotion of childhood growth and nutrition (Ruales *et al.* 2002). However, clinical trials have not yet assessed the potential adaptogenic effects of quinoa products, such as enhancement of mental performance under stress or improved recovery following intense physical activity. Since numerous in vitro and in vivo studies have demonstrated the stress-reducing and performance-enhancing properties of quinoa's phytochemical constituents, (Kokaska *et al.* 2009), ongoing investigations may show quinoa to be an effective adaptogen.

References

Adler, J.H., and Grebenok, R.J. (1995) Biosynthesis and distribution of insect-molting hormones in plants--a review. *Lipids*, **30** (3), 257–262.

Arthur, S.T., Zwetsloot, K.A., Lawrence, M.M. *et al.* (2014) *Ajuga turkestanica* increases Notch and Wnt signaling in aged skeletal muscle. *European Review for Medical and Pharmacological Sciences*, **18**, 2584–2592.

Bhargava, A., and Srivastava, S. (2013) *Quinoa: Botany, Production and Uses*. CABI, Boston.

Bhargava, A., Shukla, S., and Ohri, D. (2006) *Chenopodium quinoa* – an Indian perspective. *Industrial Crops and Products*, **23**, 73–87.

Brekhman, I.I., and Dardymov, I.V. (1968) New substances of plant origin which increase nonspecific resistance. *Annual Review of Pharmacology*, **8**, 419–430.

Cheng, D.M., Kutzler, L.W., Boler, D.D., Drnevich, J., Killefer, J., and Lila, M.A. (2013) Continuous infusion of 20-hydroxyecdysone increased mass of triceps brachii in C57BL/6 mice. *Phytotherapy Research*, **27** (1), 107–111.

Cusack, D.F. (1984) Quinua: Grain of the Incas. *Ecologist*, **14** (1), 21–31.

De Carvalho, F.G., Ovidio, P.P., Padovan, G.J., Jordao Jr., A.A., Marchini, J.S., and Navarro, A.M. (2014) Metabolic parameters of postmenopausal women after quinoa or corn flakes intake—a prospective and double-blind study. *International Journal of Food Sciences and Nutrition*, **65** (3), 380–385.

Dillehay, T.D., Rossen, J., Andres, T.C., and Williams, D.E. (2007) Preceramic adoption of peanut, squash, and cotton in northern Peru. *Science*, **316**, 1890–1893.

Dinan, L. (2009) The Karlson Lecture. Phytoecdysteroids: What use are they? *Archives of Insect Biochemistry and Physiology*, **72** (3): 126–141.

Dinan, L., and Lafont, R. (2006) Effects and applications of arthropod steroid hormones (ecdysteroids) in mammals. *Journal of Endocrinology*, **191**, 1–8.

Estrada, A., Redmond, M.J., and Laarveld, B. (1997) *Quinoa Saponin Compositions and Methods of Use*. US Patent 5688772.

Food and Agriculture Organization of the United Nations (2011) Quinoa: An ancient crop to contribute to world food security, http://www.fao.org/docrep/017/aq287e/aq287e.pdf. (accessed 8 August 2014).

Foucault, A.S., Mathé V., Lafont R. *et al.* (2011) Quinoa extract enriched in 20-hydroxyecdysone protects mice from diet-induced obesity and modulates adipokines expression. *Obesity*, **20**, 270–277.

Fuentes, F., Martinez, E.A., Hinrichsen, P.V., Jellen, E.N., and Maughan, P.J. (2009) Assessment of genetic diversity patterns in Chilean quinoa (*Chenopodium quinoa* Willd.) germplasm using multiplex fluorescent microsatellite markers. *Conservation Genetics*, **10**, 369–377.

Gomez-Caravaca, A.M., Segura-Carretero, A., Fernandez-Gutierrez, A., and Caboni, M.F. (2011) Simultaneous determination of phenolic compounds and saponins in quinoa (*Chenopodium quinoa* Willd) by a liquid chromatography-diode array detection-electrospray ionization-time-of-flight mass spectrometry methodology. *Journal of Agricultural and Food Chemistry*, **59**, 10815–10825.

Gorelick-Feldman, J. Maclean, D., Ilic, N. *et al.* (2008) Phytoecdysteroids increase protein synthesis in skeletal muscle cells. *Journal of Agricultural and Food Chemistry*, **56** (10), 3532–3537.

Graf, B.L., Cheng D.M., Esposito, D. *et al.* (2014) Compounds leached from quinoa seeds inhibit matrix metalloproteinase activity and intracellular reactive oxygen species. *International Journal of Cosmetic Science*, **37** (2), 212–221.

Graf, B.L., Poulev, A., Kuhn, P., Grace, M., Lila, M.A., and Raskin, I. (2014b) Quinoa seeds leach phytochemicals and other compounds with anti-diabetic properties. *Food Chemistry*, **163**, 178–185.

Graf, B.L., Rojo, L.E., Delatorre-Herrera, J., Poulev, A., Calfio, C., and Raskin, I. (2016) Phytoecdysteroids and flavonoid glycosides among Chilean and commercial sources of *Chenopodium quinoa*: variation and correlation to physicochemical characteristics. *Journal of the Science of Food and Agriculture*, **96** (2), 633–643.

Hellin, J., and Higman, S. (2003) *Feeding the Market: South American Farmers, Trade and Globalization*. Kumarian Press, Inc., Bloomfield.

Klein, R. (2004) *Phylogenetic and Phytochemical Characteristics of Plant Species with Adaptogenic Properties*. MS thesis, Montana State University.

Kokoska, L., and Janovska, D. (2009) Chemistry and pharmacology of *Rhaponticum carthamoides*: A review. *Phytochemistry*, **70**, 842–855.

Kumpun, S., Maria, A., Crouzet, S., Evrard-Todeschi, N., Girault, J.-P., and Lafont, R. (2011) Ecdysteroids from *Chenopodium quinoa* Willd., an ancient Andean crop of high nutritional value. *Food Chemistry*, **125** (4), 1226–1234.

Lafont, R. (1998) Phytoecdysteroids in world flora: Diversity, distribution, biosynthesis and evolution. *Russian Journal of Plant Physiology*, **45** (3), 276–295.

Martindale, W. (1894) *Coca and Cocaine: Their History, Medical and Economic Uses, and Medicinal Preparations*, 3rd edition. H. K. Lewis, London.

Msika, P. (2012) *Composition Containing a Quinoa Extract for Dermatological Use*. US Patent 9125879.

National Research Council (1989) *Lost crops of the Incas: Little Known Plants of the Andes with Promise for Worldwide Cultivation*. National Academies Press, Washington, D.C.

Panossian, A., and Wikman, G. (2009) Evidence-based efficacy of adaptogens in fatigue, and molecular mechanisms related to their stress-protective activity. *Current Clinical Pharmacology*, **4** (3), 198–219.

Panossian, A., and Wikman, G. (2010) Effects of adaptogens on the central nervous system and the molecular mechanisms associated with their stress—protective activity. *Pharmaceuticals*, **3** (1), 188–224.

Ruales, J., De Grijalva, Y., Lopez-Jaramillo, P., and Nair, B.M. (2002) The nutritional quality of infant food from quinoa and its effect on the plasma level of insulin-like growth factor-1 (IGF-1) in undernourished children. *International Journal of Food Sciences and Nutrition*, **53** (2), 143–154.

Russo, M., Spagnuolo, C., Tedesco, I., Bilotto, S., and Russo, G.L. (2012) The flavonoid quercetin in disease prevention and therapy: Facts and fancies. *Biochemical Pharmacology*, **83**, 6–15.

Scanlin, L., and Stone, M. (2008) *Quinoa Protein Concentrate, Production and Functionality*. US Patent 7563473.

Small, E. (2013) Quinoa—is the United Nations' featured crop of 2013 bad for biodiversity? *Biodiversity*, **14** (3), 169–179.

Syrov, V.N., and Khushbaktova, Z.A. (1996) Wound-healing effects of ecdysteroids. *Doklady Akademii Nauk Respubliki Uzbekistana*, **12**, 47–50.

Syrov, V., Khushbaktova, Z.A., and Shakhmurova, G.A. (2008) Effects of phytoecdysteroids and bemithyl on functional, metabolic, and immunobiological parameters of working capacity in experimental animals. *Eksperimental'naia i klinicheskaia farmakologiia*, **71** (5), 40–43.

Vega-Galvez, A., Miranda, M., Vergara, J., Uribe, E., Puente, L., and Martinez, E.A. (2010) Nutrition facts and functional potential of quinoa (*Chenopodium quinoa* Willd.) an ancient Andean grain: a review. *Journal of the Science of Food and Agriculture*, **90**, 2541–2547.

Veillet, S., and Lafont, R. (2012) *Use of Phytoecdysones in the Preparation of a Composition for Acting on the Metabolic Syndrome*. U.S. Patent 8236359.

Maqui (*Aristotelia chilensis*): An Ancient Mapuche Medicine with Antidiabetic Potential

(C. Jofré-Jimenez, H. Aguayo-Cumplido, C. Calfío, L. E. Rojo)

Maqui in the Mapuche Culture and Traditional Medicine

The Mapuche people are native inhabitants of Southern Chile and Argentina. They are currently the major native ethnic group in Chile with over 300,000 people concentrated in the Araucanía Region. This nation of fearless warriors kept both the mighty Inca Empire and the powerful Spanish conquistadors at bay for centuries. The Mapuche medicine system is based on sacred rituals and herbal preparations dating back hundreds of years. The federal government actively supports traditional Mapuche medicine; it has been formally introduced as part of the regular local public health care system in Chile (Ladio and Lozada 2000). For Mapuches, their traditional medicine is part of a holistic view of the universe, in which health and illness are closely linked and balanced in harmony. Thus, all human actions affect this balance (Albornoz *et al.* 2004), and concepts like health, physical wellness, or disease are dynamically linked (Grebe 2006). The "Machi" is the spiritual leader; a priestess who is able to heal, provide spiritual guidance, communicate with the deities, and ward off evil spirits. She is responsible for healing and bridging the community with the "ngen" (the guardian spirits) (Mujica *et al.* 2004). During the healing rituals, the Machi conveys religious and mythological beliefs (Grebe 1995) and combats the symptoms but not the cause of the

Figure 5.28 Maqui fruits. © 2014 Belén Jara.

diseases. In order to do this, she goes into a trance in which she deals with the disturbing spirits and finds the exact remedy to restore the spiritual balance and well-being of an ill person, especially through the use of herbal teas, poultices, and ointments (Echeverría *et al.* 2002).

According to Mapuche beliefs, all plants have medicinal properties. Their word *lawen* means both "plant" and "remedy" (CONAMA 2008). They consider maqui (*Aristotelia chilensis*, Figure 5.28) as not only a medicinal plant but also a symbol of prosperity for Mapuche people. *A. chilensis* is an endemic Chilean plant that belongs to the Elaeocarpaceae family. The fruit is a fleshy, glossy, purplish to black, round berry (Madaleno 2007). This dioecious plant blooms between the months of October to December. It can be found in the Andes between latitudes 31° and 42°, especially in Chile's central valley and even on Juan Fernandez Island (Rodríguez *et al.* 1983). The Mapuche people usually refer to the groves of maqui plants that grow in wet soil or near streams as "macales." This resilient plant, unlike many other species, is capable of growing rather easily in extremely harsh environments like burned lands and deforested areas. Perhaps this is why maqui is considered a living symbol of a resilient species (Hoffmann *et al.* 1992).

In Mapuche traditional medicine, an infusion of maqui leaves is used to heal wounds, reduce fever, treat diarrhea and stomachaches, calm sore throats and swollen tonsils, and to heal oral ulcers (Madaleno and Gurovich 2004; Villagran *et al.* 1983). Mapuche midwives give women infusions of maqui leaves to facilitate childbirth (Alarcón and Nahuelcheo 2008). Maqui berries can be eaten fresh or dry, in the form juice, jam, or a fermented liquor (locally known as *techu*). Ethnomedical records indicate that the natives from Southern Chile used maqui fruit to treat skin inflammation, intestinal disorders, sore throats, infected wounds, dietary fiber, hemorrhoids, and migraines (Molgaard *et al.* 2011; Ojeda *et al.* 2011; Píriz 2013; Schreckinger *et al.* 2010). Maqui is also used by Mapuches to make wooden souvenirs and to handcraft musical instruments. According to Mapuche traditional knowledge, carvings made out of selected maqui wood have unique pleasurable sounds (Quintriqueo *et al.* 2011).

Maqui plays an important role during Mapuche's sacred ceremonies like the *machitun* (healing ceremony) and the *nguillatun* (praying ceremony). In these rituals, branches of maqui, cinnamon leaves, and bay leaves are tied to the *rehue* (Figure 5.29) and thus offered by the Machi to the spirits. The rehue is a pole carved from "laurel" or other trees, and it is always the center of the nguillatun. It is through the rehue and the plants offerings that the Machi establishes a communication with spiritual world (Kraster 2003).

Figure 5.29 "Rehue" in Araucanía Region of Southern Chile. © 2014 Bernarda Calfío.

Several studies have described the chemistry and pharmacology of maqui (Table 5.4). Alkaloids, polyphenols, and anthocyanins have been identified as bioactive secondary metabolites in maqui leaves and fruit (Table 5.4). Maqui berry contains a low concentration of indole and quinoline alkaloids (Cespedes *et al.* 1993) and high concentrations of anthocyanins (137.6 mg/100 g fresh weight). Eight types of anthocyanins have been identified; delphinidin 3,5-O-diglucoside and delphinidin 3-O-sambubioside-5-O-glucoside are the most abundant (Escribano-bailón *et al.* 2006; Tanaka *et al.* 2013). Biochemical characterization of maqui leaves have revealed the presence of aristone, aristoteline, aristotelinine, and aristotelone, alkaloids unique to the genus *Aristotelia* (Schreckinger *et al.* 2010a; Zabel *et al.* 1980). A maqui leaf extract rich in

Table 5.4 Main phytochemical constituents reported in *Aristotelia chilensis*.

Phytochemical constituent	Plant organ	Reported bioactivity	Reference
Alkaloids			
8-Oxo-9-dehydrohobartine	Leaves	NBR[*]	Cespedes *et al.* 1990
8-Oxo-9-dehydromakomakine	Leaves	NBR	
Aristone	Leaves	NBR	Bittner *et al.* 1978
Aristotelinine	Leaves	NBR	
Aristotelone	Leaves	NBR	Bhakuni *et al.* 1976
Aristoteline	Leaves	NBR	
Phenolics			
Delphinidin-3-O-sambubioside-5-O-glucoside	Fruit	Antidiabetic Antioxidant Anti-apoptotic	Céspedes *et al.* 2010; Escribano-bailón *et al.* 2006; Gironés-Vilaplana *et al.* 2012; Rojo *et al.* 2012; Tanaka *et al.* 2013
Delphinidin-3,5-O-diglucoside	Fruit	Antioxidant Anti-apoptotic	
Cyanidin-3-sambubioside-5-glucoside	Fruit	Antioxidant	Escribano-bailón *et al.* 2006; Céspedes *et al.* 2010; Gironés-Vilaplana *et al.* 2012
Cyanidin-3-sambubioside-5-glucoside	Fruit	Antioxidant	
Cyanidin-3,5-diglucoside	Fruit	Antioxidant	
Delphinidin-3-glucoside	Fruit	Antioxidant	
Delphinidin-3-sambubioside	Fruit	Antioxidant	Escribano-bailón *et al.* 2006; Céspedes *et al.* 2010
Cyanidin-3-sambubioside	Fruit	Antioxidant	
Cyanidin-3-glucoside	Fruit	Antidiabetic Antioxidant Anticancer	Escribano-bailón *et al.* 2006; Gironés-Vilaplana *et al.* 2012; Marczylo *et al.* 2009

[*] NBR: No bioactivity reported.

alkaloids showed antiviral and nematicidal activity (Schreckinger *et al.* 2010a). Leaf extracts also induce alterations in the morphology of human erythrocytes from the normal discoid shape to an echinocytic form, an activity attributed to flavonoids (Suwalsky *et al.* 2008).

Maqui berry displays powerful antioxidant capacity, and protects against low-density lipoprotein oxidation and oxidative damage in cultured endothelial cells, suggesting an anti-atherogenic effect (Miranda-Rottmann *et al.* 2002). Polyphenol extracts from maqui berry have shown anti-inflammatory (Céspedes *et al.* 2010) and anti-adipogenenic effects (Schreckinger *et al.* 2010b). A crude extract from maqui berry was capable of inhibiting the activity of α-glucosidase and α-amylase, enzymes responsible for the intestinal degradation of complex carbohydrates into glucose (Rubilar *et al.* 2011). The same study suggested that catechin, epicatechin, quercetin, and kaempferol are responsible for this enzymatic inhibition. A recent study on diet-induced obese hyperglycemic mice suggested that delphinidin 3-sambubioside-5-glucoside (DSG) is the main bioactive anthocyanin responsible for the antidiabetic effect of maqui berry (Rojo *et al.* 2012). DSG displayed insulin-like effects in muscle and liver cells. Fuentes *et al.* (2013) proposed that maqui berry's hypoglycemic effect is due to the synergistic effect of anthocyanins, quercetin, and rutin. Collectively, these studies suggest that maqui polyphenols might have a prophylactic effect against metabolic syndrome by ameliorating hyperglycemia and hyperlipidemia. Although no ethnomedical records are available on maqui as an antidiabetic fruit, and the biochemical/pharmacological results have not yet been confirmed at the clinical level, there is one report that Mapuche children have a lower incidence of insulin-dependent diabetes than their Caucasian counterparts (Larenas *et al.* 1996). This epidemiological observation might be explained, in part, by a regular intake of maqui fruit. It seems the story of maqui is still unfolding, as much of its ethnomedicinal uses (anti-inflammatory, digestive disorders, fever and skin injuries) remain to be studied by modern science.

Commercial uses

Maqui-derived products, such as beverages, juice concentrates, and capsules, have gained significant popularity in the food and nutraceuticals industry. Several preclinical studies have shown that anthocyanins and other polyphenols from maqui may prevent symptoms of chronic non-communicable diseases, such as obesity, cancer, cardiovascular, and neurodegenerative disorders (Céspedes *et al.* 2008; Prior and Gu 2005; Sasaki *et al.* 2007).

Initially, due to its particularly powerful antioxidant capacity and high anthocyanin content, maqui was seen as a novel superfruit, but current knowledge on maqui medicinal properties goes far beyond the antioxidant effect. Maqui's antioxidant capacity is superior to that of other common berry fruits, like acai, blueberries, strawberries, and cranberries (McDougall and Stewart 2012). The leaves of this plant have drawn attention of medical researchers, and in a recent study a nutritional supplement was prepared by micro-encapsulation of maqui leaf extracts (Vidal *et al.* 2013) The stability issue of anthocyanins is still a challenge, especially for liquid formulations. Recent studies showed lemon juice can help stabilize maqui polyphenols (Gironés-Vilaplana *et al.* 2012; 2014). The colorful anthocyanins from maqui have been used to enhance the color of red wine and for the production of organic food compatible dyes (Misle *et al.* 2011).

As metabolic syndrome is increasing worldwide, it has been suggested that anthocyanins from maqui berries could be used as a supplement to ameliorate type 2 diabetic conditions. According to a recent study, DSG, a distinctive anthocyanin of maqui berry has insulin-like properties in vivo and it can increase insulin sensitivity in muscle and liver cells (Rojo *et al.* 2012). This antidiabetic effect is seemingly potentiated by the other anthocyanins in maqui. This finding illustrates antidiabetic potential of maqui, stimulating interest for further research.

This work was funded by The Chilean National Commission for Scientific and Technological Research (CONICYT) National Scientific and Technological Development Fund (FONDECYT) grant number 11140915.

References

Alarcón, A., and Nahuelcheo, Y. (2008) Creencias sobre el embarazo, parto y puerperio en la mujer mapuche: conversaciones privadas. *Chungara, Revista de Antropología Chilena*, **40** (2), 193–202.

Albornoz, A., Farías, V., Montero, G., and Negri, A. (2004) *Introducción a la complejidad herbolaria de la medicina tradicional mapuche. Pampa-Patagonia Argentina: análisis multidisciplinario*. University of Siena, Siena.

Bhakuni, D., Silva, M., Matun, S., and Sammes, P. (1976) Aristoteline and aristotelone, an usual indole alkaloid from *A. chilensis. Phytochemistry*, **15** (4), 574–575.

Bittner, M., Silva, M., Gopalakrishna, E.M., Watson, W., and Zabel, V. (1978) New alkaloids from *Aristotelia chilensis* (Mol.) Stuntz. *Journal of the Chemical Society, Chemical Communications*, **2**, 79–80.

Cespedes, C., Jakupov, J., Silva, M., and Watson, W. (1990) Indole alkaloids from *Aristotelia chilensis. Phytochemistry*, **29** (4), 1354–1356.

Céspedes, C.L., Valdez-Morales, M., Avila, J.G., El-Hafidi, M., Alarcón, J., and Paredes-López, O. (2010) Phytochemical profile and the antioxidant activity of Chilean wild black-berry fruits, *Aristotelia chilensis* (Mol) Stuntz (Elaeocarpaceae). *Food Chemistry*, **119** (3), 886–895.

Céspedes, C.L., El-Hafidi, M., Pavon, N., and Alarcon, J. (2008) Antioxidant and cardioprotective activities of phenolic extracts from fruits of Chilean blackberry *Aristotelia chilensis* (Elaeocarpaceae), maqui. *Food Chemistry*, **107** (2), 820–829.

Cespedes, C., Jakupov, J., Silva, M., and Tsicwritzis, F. (1993) A quinoline alkaloid from *Aristotelia chilensis. Phytochemistry*, **34** (3), 881–882.

Comission Nacional Del Medio Ambiente (2008) *Biodiversidad de Chile, Patrimonio y Desafíos*, 2nd edition. Ocho Libros Editores Ltda, Santiago.

Echeverría, R., González, P., Sánchez, A., and Toro, P. (2002) *Imaginario social de salud pehuenche de la comunidad de Callaqui en el Alto Bío-Bío*. Departamento de Psicología, Universidad de Concepción, Concepción.

Escribano-bailón, M., Alcalde-eon, C., Muñoz, O., Rivas-Gonzal, J., and Santos-Buelga, C. (2006) Anthocyanins in berries of maqui [*Aristotelia chilensis* (Mol.) Stuntz]. *Phytochemical Analysis*, **17** (1), 8–14.

Fuentes, O., Fuentes, M., Badilla, S., and Troncoso, F. (2013) Maqui (*Aristotelia chilensis*) and rutin (quercetin-3-O-rutinoside) protects against the functional impairment of the endothelium-dependent vasorelaxation caused by a reduction of nitric oxide availability in diabetes. *Boletín Latinoamericano y del Caribe de Plantas Medicinales y Aromáticas*, **12** (3), 220–229.

Gironés-Vilaplana, A., Mena, P., García-Viguera, C., Moreno, and Diego A. (2012) A novel beverage rich in antioxidant phenolics: maqui berry (*Aristotelia chilensis*) and lemon juice. *LWT—Food Science and Technology*, **47** (2), 279–286.

Gironés -Vilaplana, A., Mena, P., Moreno, D.A., and Garcia-Viguera, C. (2014) Evaluation of sensorial, phytochemical and biological properties of new isotonic beverages enriched with lemon and berries during shelf life. *Journal of the Science of Food and Agriculture*, **94** (6), 1090–1100.

Grebe, M. (1995) Etnociencia, creencias y simbolismo en la herbolaria chamánica mapuche. *Enfoques en atención primaria de Salud*, **9** (2), 6–10.

Grebe, M. (2006) *Culturas Indígenas de Chile, un estudio preliminar*. Pehuén Editores Ltda, Santiago.

Hoffmann, A., Farga, C., Lastra, J., and Veghazi, E. (1992) *Plantas medicinales de uso común en Chile*. Ediciones Fundacion Claudio Gay, Santiago.

Kraster, A. (2003) El uso de sistema de salud tradicional en la población Mapuche: Comportamiento y Percepción: Mapuche, http://www.mapuche.nl/(accessed 5 January 2016).

Ladio, A. H., and Lozada, M. (2000) Edible wild plant use in a Mapuche community of Northwestern Patagonia. *Human Ecology*, **28** (1), 53–71.

Larenas, G., Montecinos, A., Manosalva, M., Barthou, M., and Vidal, T. (1996) Incidence of insulin-dependent diabetes mellitus in the IX region of Chile: ethnic differences. *Diabetes Research and Clinical Practice*, **34**, (S1)47–51.

Madaleno, I.M. (2007) Etno-Farmacología en Iberoamérica, una alternativa a la globalización de las prácticas de cura. *Cuadernos Geográficos*, **41** (2), 61–95.

Madaleno, I.M., and Gurovich, A. (2004) "Urban versus rural" no longer matches reality: endurance of an early public agro-residential development in peri-urban Santiago, Chile. *Cities*, **21** (6), 513–526.

Marczylo, T.H., Cooke, D., Brown, K., Steward, W.P., and Gescher, A.J. (2009) Pharmacokinetics and metabolism of the putative cancer chemopreventive agent cyanidin-3-glucoside in mice. *Cancer Chemotherapy and Pharmacology*, **64** (6), 1261–1268.

McDougall, G.J., and Stewart, D. (2012) *Berries and Health: A Review of the Evidence*. The James Hutton Institute, Dundee.

Miranda-Rottmann, S., Aapillaga, A., Perez, D., Vasquez, L., Martinez, A., and Leighton, F. (2002) Juice and phenolic fractions of the berry *Aristotelia chilensis* inhibit LDL oxidation in vitro and protect human endothelial cells against oxidative stress. *Journal of Agricultural and Food Chemistry*, **50** (26), 7542–7547.

Misle, E., Garrido, E., Contardo, H., and González, W. (2011) Maqui [*Aristotelia chilensis* (Mol.) Stuntz]-the amazing Chilean tree- A review. *Journal of Agricultural Science and Technology B*, **1**, 473–482.

Molgaard, P., Holler, J.G., Asar, B. *et al.* (2011) Antimicrobial evaluation of Huilliche plant medicine used to treat wounds. *Journal of Ethnopharmacology*, **138** (1), 219–227.

Mujica, A., Pérez, M.V., Gonzales, C., and Simon, J. (2004) Conceptos de enfermedad y sanación en la cosmovisión Mapuche e impacto de la cultura occidental. *Ciencia y enfermería*, **10** (1), 9–16.

Ojeda, J., Jara, E., Molina, L. *et al.* (2011) Effects of *Aristotelia chilensis* berry juice on cyclooxygenase 2 expression, NF-κB, NFAT, ERK1/2 and PI3K/Akt activation in colon cancer cells. *Boletín Latinoamericano y del Caribe de Plantas Medicinales y Aromáticas*, **10** (6), 543–552.

Píriz, V.P. (2013) *Patrimonio natural y conocimiento tradicional mapuche en la Reserva Nacional Mocho-Choshuenco (Región de Los Ríos)*. Universidad Austral de Chile, Valvida.

Prior, R.L., and Gu, L. (2005) Occurrence and biological significance of proanthocyanidins in the American diet. *Phytochemistry*, **66** (18), 2264–2280.

Quintriqueo, S., Gutiérrez, M., and Contreras, Á. (2011) Conocimientos sobre colorantes vegetales Contenidos para la educación intercultural en ciencias. *Pefiles educativos*, **34** (138), 109–123.

Rodríguez, R., Matthei, O., and Quezada, M. (1983) *Flora árborea de Chile*. Universidad de Concepción, Concepción.

Rojo, L.E., Ribnicky, D., Logendra, S. *et al.* (2012) In vitro and in vivo anti-diabetic effects of anthocyanins from maqui berry (*Aristotelia chilensis*). *Food Chemistry*, **131** (2), 387–396.

Rubilar, M., Jara, C., Poo, Y. *et al.* (2011) Extracts of maqui (*Aristotelia chilensis*) and murta (*Ugni molinae* Turcz.): sources of antioxidant compounds and α-glucosidase/α-amylase inhibitors. *Journal of Agricultural and Food Chemistry*, **59** (3), 1630–1637.

Sasaki, R., Nishimura, N., Hoshino, H. *et al.* (2007) Cyanidin 3-glucoside ameliorates hyperglycemia and insulin sensitivity due to downregulation of retinol binding protein 4 expression in diabetic mice. *Biochemical Pharmacology*, **74** (11), 1619–1627.

Schreckinger, M.E., Lotton, J., Lila, M.A., and Gonzalez, E. (2010a) Berries from South America: a comprehensive review on chemistry, health potential, and commercialization. *Journal of Medicinal Food*, **13** (2), 233–246.

Schreckinger, M.E., Wang, J., Yousef, G., Lila, M.A., and Gonzalez de Mejia, E. (2010b) Antioxidant capacity and in vitro inhibition of adipogenesis and inflammation by phenolic extracts of *Vaccinium floribundum* and *Aristotelia chilensis*. *Journal of Agricultural and Food Chemistry*, **58** (16), 8966–8976.

Suwalsky, M., Vargas, P., Avello, M., Villena, F., and Sotomayor, C.P. (2008). Human erythrocytes are affected in vitro by flavonoids of *Aristotelia chilensis* (maqui) leaves. *International Journal of Pharmaceutics*, **363**, 85–90.

Tanaka, J., Kadekaru, T., Ogawa, K., Hitoe, S., Shimoda, H., and Hara, H. (2013) Maqui berry (*Aristotelia chilensis*) and the constituent delphinidin glycoside inhibit photoreceptor cell death induced by visible light. *Food Chemistry*, **139** (1–4), 129–37.

Vidal, L., Avello, M., Loyola, C. *et al.* (2013) Microencapsulation of maqui (*Aristotelia chilensis* [Molina] Stuntz) leaf extracts to preserve and control antioxidant properties. *Chilean Journal of Agricultural Research*, **73** (1), 17–23.

Villagran, C., Meza, I., Silva, E., and Vera, N. (1983) *Nombres folklóricos y usos de la flora de la Isla de Quinchao, Chiloé*. Ministerio de Educación Pública, Museo de Historia Natural, Santiago.

Zabel, V., Watson, W.H., Bittner, M., and Silva, M. (1980) Structure of aristone, a unique indole alkaloid from *Aristotelia chilensis* (mol.) stuntz, by X-ray crystallographic analysis. *Journal of the Chemical Society, Perkin Transactions* **1**, 2842–2844.

Betalains from *Chenopodium quinoa:* Andean Natural Dyes with Industrial Uses beyond Food and Medicine

(K. Cubillos-Roble, B. L. Graf, A. Troncoso-Fonseca, J. Delatorre-Herrera, L. E. Rojo)

History and Coloristic Properties of Quinoa

Historical records indicate that the natives of South America domesticated quinoa (*Chenopodium quinoa* Willd.) over 5000 years ago (Sepúlveda *et al.* 2004). When the Native Americans settled in the Andes, quinoa was baptized with the Quechua word *Chisiya mama* or "the mother grain." Archaeologists believe the Tiwanaku culture, the most prominent Pre-Incan civilization, based its diet on the quinoa grain (Kolata 2009), as they recognized its extraordinary nutritional traits (Lamothe *et al.* 2015). Chipaya people, one of the earliest cultures of the Andes (current Chile, Peru, and Bolivia), also attributed a magical force to the quinoa plant, as they used it in religious ceremonies like the mortuary ritual (Esatbeyoglu *et al.* 2014). In mortuary rituals, mourners prepared bags with cooked and raw quinoa for the deceased to eat during their journey through the spiritual world (Acosta Veizaga 2001).

In addition to its nutritional value (FAO 2011), quinoa is rich in betalains, a class of natural pigments found in the seeds and aerial parts of the plant (Tang *et al.* 2015)· In Bolivia, the Spanish word *mojuelo* is the name for a betalain-rich by-product of quinoa processing. Mojuelo is seen as a source of colorants for foodstuff and llama wool. Historical records indicate that natural colorants were common in Andean pre-Hispanic cultures, but as commonly happens in human history, the social interaction between Spanish conquerors and Andean natives resulted in the introduction of new colorants like carminic acid, beet juice, beet powder (dehydrated beets), and beta carotene and the loss of local natural pigments, like those from quinoa seeds (Cassman 1990).

Although betalains are typically associated with beetroots (Nemzer *et al.* 2011), these natural pigments are also present in other members of Caryophyllales, including certain varieties of quinoa. Betalains impart an attractive red-purple color to the fruits/seeds, leaves, and inflorescences of this Andean crop (Figure 5.30). The variation in betalain content in the pericarp among quinoa varieties gives the fruits/seeds their characteristic variety of colors (Figure 5.31). Betalains are water-soluble nitrogen-containing compounds derived from tyrosine. To date, 55 different chemical structures of betalains have been described. They are found in 13 plant families of the order Caryophyllales (Castellanos-Santiago and Yahia 2008), including Amaranthaceae. These natural dyes are divided into two primary subgroups (Figure 5.32A–C) that differ in their pigmenting properties: betacyanins (red-violet) and betaxanthins (yellow-orange) (Strack *et al.* 2003). Other common food sources of betacyanins and betaxanthins are Swiss chard (*Beta vulgaris* L. ssp. Cicla) and cactus fruit [*Opuntia ficus-indica* (L.) Mill.] (Tang *et al.* 2015). The best known betacyanin is betanin, which gives beets and cactus fruits their

Figure 5.30 Variety of colors of quinoa fruits and vegetative parts. © 2014 David Wu.

Figure 5.31 Variation in quinoa seed color. © 2007 Michael Hermann, http://www.cropsforthe future.org/.

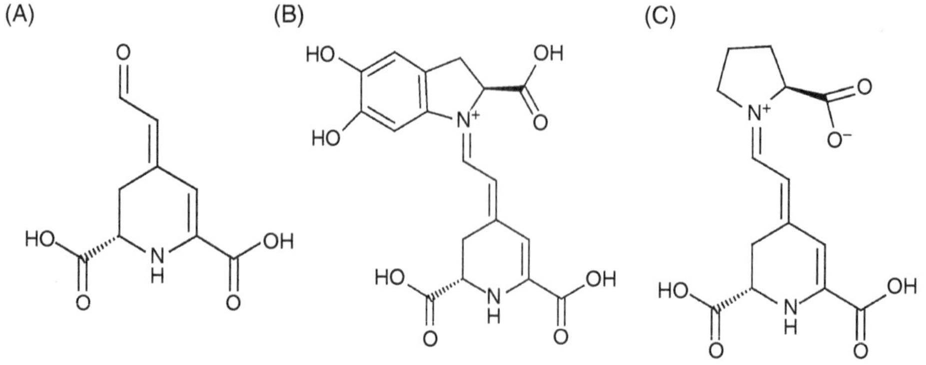

(A)   (B)   (C)

Figure 5.32 The structure scheme shows (A) betalamic acid, the chromophore and precursor of all betalains; (B) betanidin, the aglycone of most of the betacyanins; and (C) indicaxanthin, a proline-containing betaxanthin.

typical red-violet color. However, one recent study has reported betanin and isobetanin to be the most abundant betalains in three different quinoa genotypes (Tang *et al.* 2015). Due to the color of betacyanins, which absorb visible light in the range of 536–538 nm, they are often mistaken for anthocyanins, a group of polyphenols with maximum UV absorbance of 520 nm (Tang *et al.* 2015). However, betacyanins are structurally distinct from anthocyanins (Figure 5.32A–C). Betacyanins are biosynthesized from tyrosine by the condensation of betalamic acid with a dihydroxyphenylalanine (DOPA) derivative, while betaxanthins are formed from the condensation of betalamic acid with an amino acid or amino acid derivative. Betalains are normally extracted with acidified aqueous methanol (Tang *et al.* 2015) and are stable between pH 3 and 7, which makes them attractive as food colorants. Surprisingly, rigorous chemical characterization of beta-lains in quinoa is still lacking, as well as an understanding of genetic and environmental effects on betalain production in quinoa.

Bolivia and Peru are presently the major global producers of quinoa; Bolivia is the world-leading exporter of raw quinoa and quinoa-based products (FAO 2013). It has the largest collection of ecotypes and a seed bank with over 3000 varieties of quinoa seeds. Although Chile and the United States are small producers of quinoa seeds, American and Chilean researchers are developing innovations in order to increase the value of quinoa seeds beyond the commodity prices (FAO 2011; Graf *et al.* 2014).

Colorful betalains from quinoa provide a variety of shelf-stable, bright colorants with multifaceted applications (Moreno *et al.* 2008). For example, betanin can be used for coloring dairy products such as yogurt and ice cream, as well as sauces and soups. Color is among the key features for food attractiveness. The right color in foodstuffs can increase their acceptance and market price. In fact, higher prices are not a deterrent to consumers when purchasing more healthy-looking products. Food color can positively influence the hedonic and sensory experience of food. Thus, the right combination of flavor and color can positively reinforce the intake of certain edible products. According to the regulation on food additives, betanin is permitted as a natural red food colorant (E162) in the European Union (Moreno *et al.* 2008), and as a component of beet juice by the FDA. Moreover, betanin is used as a colorant in cosmetics and pharmaceutical formulations due to the safe toxicity profile (Esatbeyoglu *et al.* 2014).

The industrial processing of quinoa seeds for food applications yields significant amounts of *mojuelo* in Perú and Bolivia (Acuña Soliz 2006) The production of mojuelo has increased dramatically due to the high and increasing international demand for quinoa, providing new opportunities for technology-based business among quinoa farmers in Chile, Peru, and Bolivia (Acuña Soliz 2006).

A number of Chilean and Peruvian quinoa varieties are particularly rich in betalains, and the stems and leaves of these quinoa varieties can be used to obtain natural dyes. The coloristic properties of quinoa betalains vary depending on the specific compound, plant organ, and method used for extraction. Currently, Peruvian women are using quinoa betalains to stain fabrics, soaps, ice cream, and modeling clay (FAO 2011).

In conclusion, the use of colorful ecotypes of *Chenopodium quinoa*, not only as a staple crop but also as a source of natural pigments with broad industrial applications, is drawing the attention of entrepreneurs, farmers, and consumers in Andean countries. Further development of betalain extraction/concentrating technologies may offer new sources of income for Andean quinoa farmers from Peru, Bolivia, Ecuador, and Chile, which along with the sustainable production of "Mother Grain" might increase profitability of quinoa in the coming years.

This work was funded by The Chilean National Commission for Scientific and Technological Research (CONICYT) National Scientific and Technological Development Fund (FONDECYT) grant number 11140915.

References

Acosta Veizaga, O. (2001) La muerte en el contexto uru: El caso chipaya. *Chungará (Arica)*, **33**, 259–270.

Acuña Soliz, Y.J. (2006) *Extracción del colorante del mojuelo de la quinua*, PhD Thesis, Universidad Técnica de Oruro, Bolivia. Available from: Banco de Tesis. (accessed 6 January, 2016).

Cassman, V. (1990) Natural dye research in the South Central Andes. *WAAC Newsletter*, **12** (2), 2–3.

Castellanos-Santiago, E., and Yahia, E.M. (2008) Identification and quantification of betalains from the fruits of 10 Mexican prickly pear cultivars by high-performance liquid chromatography and electrospray ionization mass spectrometry. *Journal of Agricultural and Food Chemistry*, **56**, 5758–5764.

Graf, B.L., Poulev, A., Kuhn, P., Grace, M.H., Lila, M.A., and Raskin, I (2014) Quinoa seeds leach phytoecdysteroids and other compounds with anti-diabetic properties. *Food Chemistry*, **163**, 178–185.

Esatbeyoglu, T., Wagner, A.E., Schini-Kerth, V.B., and Rimbach, G. (2014) Betanin-A food colorant with biological activity. *Molecular Nutrition & Food Research*, **59** (1), 36–47.

Food and Agriculture Organization of the United Nations (2001) *Quinoa: An Ancient Crop to Contribute to World Food Security*. Food and Agriculture Organization of the United Nations (FAO), Regional Office for Latin America and the Caribbean, Santiago.

Food and Agriculture Organization of the United Nations (2013). FAOSTAT database collections. Food and Agriculture Organization of the United Nations, Rome, http://faostat.fao.org (accessed 6 January, 2016).

Kolata, A.L. (2009) *Quinua: Producción, Consumo y Valor Social en el Contexto Histórico*. Department of Anthropology, University of Chicago, Chicago.

Lamothe, L.M., Srichuwong, S., Reuhs, B.L., and Hamaker BR. (2015) Quinoa (*Chenopodium quinoa* W.) and amaranth (*Amaranthus caudatus* L.) provide dietary fibres high in pectic substances and xyloglucans. *Food Chemistry*, **167**, 490–496.

Moreno, D.A., García-Viguera, C., Gil, J.I., and Gil-Izquierdo, A. (2008) Betalains in the era of global agri-food science, technology and nutritional health. *Phytochemistry Reviews*, 7 (2), 261–280.

Nemzer, B., Pietrzkowski, Z., Spórna, A. *et al.* (2011) Betalainic and nutritional profiles of pigment-enriched red beet root (*Beta vulgaris* L.) dried extracts. *Food Chemistry*, **127** (1), 42–53.

Sepúlveda, J., Thomet, M., Palazuelos, P., and Mujica, M.A. (2004) *La kinwa mapuche recuperacion de un cultivo para la alimentacion*, Imprenta Andalién, Temuko.

Strack, D., Vogt, T., and Willibald, S. (2003) Recent advances in betalain research. *Phytochemistry*, **62**, 247–269.

Tang, Y., Li, X., Zhang, B., Chen, P.X., Liu, R., and Tsao, R. (2015) Characterisation of phenolics, betanins and antioxidant activities in seeds of three *Chenopodium quinoa* Willd. genotypes. *Food Chemistry*, **166**, 380–388.

6

Asia

P. Li, W. Gu, C. Long, B. M. Schmidt, S. S. Ningthoujam, D. S. Ningombam,
A. D. Talukdar, M. D. Choudhury, K. S. Potsangbam, H. Singh, S. Khatoon, and
M. Isman

© 2010 Ilya Raskin

Ethnobotany: A Phytochemical Perspective, First Edition. Edited by B. M. Schmidt and D. M. Klaser Cheng.
© 2017 John Wiley & Sons Ltd. Published 2017 by John Wiley & Sons Ltd.

Introduction

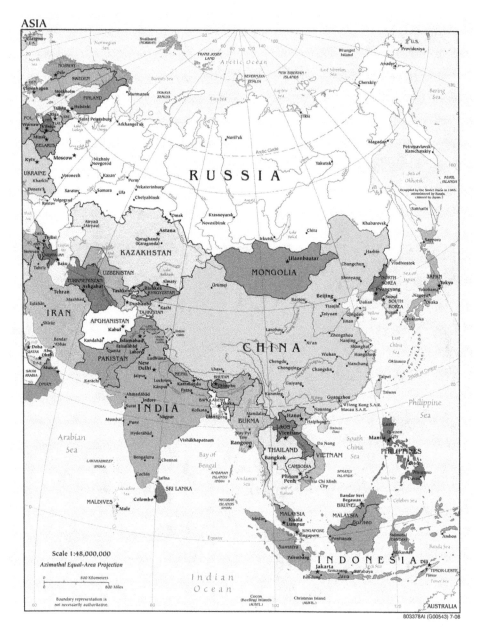

Figure 6.1 Political map of Asia (courtesy of United States of America Central Intelligence Agency).

Asia is the largest continent on earth, with about 60% of the world's population. The western border of Asia starts at the Red Sea and runs north through the Black Sea and Ural Mountains (Figure 6.1). To the southeast, the Malay Peninsula and Indonesia separate Asia from Oceania. Terrestrial ecosystems include deciduous and coniferous forest, rainforest, grassland, savanna, desert, mangrove, alpine, and tundra. With such a vast land area, it contains some of the earth's most extreme climates from the cold Arctic winds in Siberia to the dry Gobi desert and hot monsoon rains in the south.

Asia is the home to two major systems of traditional medicine, traditional Chinese medicine (TCM) and Ayurveda, not to mention Kampo, Unani, and many more local traditions. Some medicinal plants are unique to specific areas or indigenous people, while others are used universally. One reason for similarities in herbal medicines may be due to similarities of species from one country to the next. Another explanation is that people traveling along trade routes spread ethnobotanical knowledge throughout the region. For instance, plants that hold high importance in TCM such as ginseng (*Panax ginseng* C.A.Mey.) and ginkgo (*Ginkgo biloba* L.) also appear in Vietnamese monographs (Van Sam 2008). And plants important in Ayurveda like *Cassia* spp. and henna (*Lawsonia inermis* L.) appear in pharmacopeias in countries highly influenced by Indian traders, like Indonesia and Thailand.

For the purposes of ethnobotany, Asia can be separated into five somewhat distinct regions: Central Asia, Western Asia (Asia Minor, Iran, and Transcaucasia), South Asia, Southeast Asia, and East Asia. There is considerable overlap between regions, reflecting the long history of interaction (e.g., trade, war, and migration) between different Asian cultures.

Central Asia

Central Asia (i.e., Kazakhstan, Kyrgyzstan, Tajikistan, Turkmenistan, and Uzbekistan) has classically been a crossroads for nomadic people and traders moving goods across the Silk Road traversing the treeless Eurasian Steppe. Neighboring cultures left their mark on the region, especially Persian, Turkish, Islamic, Chinese, and Tibetan people. In modern times, the area was part of the Soviet Union and is still home to thousands of ethnic Russians. Much of the region is not suitable for agriculture, especially high mountain regions (Figure 6.2) and the dry Gobi desert. This contributed to the prevailing nomadic culture where raising livestock was more common than growing crops. Central Asia may not receive as much attention from the ethnobotanical community as some of its neighbors, but it is the center of diversity for some important economic crops listed in Table 6.1.

A number of medicinal plants are used in folk medicine in Central Asia, including well-known herbs like Saint John's wort (*Hypericum perforatum* L.), hyssop (*Hyssopus seravschanicus* Pazij), feverfew (*Tanacetum parthenium* L. Schultz-Bip.), and lemon balm (*Melissa officinalis* L.) (Figure 6.3). Several native *Artemisia* spp. have been used to treat a range of disorders (see Artemisia case study in the section titled "Artemisia Species and Human Health" in Chapter 6). Wild mint species *Mentha longifolia* (L.) Huds. is used as a gargle for respiratory conditions, liver disorders, and to improve appetite. It has also been reported as a snake and insect repellent. Herbs like dill (*Anethum graveolens* L.), basil (*Ocimum basilicum* L.), clary sage (*Salvia sclarea* L.),

Figure 6.2 A farmer dries his rice in the high mountains of Bhutan. © 2010 Ilya Raskin.

Table 6.1 Crops that originated in Central Asia.

Fruits and vegetables		Grains and legumes	
Almond	*Prunus dulcis* (Mill.) D.A.Webb, Rosaceae	Alfalfa	*Medicago sativa* L., Fabaceae
Apple	*Malus pumila* Mill., Rosaceae	Wheat	several *Triticum* spp., Poaceae
Pear	*Pyrus communis* L., Rosaceae	Mustard	*Brassica juncea* (L.) Czern., Brassicaceae
Pistachio	*Pistacia vera* L., Anacardiaceae	Lentil	*Lens culinaris* Medik., Fabaceae
Quince	*Cydonia oblonga* Mill., Rosaceae	Buckwheat	*Fagopyrum esculentum*, Polygonaceae
Garlic	*Allium sativum* L., Amaryllidaceae	**Fiber, medicine, and dyes**	
Pea	*Pisum sativum* L., Fabaceae	Hemp	*Cannabis sativa* L., Cannabaceae
Onion	*Allium cepa* L., Amaryllidaceae	Woad	*Isatis tinctoria* L., Brassicaceae
Spinach	*Spinacia oleracea* L., Amaranthaceae		
Walnut	*Juglans regia* L., Juglandaceae		

blue mint (*Ziziphora clinopodioides* Lam.), black cumin (*Bunium persicum* B. Fedtsch), and *Shnk* (*Galagania fragrantissima* Lypsky) are used as both food and medicine. Perhaps the most well-known traditional medicine and psychoactive plant from the area is marijuana. Marijuana and hemp are strains of the same species, *Cannabis sativa* L. The Chinese may have first domesticated *C. sativa* for hemp fiber and oil

Figure 6.3 A woman sells medicinal herbs at a market in Central Asia. © 2008 Slavik Dushenkov.

production (see the section titled "Cannabaceae: Hemp Family" in Chapter 1). But wild *Cannabis* species first evolved somewhere in Central Asia, from Northern India to Western China.

Western Asia

Western Asia includes countries east of the Red Sea on the Arabian Peninsula (Israel, Jordan, Saudi Arabia, Yemen, etc.), north to the Caucus Mountains (Transcaucasia), with Iran as its Eastern border. This same region is often called the Middle East or Near East, definitions that frequently include Egypt. The climate in the region is mostly arid and semi-arid, with seasonably changing winds. Geography includes mountains, large plateaus, and desert basins.

The Achaemenid Persian Empire ruled the region from 550 BCE, a time during which plants were largely equated with medicine. A translated version of Dioscorides' *Materia Medica* (c. 50–70 CE) was the original source of knowledge for many botanical scholars in the region. Abū Ḥanīfa Dīnawarī is considered the founder of Arabic Botany. His *Kitab al-Nabat* (Book of Plants) described the growth, fruiting, and flowering of hundreds of plants, along with their agricultural applications. Persian physician Razes (c. 865–925 CE) and Persian scholar Avicenna (c. 980–1037 CE) published books containing extraction and preparation methods for dozens of herbal medicines. Many modern plant names are derived from Persian plants names such as alfalfa, pomegranate, wheat, and cumin.

Dozens of economically valuable plants originated in Western Asia, including several species of wheat and other cereals, legumes, and oil seed crops (Table 6.2). Domesticated

Table 6.2 Crops that originated in Western Asia.

Fruits and vegetables		Grains and legumes	
Artichoke	*Cynara scolymus* L., Asteraceae	Barley	*Hordeum vulgare* L., Poaceae
Beets	*Beta* spp., Chenopodiaceae	Bitter vetch	*Vicia ervilia* (L.) Willd., Fabaceae
Cabbage	*Brassica oleracea* L. Brassicaceae	Chickpea	*Cicer* spp., Fabaceae
Carrot	*Daucus carota* L., Apiaceae	Faba bean	*Vicia faba* L., Fabaceae
Citron	*Citrus medica* L. Rutaceae	Flax	*Linum usitatissimum* L., Linaceae
Cornelian cherry	*Cornus mas* L., Cornaceae	Goatgrass	*Aegilops* spp., Poaceae
Date palm	*Phoenix dactylifera* L., Arecaceae	Lentil	*Lens culinaris* Medik., Fabaceae
Fig	*Ficus carica* L., Moraceae	Oats	*Avena sterilis* L., Poaceae
Grape	*Vitis vinifera* L., Vitaceae	Pea	*Pisum sativum* L., Fabaceae
Hazelnut	*Corylus avellana* L., Betulaceae	Rye	*Secale anatolicum* Boiss., Poaceae
Lettuce	*Lactuca serriola* L., Asteraceae	Wheat	Several *Triticum* spp., Poaceae
Muskmelon	*Cucumis melo* L., Cucurbitaceae		
		Medicine and spices	
Olive	*Olea europaea* L., Oleaceae	Anise	*Pimpinella anisum* L., Apiaceae
Parsley	*Petroselinum crispum* (Mill.) Fuss, Apiaceae	Dill	*Anethum graveolens* L., Apiaceae
Parsnip	*Pastinaca sativa* L. Apiaceae	Cumin	*Cuminum cyminum* L., Apiaceae
Radish	*Raphanus sativus* L. Brassicaceae	Licorice	*Glycyrrhiza glabra* L., Fabaceae
Rocket	*Eruca vesicaria* (L.) Cav. Brassicaceae	Opium poppy	*Papaver somniferum* L., Papaveraceae
Sour cherry	*Prunus cerasus* L., Rosaceae	Saffron	*Crocus sativus* L., Iridaceae

einkorn wheat (*Triticum boeoticum* Boiss.) and emmer wheat or farro (*Triticum dicoccum* varieties) first appeared in Turkey, c. 9000 BCE. Domesticated wheat reached Greece, Cyprus, and India by 6500 BCE followed by Egypt, Germany, and Spain. Several oil crops originated in the region including *Brassica napus* L. (rapeseed oil; Figure 6.4), *Carthamus tinctorius* L. (safflower oil), *Brassica nigra* K.Koch (black mustard seed oil), *Sinapis alba* L. (white mustard seed oil), and *Olea europaea* L. (olive oil). Tragacanth gum comes from the sap of the woody shrub *Astracantha gummifera* (Labill.) Podlech

Figure 6.4 Field of rapeseed (*Brassica napus*) in bloom. © 2013 Joanna Ziel.

is used as a thickener and emulsifier in food preparation. Frankincense is an aromatic resin from *Boswellia* spp. trees in the family Burseraceae. It has been traded since ancient times as a perfume and traditional medicine. Its Biblical partner, myrrh, is another aromatic resin used as perfume. It comes from other members of the Burseraceae family, *Commiphora* spp.

Perhaps one of the most important sources of plant-based drugs today is the opium poppy (Figure 6.5), first cultivated c. 3400 BCE by the Sumerians in ancient Mesopotamia. They called the plant *Hul Gil*, "joy plant," in reference to its pleasant analgesic and euphoric properties. Cultivation spread to North Africa, Central Asia along the Silk Road, and eventually to China. For more information on the opium poppy and its phytochemistry, see the section titled "Papaveraceae: Poppy Family" in Chapter 1.

South Asia

Modern-day India, Pakistan, and Bangladesh form the bulk of South Asia, along with the smaller countries Nepal, Bhutan, Sri Lanka, and the Maldives. The Indian subcontinent is often used synonymously with South Asia. The climate is tropical monsoon in the south and temperate to the north. It is thought that Andamanese people from the Andaman Islands in the Bay of Bengal and/or Indo-Aryan migrants from Western and Central Asia were the first inhabitants of the region. The Indus Valley Civilization (i.e., Harappan Civilization, c. 3300–1300 BCE) raised domesticated livestock and grew crops that most likely originated in neighboring regions such as wheat, barley, peas, lentils, flax, mustard, sesame, and millet. Compared to other centers of diversity, few

Figure 6.5 A US Marine greets opium farmers in their poppy field in Helmand province, Afghanistan (US Marine Corps photo by Cpl. Marco Mancha).

row crops originated in South Asia (Table 6.3); most were imported from Africa or other parts of Asia. Tree cotton (*Gossypium arboreum*) is one exception. It was first cultivated in present-day Pakistan and spread to the Indus Valley by c. 3300 BCE. Archeological evidence suggests cotton weaving may have originated in the Indus Valley before spreading to the rest of South Asia. Several native fruits and spices first collected by indigenous people thousands of years ago are important commodities in the global economy (Table 6.3). The Western Ghats forest along the Malabar Coast is the center of origin for black pepper (*Piper nigrum* L.), turmeric (*Curcuma longa* L.), and small cardamom (*Elettaria cardamomum* Maton.). The eastern Himalayan region is the center of diversity for large cardamom (*Amomum* spp.), *Citrus* spp., and a second center of diversity for turmeric.

Ayurveda is an Indian system of traditional medicine based on the Atharva Veda, an ancient Hindu Sanskrit text. This fourth veda, written c. 1200–1000 BCE, contains medical treatises, spiritual verse (hymns, incantations, spells, prayers, and mantras), and herbal remedies. Many of the herbal remedies are in the form of spiritual verse. Other texts followed the Atharva Veda including the Caraka, Sushruta, and Bhela samhitas, building onto the foundation of medical knowledge. Basic Ayurvedic principles follow the logic that health is based on our environment, mind, body, and spirit, which are all governed by the forces (*doshas*) of wind (*vata*), fire (*pitta*), and earth (*kapha*). Each person tends to have one dominant *dosha*, which can become imbalanced. For each *dosha*, there is a list of herbal remedies (Table 6.4).

From 600 BCE to 400 CE, 16 great kingdoms called *Mahājanapadas* were formed in India. Around the same time, Siddhartha Gautama founded Buddhism in northeastern

Table 6.3 Crops that originated in South Asia.

Fruits and vegetables		Spices, dyes, and wood	
Assam tea	*Camellia sinensis* var. *assamica*, Theaceae	Black pepper	*Piper nigrum* L., Piperaceae
Cucumber	*Cucumis sativus* L., Cucurbitaceae	Cardamom	*Elettaria* and *Amomum* spp., Zingiberaceae
Mango	*Mangifera indica* L., Anacardiaceae	Cinnamon	*Cinnamomum verum* J.Presl, Lauraceae
Orange	*Citrus sinensis* (L.) Osbeck, Rutaceae	Ginger	*Zingiber officinale* Roscoe, Zingiberaceae
Taro	*Colocasia esculenta* (L.) Schott, Araceae	Henna	*Lawsonia inermis* L., Lythraceae
Grains and legumes		Indigo	*Indigofera tinctoria* L., Fabaceae
Mung bean	*Vigna radiata* (L.) R.Wilczek, Fabaceae	Sandalwood	*Santalum album* L., Santalaceae
Fiber		Turmeric	*Curcuma longa* L., Zingiberaceae
Tree cotton	*Gossypium arboreum* L., Malvaceae		

India. Siddhartha had a close relationship with plants. He was reportedly born under a sorrowless tree [*Saraca asoca* (Roxb.) Willd.], became Buddha under a fig tree (*Ficus religiosa* L.), and died under a shala tree (*Shorea robusta* Gaertn.). Buddhists used special wood to construct their temples including teak (*Tectona grandis* L.f.), *Chukrasia tabularis* A.Juss., Malay bushbeech (*Gmelina arborea* Roxb.), and *Magnolia baillonii* Pierre. They also planted temple gardens of holy plants consisting of five Buddhist trees and six Buddhist flowers.

By c. 320–550 CE, the Gupta Dynasty covered most of the subcontinent, bringing peace to the region. This Golden Age of India formed Hindu culture and brought about many scientific advances including the refinement and crystallization of sugar from sugarcane. Two species of *Saccharum* were involved in the development of today's modern cultivars, *S. officinarum* and *S. spontaneum*. *S. spontaneum* is indigenous to India, while *S. officinarum* originated in Polynesia and Southeast Asia. Sugar crystals were known as *khanda* in India, the source of the English word "candy." Hindu culture also made use of intoxicating plants such as *Datura metel* and *Cannabis sativa* for religious and ceremonial purposes.

By c. 1200 CE, Muslim invaders conquered and destroyed many of the Buddhist temples and Hindu kingdoms. Invasion by Mongols in 1397 fractured the Indian cultural landscape yet again, but led to one of the longest-lived empires in human history, the Mughal Empire. Mughal emperors claimed to be direct descendants of Genghis Khan, founder of the Mongol Empire. Meitei Pangals are members of the Muslim community who settled in the northwest state of Manipur from the sixteenth to seventeenth centuries. Tribes in Manipur belong to two generic groups: Naga or Kukis. Most of them

Table 6.4 Common Ayurvedic medicinal herbs for different *doshas* (Dass, 2013).

	Medicine	Common name	Latin name
Pitta herbs	Amalaki	Emblic myrobalan or Indian gooseberry	*Phyllanthus emblica* L.
	Chandan	Sandalwood	*Santalum album* L.
	Draksha	Grape	*Vitis vinifera* L.
	Guduchi	Heart-leaf moonseed	*Tinospora sinensis* (Lour.) Merr.
	Ushira	Vetiver grass	*Chrysopogon zizanioides* (L.) Roberty
	Yashtimadhu	Licorice	*Glycyrrhiza glabra* L.
Kapha herbs	Ajmoda	Celery seed	*Apium graveolens* L.
	Haridra	Turmeric	*Curcuma longa* L.
	Haritaki	Yellow myrobalan	*Terminalia bellirica* (Gaertn.) Roxb.
	Lavang	Cloves	*Syzygium aromaticum* (L.) Merr. & L.M.Perry
	Putiha	Mint	*Mentha spicata* L.
	Sunthi	Dried ginger	*Zingiber officinale* Roscoe
Vata herbs	Ashwagandha	Winter cherry	*Withania somnifera* (L.) Dunal
	Kantakari	Wild eggplant	*Solanum virginianum* L.
	Methika	Fenugreek	Trigonella foenum-graecum L.
	Rasona	Garlic	*Allium sativum* L.
	Triphala	Myrobalan mix	*Phyllanthus emblica* L., *Terminalia bellirica* (Gaertn.) Roxb., and *Terminalia chebula* Retz.

practice Christianity, but still retain their pre-Christian culture and tradition. The Meitei Pangal traditional medicine is influenced by Unani medicine, indigenous folk medicine, and other Indian medicine systems, reflecting the cultural diversity of the region (see jaundice case study in the section titled "Traditional Treatment of Jaundice in Manipur, Northeast India" in this chapter). Unani medicine is a traditional system of medicine practiced by Muslims in South Asia. It is based on Greco-Arab medicine, especially the teachings of Aesculapius, Hippocrates, and Ibn Sina's *Canon of Medicine*. In this system, the body is composed of four elements (earth, air, water, and fire) and four humors (blood, phlegm, yellow bile, and black bile). Each element and humor has its own temperament: cold, hot, wet, and/or dry. As with Ayurveda, there are numerous herbal medicines for any type of medical condition.

The Nepali community did not arrive in Manipur until after World War II. Their traditional medicine systems are different from all other communities in Manipur. Some follow Tibetan medicine, largely based on the balance of five basic elements: earth, water, fire, air, and space. Tibetan healers may prescribe herbal medicines and/or changes in diet along with massage, moxa, cupping, saunas, emetics, or gold needle therapy. The Nepal government recognizes 30 medicinal plant species as "priority herbs" for research, development, and cultivation.

Shortly after the Meitei Pangals were settling in northern India, Europeans were setting up trading posts in the south. By 1769, the British East India Company fought off all other European forces and controlled all European trade out of India. The British *Raj* (Hindu word for "rule") lasted from 1858 to 1947. Under the Raj, the British greatly improved Indian infrastructure and ratcheted up agricultural production of crops like jute, tea, and indigo (see Indigo case study in the section titled "Indigo: The Devil's Dye and the American Revolution" in Chapter 7), frequently using Indians for slave labor.

Southeast Asia

Southeast Asia is defined as the 11 countries that range from China to India. It can be further divided into the mainland peninsula of Southeast Asia, called Indochina, and the island or maritime region. Mainland Southeast Asia consists of small mountain ranges and foothills with long rivers that begin in the highlands of China and India. The rivers flow into lowland plains that are separated by hills and mountainous terrain. The island region is comprised of several large (e.g., Borneo, Sumatra, and Java) and numerous small islands. Nearly all of Southeast Asia has a tropical climate; therefore, temperatures are generally hot and humid with abundant rainfall. The rainforest of Southeast Asia is second in size only to the Amazon rainforest. It is an area of high bio-diversity and endemic species. About 27% of the world's mangrove forests are located in Southeast Asia (FAO 2006).

According to archeological evidence, humans have inhabited Southeast Asia for at least 40,000 years. Tribal groups from China traveled down the long rivers, bringing their language, religion, and culture with them. Native people migrated from Australia and New Guinea to the Philippines and Malay-Indonesian archipelago. *Orang laut*, or sea people, lived exclusively on boats. Others thrived in the deep jungles, including infamous "headhunter" groups. Around 2000 years ago, the area experienced cultural changes coming from China and India. The Chinese expanded their empire down the Yangtze River into Vietnam. They brought along Confucian philosophy, Buddhism, and Taoism. The rest of mainland and island Southeast Asia were highly influenced by Indian culture, as a result of trade across the Bay of Bengal.

Many products from the rainforest and ocean are unique to the region, which histori-cally led to intense international trade. Cloves [*Syzygium aromaticum* (L.) Merr. & L.M.Perry], mace (arils of *Myristica fragrans* Houtt.), and nutmeg (seed of *M. fragrans*) were once only found on small islands in Indonesia called the Spice Islands. The wild progenitor of rice (*Oryza sativa* L.) originated somewhere on the South Asia plate. Wild species can be found everywhere in Southeast Asia. The ecogeographic race *indica* is grown commercially throughout the region (Figure 6.6A–B). *Sinica* and *javanica* are grown to a lesser extent. After rice, other important economic crops of the region include (in descending order of 2013 production value) palm oil, natural rubber, cassava, sugarcane, bananas, and fresh vegetables (FAOSTAT 2013). Coffee is also a major product for some countries. Vietnam is the second largest producer of coffee in the world, followed by Indonesia (FAOSTAT 2013). The oil palm (*Elaeis guineensis*) may have originated in West Africa, but in 2013, Indonesia and Malaysia produced 85% of the world's palm oil (FAOSTAT 2013). Table 6.5 lists several important crops that originated in Southeast Asia.

(A)

(B)

Figure 6.6 (A) Farmers in the distance tend to their rice fields near Nha Trang, Vietnam; (B) a load of rice travels down the Mekong.

Table 6.5 Crops that originated in Southeast Asia.

Fruits and vegetables		Grains, oils, and latex	
Breadfruit	*Artocarpus altilis* Fosb., Moraceae	Candlenut	*Aleurites moluccanus* (L.) Willd., Euphorbiaceae
Coconut palm	*Cocos nucifera* L., Arecaceae	Rice	*Oryza sativa* L., Poaceae
Cucumber	*Cucumis sativus* L., Cucurbitaceae	Rubber	*Ficus elastica* Roxb. ex Hornem., Moraceae
Durian	*Durio zibethinus* L., Malvaceae	**Herbs, spices, and aromas**	
Mangosteen	*Garcinia* × *mangostana* L., Clusiaceae	Cloves	*Syzygium aromaticum* (L.) Merr. & L.M.Perry, Myrtaceae
Rambutan	*Nephelium lappaceum* L., Sapindaceae	Nutmeg, mace	*Myristica fragrans* Houtt., Myristicaceae
Starfruit	*Averrhoa carambola* L., Oxalidaceae	Ylang-ylang	*Cananga odorata* (Lam.) Hook.f. & Thomson, Annonaceae
Taro	*Colocasia esculenta* (L.) Schott, Araceae		
Stimulants		**Sweeteners**	
Betel	*Areca catechu* L., Arecaceae	Sugar palm	*Arenga pinnata* (Wurmb) Merr., Arecaceae
Kratom	*Mitragyna speciosa* (Korth) Havil. Rubiaceae	Sugarcane	*Saccharum officinarum* L. Poaceae

Several native plants are used for religious purposes in the region. For example, holy basil and Thai holy basil (*Ocimum tenuiflorum* L. cultivars) are used in Hinduism for worship of the God Vishnu and other deities. The talipot palm (*Corypha umbraculifera* L.), sacred fig (*Ficus religiosa* L.), council tree (*F. altissima* Blume), betel (*Areca catechu* L.), and Palmyra palm (*Borassus* spp.) or coconut palm are five trees that Buddhist monks must plant in front of their monasteries. In addition, they plant water lily [*Crinum asiaticum* var. *sinicum* (Roxb. ex Herb.) Baker], white ginger lily (*Hedychium coronarium* J.Koenig), champak (*Michelia champaca* L.), plumeria (*Plumeria rubra* L.), and Chinese dwarf banana [*Ensete lasiocarpum* (Franch.) Cheesman] (Kleinmeyer 2004). The lotus (*Nelumbo nucifera* Gaertn.), or water lily, is considered sacred throughout the region. It is a divine symbol in Hinduism, Buddhism, and Confucianism.

Wild plants are the basis of traditional medicine throughout Southeast Asia. Many people living in rural areas still rely on traditional herbs as their primary source of medicine since access to healthcare is limited both by economics and availability. Some native herbs have been developed into pharmaceuticals. Reserpine is an alkaloid antipsychotic and antihypertensive drug derived from Snakeroot [*Rauvolfia serpentina* (L.) Benth. ex Kurz] (Monachino 1954). Himalayan Yew (*Taxus wallichiana* Zucc.) is a source of the diterpene cancer drug paclitaxel (Miller 2011). Elephant's Foot (*Dioscorea deltoidea* Wall. ex Griseb.) is a source of the steroid sapogenin diosgenin, a compound originally discovered in Mexican yams and used to make oral contraceptives.

East Asia

Most people in East Asia are part of the Sinosphere, regions influenced by the culture of China. Countries include China, Korea, Taiwan, Japan, and Mongolia. Humans may have inhabited East Asia for more than 40,000 years, but the earliest historical records describe the Xia dynasty of China (c. 2100–1600 BCE). Several other dynasties followed (see timeline; Figure 6.7) with significant cultural, religious, and ethnobotanical developments. Herbal medicines were likely used since the Neolithic period, based on archeological evidence. Mythical emperor Shen Nong (c. 2800 BCE) was said to have discovered berberine, invented tea, and taught people how to use plants for food and medicine. Tea consumption and TCM as it stands today have their origins in the Han Dynasty (for more information on tea production, see the section titled "Theaceae: Tea Family" in Chapter 1). Through centuries of war and trade with their neighbors, Chinese culture spread throughout the region.

Several important crops originated in East Asia, including Chinese varieties of cabbage used for the Korean dish kimchi and wasabi [*Eutrema japonicum* (Miq.) Koidz] (Figure 6.8). Ramie [*Boehmeria nivea* (L.) Gaudich.] fibers were some of the earliest fibers used to make clothing and mummy wrappings. In China, pairs of walnuts rotated in the palm of the hand are thought to increase blood circulation. Particularly symmetrical pairs of walnuts are valued as a status symbol. Lacquerware handicrafts produced throughout East and Southeast Asia are made with the urushiol-based sap of the Chinese lacquer tree, *Toxicodendron verniciflluum* (Stokes) F.A. Barkley (Figure 6.9). Tung oil, used mostly as furniture finish, comes from the nuts of Chinese native *Vernicia*

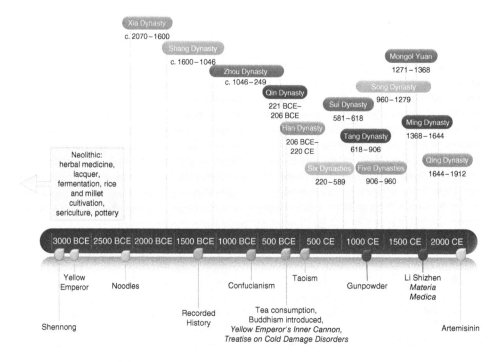

Figure 6.7 Timeline of Chinese history including dynasties and important historical developments.

Figure 6.8 Wasabi for sale in Japan. True wasabi comes from the enlarged stems of *Wasabia japonica* and is not generally available outside Japan. © 2015 Antonio Rubio.

Figure 6.9 A woman in Cambodia working on a lacquered vase. © 2006 Marc A. Garrett.

fordii (Hemsl.) Airy Shaw. Several species of citrus can be traced back to China and northeastern India. Buddha's hand (Figure 6.10), an ornamental variety of citron (*Citrus medica* L.), is a particularly unusual variety of citrus as the carpels are segmented to form finger-like appendages. The cultivars range from more open hand to closed hand forms. They are used primarily for their fragrance from the compound β-ionone (Shiota 1990) and as offerings in temples. They have no juice or pulp, but their fragrant rind can be used to flavor dishes or made into a bitter tonic as medicine.

Figure 6.10 Buddha's hand, a variety of citron (*Citrus medica* L.)

Table 6.6 Crops that originated in East Asia.

Fruits and vegetables		Grains, legumes, and fiber	
Apricot	*Prunus armeniaca* L., Rosaceae	Broomcorn millet	*Panicum miliaceum* L., Poaceae
Chinese or napa cabbage	*Brassica rapa* L., Brassicaceae	Buckwheat	*Fagopyrum esculentum,* Polygonaceae
Chinese yam	*Dioscorea polystachya* Turcz., Dioscoreaceae	Japanese barnyard millet	*Echinochloa* spp., Poaceae
Litchi	*Litchi chinensis* Sonn., Sapindaceae	Ramie	*Boehmeria nivea* (L.) Gaudich., Urticaceae
Peach	*Prunus persica* (L.) Batsch, Rosaceae	Soybean	*Glycine max* (L.) Merr., Fabaceae
Pear	*Pyrus communis* L., Rosaceae	**Stimulants**	
Wasabi	*Eutrema japonicum* (Miq.) Koidz., Brassicaceae	Tea	*Camellia sinensis* (L.) Kuntze, Theaceae
Water chestnut	*Eleocharis dulcis* (Burm.f.) Trin. ex Hensch.		

Three major religions in the region, Confucianism, Buddhism, and Taoism, helped shape Chinese culture and their relationship with plants. Chinese philosopher Confucius (551–479 BCE) developed Confucianism during the Zhou Dynasty. The Four Gentleman in Confucianism are the plum, orchid, bamboo, and chrysanthemum. These iconic plants are common subjects in Chinese art. Buddhism was introduced to China from Nepal during the Han dynasty (206 BCE–220 BCE); Taoism developed at the end of the second century CE. By the fourth century CE, Buddhism had spread to Korea and by 600 CE, it spread to Japan. Along with the knowledge of new religions, Japanese diplomats and students returning from China brought back miniature plants, a love affair that developed into the Japanese art form known as bonsai. During the twelfth century, the iconic Japanese tea ceremony and flower arranging became popular with the privileged class.

Shamanistic medicine began during the Shang Dynasty. The Shang believed illness was caused by the supernatural, such as curses and demons. Shamans could speak to deceased ancestors, asking the God Shang Ti for help to expel demons. TCM may not be the oldest system of traditional medicine, but it is probably the most well known. Kampo (Japanese) and Korean traditional medicines are both based on TCM, but adapted to each culture. TCM was likely introduced with Buddhism to Korea and Japan c. 538 CE. Two publications during the Han Dynasty, *Yellow Emperor's Inner Cannon* and *Treatise on Cold Damage Disorders*, formed the basis of most TCM concepts. Herbalism plays a significant role in TCM. *Panax ginseng* C.A.Mey. root can be found in nearly every Asian pharmacopeia, but it originated in China. It is described as sweet and cold, often prescribed in herbal combinations to treat the digestive system and "cold" disorders. *Ma huang* (*Ephedra sinica* Stapf) was prescribed for common cold, coughs, asthma, headaches, and allergies. It has stimulant properties due to the presence of ephedrine alkaloids. Aconite (monk's hood, *Aconitum* sp.) was a "hot" herb prescribed for cold disorders such as arthritis. The presence of cardio- and neurotoxic aconitine and other *Aconitum* alkaloids makes consumption dangerous. Licorice (*Glycyrrhiza glabra* L.) is widely used in TCM to sweeten herbal formulas (perhaps to increase palatability) and to replenish the *qi* (vital energy). There are hundreds of TCM herbs still in use today, comprising thousands of formulations. The antimalarial drug artemisinin is based on the use of *Artemisia annua* L. for fevers in TCM. A team of Chinese scientists lead by Youyou Tu discovered artemisinin in 1971.

References and Additional Reading

Cooper, R. (2015) Re-discovering ancient wheat varieties as functional foods. *Journal of Traditional and Complementary Medicine*, **5** (3), 138–143.

Dass, V. (2013) *Ayurvedic Herbology – East & West: The Practical Guide to Ayurvedic Herbal Medicine*. Lotus Press, Racine WI.

Eisenman, S.W., Zaurov, D.E., and Struwe, L. (eds) (2013) *Medicinal Plants of Central Asia: Uzbekistan and Kyrgyzstan*. Springer, New York.

Harris, D.R. (1996) *The Origins and Spread of Agriculture and Pastoralism In Eurasia: Crops, Fields, Flocks And Herds*. Routledge, New York.

Larson, G., Albarella, U., Dobney, K. *et al.* (2007) Ancient DNA, pig domestication, and the spread of the Neolithic into Europe. *Proceedings of the National Academy of Sciences*, **104** (39), 15276–15281.

Lev-Yadun, S., Gopher, A., and Abbo, S. (2000) The cradle of agriculture. *Science*, **288** (5471), 1602–1603.

Murphey, R. (2013) *A History of Asia*, 7th edition. Pearson, New York.

Park, H.-L., Lee, H.-S., Shin, B.-C. *et al.* (2012) Traditional medicine in China, Korea, and Japan: a brief introduction and comparison. *Evidence-Based Complementary and Alternative Medicine*, available from: <http://www.hindawi.com/journals/ecam/2012/429103/> [accessed 21 October, 2015].

Sharafzadeh, S., and Alizadeh, O. (2012) Some medicinal plants cultivated in Iran. *Journal of Applied Pharmaceutical Science*, **2** (1), 134–137.

Shiota, H. (1990) Volatile components in the peel oil from fingered citron (*Citrus medica* L. var. sarcodactylis Swingle). *Flavour and Fragrance Journal*, **5** (1), 33–37.

Zohary, D., Hopf, M., and Weiss, E. (2012) *Domestication of Plants in the Old World: The Origin and Spread of Domesticated Plants in Southwest Asia, Europe, and the Mediterranean Basin*, 4th edition. Oxford University Press, Oxford.

Ethnobotany of Dai People's Festival Cake in Southwest China

(P. Li, W. Gu, C. Long)

Ethnobotany

The New Year celebration of the Dai Calendar is an important traditional festival, called Songkran or Water-Splashing Festival, for Dai people, who generally reside in Southern Yunnan, southwest China. The festival usually takes place in mid-April each year and involves three days of celebrations that include sincere, yet light-hearted religious rituals that invariably end in merrymaking, where everyone ends up getting splashed, sprayed, or doused with water (China Highlights 2015).

Khaonuosuo, the traditional festival cake, is usually prepared by each family and eaten at a local community gathering in celebration of the Dai Calendar New Year. As the Dai people explain, *Khaunuosuo* should only be prepared with a perfect combination of color, aroma, and taste. *Gmelina arborea* Roxb. (Lamiaceae) flowers will be dried and finely ground, and then mixed with other ingredients including glutinous rice flour, brown sugar, sesame, and peanut. The contents are all mixed with water and wrapped with soaked banana leaves (*Musa* spp.), or other leaves of tiger grass (*Thysanolaena latifolia*), *Phrynium*, and bamboo, made into rectangular pieces 10 cm long and 5 cm wide. After steaming for about an hour, the *Khaonuosuo* is ready (Figure 6.11). The festival cake carries significant meaning for the Dai people, symbolizing auspiciousness with the health and safety that will accompany them throughout the New Year. A Dai proverb says, "one can live an extra year when they eat the *Khaonuosuo* in the traditional New Year Festival."

Interestingly, *G. arborea* flowers (Figure 6.12) are used as coloring and flavoring ingredients for festival cakes in the Water-Splashing Festival by the Dai people. *G. arborea*,

Figure 6.11 Traditional Dai people's festival cake.

Figure 6.12 *Gmelina arborea* flowers © 2016 Nilgiri Biosphere Reserve, Keystone Foundation.

called *Maisuo* by the local Dai people, is native to southern and southeastern Asia, including Bangladesh, Bhutan, China, India, Indonesia, Laos, Malaysia, Myanmar, Nepal, Philippines, Sri Lanka, Thailand, and Vietnam (Chen and Gilbert 1994; Pei and Chen 1982). It has gained prominence because of its widespread economic importance, its uses for timber and medicine, and its ecological function. The Dai people in China have considered *G. arborea* as an edible plant since ancient times and have often used it to make delicious cakes. The climate of Southern Yunnan is highly humid and hot. Especially, during the celebration of the Dai's New Year, it is the hottest time of the year. While many other foods quickly deteriorate in such conditions, the traditional *Khaonuosuo* is preserved for a couple of days at room temperature.

The Dai people have traditionally used *G. arborea* flowers as a food additive for coloring and spice, but it is possible that the flowers have an antimicrobial function in the cakes. The results of a pharmacological study on *G. arborea* flowers provided evidence to support this antimicrobial activity hypothesis and offer a theoretical basis for the development and utilization of edible plants from the traditional knowledge of the Dai people.

Chemical Components and Bioactivity

G. arborea has a long history of medicinal use in tropical Asian civilizations (Kaswala *et al.* 2012). The heartwood, leaves, and bark have been used to treat liver diseases, inflammation, injury, and eczema. Previous phytochemical studies have isolated and identified constituents that are responsible for the medicinal importance of the plant including flavonoids, lignans, iridoid glycosides, and alkaloids in the aerial parts of *G. arborea* (Gu *et al.* 2013).

Researchers have also found that different parts of *G. arborea* show significant antimicrobial activities. Kawamura and colleagues reported that the lignans isolated from the heartwood of *G. arborea* possessed antifungal activity (*Trametes versicolor* and *Fomitopsis palustris*) (Kawamura *et al.* 2004). In vitro studies of the crude extracts of leaves and stem bark of *G. arborea* demonstrated antimicrobial activity against the growth of *Escherichia coli*, *Klebsiella pneumonieae*, *Proteus mirabilis*, *Shigella dysenteriea*, and *Salmonella typhi* (EI-Mahmood *et al.* 2010), bacteria that may cause

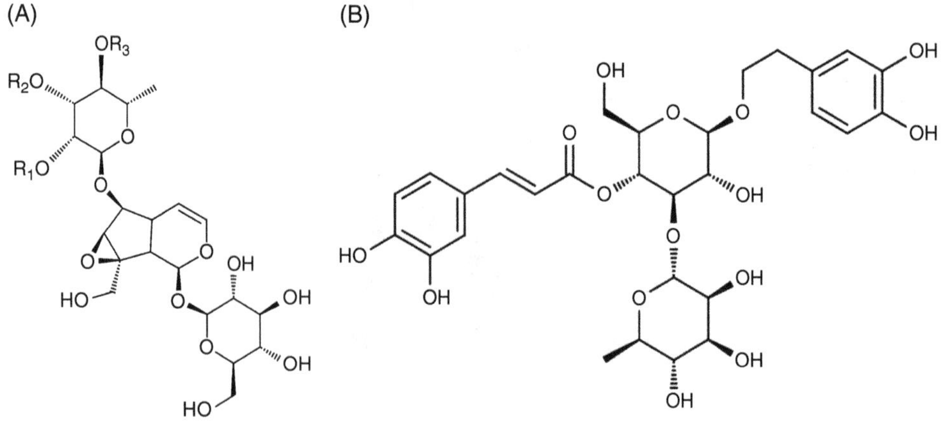

Figure 6.13 Structures of compounds isolated from *G. arborea* flowers: acylated iridoid glycosides (A) and verbascoside (B).

	R1	R2	R3
1	*trans*-cinnamoyl	isovaleryl	H
2	*trans-p*-hydroxycinnamnoyl	*trans-p*-hydroxycinnamnoyl	H
3	H	H	*trans*-cinnamoyl
4	H	*trans*-cinnamoyl	*trans*-cinnamoyl
5	*trans*-cinnamoyl	*trans*-cinnamoyl	H
6	*trans*-cinnamoyl	H	H
7	H	*trans-p*-hydroxycinnamnoyl	H
8	*trans-p*-hydroxycinnamnoyl	H	H
9	H	*trans*-cinnamoyl	H

diarrhea, dysentery, and stomach abnormalities. Investigations into the volatile constituents of *G. arborea* flowers identified 37 compounds by GC-MS, but the crude volatile constituents did not show any activity on *Staphylococcus aureus, Bacillus cereus, Bacillus cereus, Escherichia coli,* or *Paecilomyces lilacinus* (Wang *et al.* 2010).

Our study focused on the edible flowers of *G. arborea*, on the basis of our previous ethnobotanical observations and traditional knowledge of the Dai people. The methanolic extract of *G. arborea* flowers was analyzed by antimicrobial assays, and the results showed moderate antibacterial activities against *Staphylococcus aureus,* which is on the primary causes of food spoilage, supporting its traditional use.

Further phytochemical investigation yielded 15 compounds, including 9 acylated iridoid glycosides (Figure 6.13 A 1–9), 5 acylated rhamnopyranoses, and verbascoside (Figure 6.13 B) from *G. arborea* flowers. A study on cytoprotective activity revealed that 6-O-α-L-(2″,3″-di-O-*trans*-p-hydroxycinnamoyl)rhamnopyranosylcatalpol (2) possessed hepatoprotective activity with EC_{50} values of 42.45 μM, which indicated a new functional benefit of *G. arborea* flowers (Gu *et al.* 2013).

Modern Perspective

So far, attention has focused on utilization of *G. arborea* timber, while phytochemical and pharmacological investigations are still lacking, especially on the flowers of the plant. Extracts and isolated constituents from the flowers of *G. arborea* demonstrated antibacterial activities and, for certain compounds, potent cytoprotective effects. The application of *G. arborea* not only shows the wisdom of the Dai people, but also provides a potentially novel approach for the use of a natural antiseptic in food industries. Investigation of the traditional use of *G. arborea* flowers has shown that its potency and apparent safety profile make it a potential natural dietary source of phytonutrients that may be important for human health.

References

China Highlights (2015) Winter Splashing Festival, http://www.chinahighlights.com/festivals/water-splashing-festival.htm

Chen, S.L., and Gilbert, M.G. (1994) Verbenaceae. In *Flora of China*, **17**, 33 (eds Z.Y. Wu and P. Raven). Science Press, Beijing, and Missouri Botanical Garden, Saint Louis.

El-Mahmood, A.M., Doughari, J.H., and Kiman, H.S. (2010) In vitro antimicrobial activity of crude leaf and stem bark extracts of *Gmelina arborea* Roxb. against some pathogenic species of *Enterobacteriaceae*. *African Journal of Pharmacy and Pharmacology*, **4**(6), 355–361.

Gu, W., Hao, X.J., Liu, H.X., Wang, Y.H., and Long, C.L. (2013) Acylated iridoid glycosides and acylated rhamnopyranoses from *Gmelina arborea* flowers. *Phytochemistry Letters*, **6** (4), 681–685.

Kaswala, R., Patel, V., Chakraborty, M., and Kamath, J.V. (2012) Phytochemical and pharmacological profile of *Gmalina arborea*: An overview. *International Research Journal of Pharmacy*, **3** (2), 61–64.

Kawamura, F., Ohara, S., and Nishida, A.(2004) Antifungal activity of constituents from the heartwood of Gmelina arborea: Part 1. Sensitive antifungal assay against Basidiomycetes. *Holzforschung*, **58** (2), 189–192.

Pei, C., and Chen, S.L. (1982) Verbenaceae. In *Flora Reipublicae Popularis Sinicae*, **65** (1), 1–49 (ed. Z.Y. Wu). Science Press, Beijing.

Wang, L.Q., Lü, J.G., Zhang, H.Q., Ouyang, J.L., and Zhang, X.Q. (2010) Volatile constituents of flowers of *Gmelina arborea*. *Flavour Fragrance Cosmetics*, **2010** (3), 6–8.

The Ethnobotany of Teeth Blackening in Southeast Asia

(B. M. Schmidt)

Teeth blackening is the practice of dyeing the teeth for religious or cultural purposes. Many tribes have abandoned the practice in recent years due to influences of Christianity and the Western ideal of white teeth (Zumbroich 2009). Southeast Asian people relied heavily on plants to blacken teeth. One of the simplest ways to blacken teeth is chewing on a plant with known coloring properties. *Piper betle* L. (known as betel leaf or *paan*; Figure 6.14A); cured seeds of *Areca catechu* L. (the areca palm, sometimes called "betel nut"; Figure 6.14B); *Epipremnum pinnatum* (L.) Engl. (various common names including dragon-tail plant); fruit of *Paederia foetida* L. (skunk vine or Chinese fever vine); and leaves or latex of *Jatropha curcas* L. have all been used, either as a single ingredient

(A)

(B)

Figure 6.14 (A) *Piper betle* leaves and (B) areca nuts for sale at a market in Hoi An, Vietnam.

Figure 6.15 Betel quid preparation in Myanmar. © 2013 Tara Shubbuck, TwoTravelaholics.com.

or mixed with other ingredients like slaked lime (calcium hydroxide) (Zumbroich 2009). The lime could come from limestone cliffs in Thailand, Laos, or Vietnam, or ground mollusk shells in island nations. While there are many ways to make a betel quid (see Patidar *et al.* 2015 for a review), typical preparation (Figure 6.15) begins with dabbing lime paste onto the betel leaf, laying thin slices of the cured areca nut on top, then folding the outer edges of the leaves across the center. The wad is placed in the mouth between the cheek and teeth and can be sucked and chewed on for hours. The process creates red saliva, "betel juice," that is spat out.

Wood tar and burn products were also used. For example, reddish resin exuded from burning sticks of *Eurya acuminata* DC. was applied to teeth. Empyreumatic oil of coconut was collected from burning coconut shells. It was either applied directly or mixed with leaf of the nipah palm (*Nypa fruticans* Wurmb) or coconut oil to make a paste that was easier to apply. Fruit tree wood was also burned and applied to teeth including durian, guava, langstat, and mangosteen. The Viet people of Vietnam (Figure 6.16) had a multistage process to blacken teeth. First, teeth were cleaned and acid etched. Next, iron sulfate, *Rhus chinensis* Mill. (Chinese gall nuts), pomegranate rind, or areca nut were applied multiple times to create an intense dark color. Empyreumatic oil of coconut was applied last to give the teeth a lacquer appearance. French colonialists thought the Viet people were using lacquer from *Toxicodendron succedaneum* (L.) Kuntze and erroneously named the practice "laquage des dents" (lacquering of teeth).

In addition to blackening properties, indigenous people believed that teeth blackening strengthened the gums and prevented tooth decay (Rooney 1995). Empyreumatic oil of coconut was included in the indigenous pharmacopeia as a treatment for toothache in the Philippines and Cambodia. Citrus leaves were chewed for toothache and other dental problems in the Indo-Malaysian archipelago. Betel leaf was believed to reduce "hot" illnesses like headache and fever (Rooney 1995). The *Areca* nut provides a relaxing effect and reduces hunger. It was also used as an antihelminthic.

Figure 6.16 A Vietnamese woman with blackened teeth.

There are different chemistries at work in each of the teeth blackening processes. The plant materials chewed for teeth blackening are not inherently colored with the exception of latex from *J. curcas*. Instead, the color develops over time with mastication and/or oxidation. The enzymatic reaction between phenols and polyphenol oxidase is likely responsible for most of the dark pigments. Polyphenol oxidase polymerizes colorless phenols to form intensely dark polyphenol oligomers like condensed tannins (Figure 6.17A). This is the same reaction that occurs in enzymatic fruit browning and creates the dark brown color associated with black tea. Reichart *et al.* (1985) studied the structural composition of the black layer on teeth from members of the P'wo Karen hill tribe in Northern Thailand. The composition was similar to that of subgingival calculus (hardened dental plaque) stained with areca nut polyphenols and covered by a protective pellicle-like layer.

A. catechu seeds contain native condensed tannins (arecatannins; Figure 6.17B) and high levels of catechin, epicatechin, and syringic acid (Zhang *et al.* 2011). Addition of slaked lime to betel quids raises the pH and causes the tannins to turn bright red (Patidar *et al.* 2015). Catechins are components of tea that polymerize to form dark pigments during fermentation. Syringic acid, a component of acai berries and red wine, can also be enzymatically polymerized. Iridoids can react with natural proteins to generate dark pigments as well. In the wood burning reactions, wood tar or exuded resin would contain a mixture of tarry residues of combustion along with naturally occurring plant pigments. The Viet method and other methods using iron relies on the formation of ferric tannates or other metallo-organic complexes to produce a dark black color.

Some of the plants used for teeth blackening have documented antihelminthic activity including *A. catechu* (Monteiro *et al.* 2011), *J. curcas* (Petti and Scully 2009),

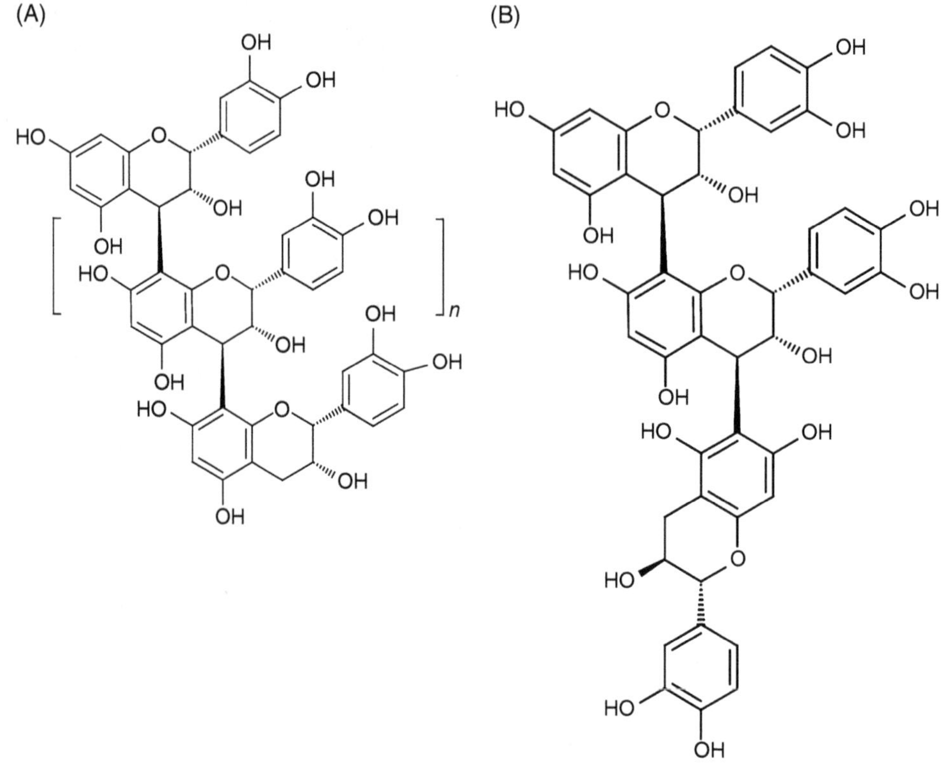

(A)

(B)

Figure 6.17 Structure of tannins that cause teeth blackening: (A) condensed tannin; (B) arecatannin B1.

and *P. foetida* (Roychoudhury *et al.* 1970), supporting traditional use. Some of the purported dental health properties of teeth blackening are also supported by scientific studies. Polyphenols, which are abundant in many teeth blackening preparations, have recently been found to have a positive effect on oral health (Sharan *et al.* 2012). The formation of a polyphenol and pellicle-like layer on blackened teeth could protect tooth enamel by reducing demineralization by bacteria-produced acids (Reichart *et al.* 1985). However, long-term use of betel quids, especially if mixed with tobacco, can increase the risk for oral leukoplakia, submucous fibrosis, and oral cancer (Jeng *et al.* 1999; Li *et al.* 2012). Betel quids without tobacco are still considered a risk factor for oral cancers and are associated with a higher risk of oral submucous fibrosis (Patidar *et al.* 2015). Adding *P. betel* inflorescences to a quid may further increase carcinogenic potential (Jeng *et al.* 1999). *A. catechu* seeds contain the pyridine alkaloids arecoline, arecaidine, guvacoline, and guvacine, with arecoline occurring at the highest level (Figure 6.18). These alkaloids are parasympathetic nervous system stimulants. In the oral cavity, they can undergo nitrosation by human cytochromes P450, giving rise to *N*-nitrosamine carcinogens such as *N*-nitrosoguvacolin (Miyazaki *et al.* 2005).

Modern Uses

Betel quids are still chewed around the globe, and so the components are readily available from international distributers. In addition to being chewed as part of a betel quid,

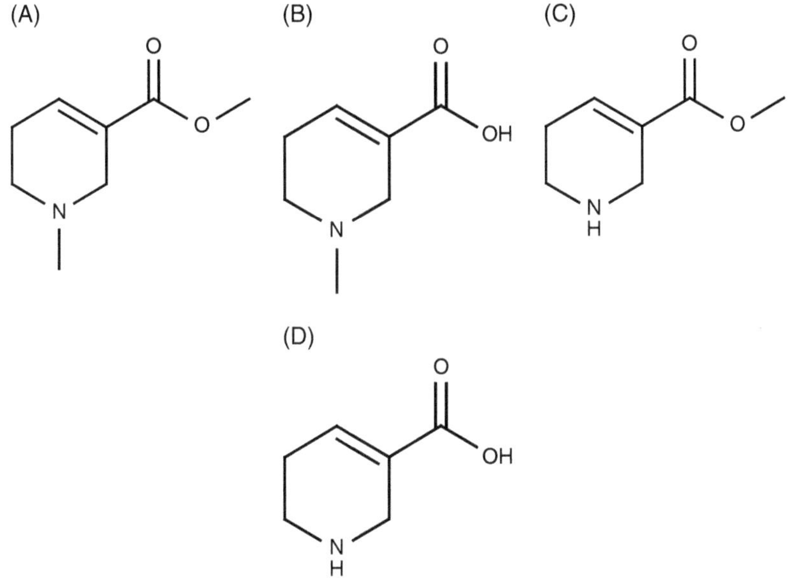

(A)   (B)   (C)

(D)

Figure 6.18 Pyridine alkaloids from areca palm (*Areca catechu* L.) seeds.

Piper betle is also used in shampoos (antifungal and pediculicidal), hair dye, toothpaste, mouthwash, and face cream (Samantaa and Agarwal 2009; Watcharawit and Soonwera 2013). *A. catechu* and plant-derived polyphenols and iridoids in general are used as food coloring and for dyeing fabrics (Watcharawit and Soonwera 2013). There is a potential for further product development as companies are working to replace petroleum-based dyes with more sustainable natural compounds.

References

Jeng, J.H., Hahn, L.J., Lin, B.R., Hsieh, C.C., Chan, C.P., and Chang, M.C. (1999) Effects of areca nut, inflorescence piper betle extracts and arecoline on cytotoxicity, total and unscheduled DNA synthesis in cultured gingival keratinocytes. *Journal of Oral Pathology and Medicine*, **28** (2), 64–71.

Li, T., Ito, A., Chen, X. *et al.* (2012) Usefulness of pumpkin seeds combined with areca nut extract in community-based treatment of human taeniasis in northwest Sichuan Province, China. *Acta Tropica*, **124** (2), 152–157.

Miyazaki, M., Sugawara, E., Yoshimura, T., Yamazaki, H., and Kamataki, T. (2005) Mutagenic activation of betel quid-specific N-nitrosamines catalyzed by human cytochrome P450 coexpressed with NADPH-cytochrome P450 reductase in *Salmonella typhimurium* YG7108. *Mutation Research*, **581** (1–2), 165–171.

Monteiro, M.V., Bevilaqua, C.M., Morais, S.M. *et al.* (2011) Anthelmintic activity of *Jatropha curcas* L. seeds on *Haemonchus contortus*. *Veterinary Parasitology*, **182** (2–4), 259–263.

Petti, S., and Scully, C. (2009) Polyphenols, oral health and disease: A review. *Journal of Dentistry*, **37** (6), 413–423.

Patidar, K.A., Parwani, R., Wanjari, S.P., and Patidar, A.P. (2015) Various terminologies associated with areca nut and tobacco chewing: A review. *Journal of Oral and Maxillofacial Pathology*, **19** (1), 69–76.

Reichart, P.A., Lenz, H., König, H., Becker, J., and Mohr, U. (1985) The black layer on the teeth of betel chewers. *Journal of Oral Pathology & Medicine*, **14** (6), 466–475.

Rooney, D.F. (1995) *Betel Chewing in South-East Asia*. Centre National de la Recherche Scientifique (CNRS), Lyon.

Roychoudhury, G.K., Chakrabarty, A.K., and Dutta, B. (1970) A preliminary observation on the effects of *Paederia foetida* on gastro-intestinal helminths in bovines. *Indian Veterinary Journal*, **47** (9), 767–769.

Samantaa, A.K., and Agarwal, P. (2009) Application of natural dyes on textiles. *Indian Journal of Fibre & Textile Research*, **34**, 384–399.

Sharan, R.N., Mehrotra, R., Choudhury, Y., and Asotra, K. (2012) Association of betel nut with carcinogenesis: revisit with a clinical perspective. *PLoS One*, 7 (8) e42759.

Watcharawit, R., and Soonwera, M. (2013) Pediculicidal effect of herbal shampoo against *Pediculus humanus capitis* in vitro. *Tropical Biomedicine*, **30** (2), 315–324.

Zhang, W.M., Wei, J., Chen, W.X., and Zhang, H.D. (2011) The chemical composition and phenolic antioxidants of Areca (*Areca catechu* L) seeds. *International Conference on Agricultural and Biosystems Engineering, Advances in Biomedical Engineering*, **1** (2), 16–22.

Zumbroich, T.J. (2009) Teeth as black as a bumble bee's wings': The ethnobotany of teeth blackening in Southeast Asia. *Ethnobotany Research & Applications*, 7, 381–398.

Artemisia Species and Human Health

(B. M. Schmidt)

Artemisia is a genus in the family Asteraceae commonly known as wormwood, mugwort, or sagebrush. With around 500 members, it is a diverse genus of herbs and shrubs that produce essential oils, sesquiterpene lactones, and polyphenols with demonstrated health benefits. Central Asia is its center of diversification, with secondary speciation areas in the Mediterranean region and the American Northwest (Valles and McArthur 2001). Historically, numerous species have been used as insecticides, food, spice, and medicine. The Egyptian Ebers Papyrus and the Indian Atharvaveda (both c. 1500 BCE) mention medicinal wormwood. Hippocrates made "Hippocratic wine" c. 400 BCE, a white wine infused with wormwood. Pliny the Elder published its formula (Naturalis Historia, 77 CE), specifying 1 lb Pontic wormwood (*A. pontica* L.) added to 40 sextarii of grape must, and boiling until reduced to one third. Dioscorides recommended *A. absinthium* L. (Figure 6.19) for stomach ailments and notes its use as a fever reducer (De Materia Medica, c. 50–70 CE). He published different recipes for wormwood wine. Culpeper described several species of wormwood in his herbal, each with their "own virtues" as bitter herbs (Culpeper 1995). He suggested common wormwood (*A. vulgaris* L.) for weakness of the stomach, gout, and gravel. Many other cultures used *Artemisia* species for various maladies. Some of the more commonly used species and ethnomedical applications are

Figure 6.19 *A. absinthium.*

listed in Table 6.7. Widespread uses include treatment for GI disorders, helminth infection, wounds, and fevers.

For centuries, *Artemisia* species have been used in "wormwood wine," a medicinal wine fortified with herbs. Wormwood wines may have been consumed as early as 1500 BCE in India and Egypt and 1250 BCE in China (Brown and Miller 2011; McGovern *et al.* 2004). Archaeobotanical evidence from the Shang and Western Zhou dynasty Changzikou Tomb suggests that medicinal *jiu* or *chang* wine was made with rice or millet, tree resins, flowers, and aromatic herbs including wormwood. Its consumption was reserved for the king (McGovern *et al.* 2004). Modern versions of these alcoholic beverages include absinthe and vermouth. It is not clear when wormwood wine made the transition from bitter medicinal tonic to enjoyable alcoholic beverage. By some accounts, the Romans offered wormwood wine as an appetite-stimulating reward to winners of chariot races (Brown and Miller 2011). It was well known throughout Europe during the Renaissance as "eisel," "alsem wine," "wermut wein," or "purl" (Brown and Miller 2011). But by the mid-1700s, its popularity was waning. In 1786, Antonio Benedetto Carpano brewed his family's recipe for wormwood wine and sold it at a shop in Torino, Italy. He named it "wermut," the German name for wormwood wine. Soon Carpano's wermut was offered in cafes around the city. After his death, it spread across Europe, and eventually became known as "vermouth."

Absinthe originated around the same time in Switzerland, but accounts of its development are inconsistent. Some say two sisters with the surname "Henriod" made the original recipe; others credit Dr. Pierre Ordinaire and Major Dubied. The recipe varies considerably, but flavorings include green anise, fennel, and wormwood (generally *A. absinthium*). French soldiers reportedly developed a taste for bitter wormwood during their time abroad fighting wars in Africa during the 1800s (Adams 2004). They used absinthe topically as an insecticide, added absinthe to drinking water as a treatment for dysentery and roundworms, and doctors recommended *Artemisia* as a replacement for quinine (Adams 2004).

Table 6.7 *Artemisia* species and their use in traditional medicine.

Species	Ethnobotanical uses[1]
A. abrotanum L.	Infusion used for GI disorders
A. absinthium L.	Dried plant used as insecticide; seed powder paste for dental pain; whole plant as general tonic; leaf powder for gastrointestinal problems and intestinal worms.
A. abyssinica Sch.Bip. ex A.Rich.	Anthelmintic, antispasmodic, antirheumatic, and antibacterial agent
A. afra Jacq. ex Willd.	Colds, coughs, diabetes, heartburn, bronchitis, and asthma
A. annua L.	Whole plant decoction used for malaria; leaves used for fever, cold, and skin disorders; leaf powder for diarrhea
A. apiacea Hance	Malaria, jaundice, and dyspepsia
A. arborescens (Vaill.) L.	Anti-inflammatory remedy
A. argyi H.Lév. & Vaniot	Conditions of the liver, spleen, and kidney
A. biennis Willd.	Topical antiseptic and internally for chest infections
A. campestris L.	Antivenin, anti-inflammatory, anti-rheumatic, and antimicrobial
A. douglasiana Besser ex Besser	Treats peptic ulcers and other GI disorders
A. dracunculus L.	Gastronomic herb used for GI problems and diabetes; antihelminthic
A. dubia Wall. ex B.D.Jacks.	Leaf powder for GI problems and intestinal worms; fresh leaf paste applied to external wounds and skin infections.
A. frigida Willd.	Leaf infusion used for colds, cough, fever, and irregular menstruation; poultice applied to wounds to stop bleeding
A. gmelinii Weber ex Stechm.	Treatment of liver diseases including hepatitis
A. herba-alba Asso.	Whole plant powder for diabetes
A. japonica Thunb.	Plant decoction used for malaria
A. judaica L.	Anthelmintic
A. maritima L. ex Hook.f.	Plant used as antiseptic and anti-inflammatory; plant decoction used for malaria; plant powder used for intestinal worms
A. nilagirica (C.B.Clarke) Pamp.	Insecticide
A. pontica L.	Flavoring for vermouth
A. princeps Pamp.	Inflammation, diarrhea, circulatory disorders; burnt plant used for wound healing
A. roxburghiana Wall. ex Besser	Plant decoction used for fever; plant powder used for intestinal worms
A. rubripes Nakai	Gastrointestinal disorders and hemostatic
A. santolinifolia Turcz. ex Krasch	Plant extract used for intestinal worms
A. scoparia Waldst. & Kit.	Used for ear pain, hepatic disorders; smoke of twigs is used for burns
A. spicigera K.Koch	Skin disorders
A. vulgaris L.	Leaf infusion used for fever, colds, epilepsy, GI inflammation, and used topically for skin disorders

[1] From Hayet *et al.* 2009; Moreman 1998; Shah 2014; Abad 2012; Swantson-Flatt *et al.* 1991; Eisenman *et al.* 2013.

During the 1950s and 1960s, there was a malaria resurgence, due in part to the emergence of chloroquine-resistant *Plasmodium falciparum*, the protozoan parasite that causes malaria. In 1967, the Chinese government set up a national project to screen traditional Chinese medicines and plant extracts for antimalarial activity. *A. annua* was one of the 640 hits in the malaria assay, but it was not immediately identified as a likely drug candidate. Initially, the *A. annua* results were difficult to reproduce. Furthermore, there was scarce Chinese ethnobotanical evidence that *Artemisia* was used specifically to treat malarial fevers (Willcox *et al.* 2011). Qinghao (*Artemisia annua*) was first mentioned in *Wu Shi Er Bing Fang* (*Prescriptions for Fifty-Two Ailments*) more than 2000 years ago. But it was not mentioned as a fever treatment until Ge Hong's *Zhou Hou Bei Ji Fang* (*Handbook of Prescriptions for Emergencies, 284–346 CE*), "A handful of qinghao immersed with 2 liters of water, wring out the juice and drink it all" (Tu 2011). *Ben Cao Gang Mu* (1596) was the first text that suggested qinghao for malarial fevers, "a treatment for hot and cold due to intermittent fever illness." In these texts, fresh qinghao leaves were pounded to expel the juice.

Tu Youyou the scientist credited with the discovery of artemisinin, suspected the active components of qinghao were destroyed by heat on the basis of Ge Hong's cold extraction process. On October 4, 1971, her research team created an *A. annua* extract 100% effective against *Plasmodium berghei* in mice and *Plasmodium cynomolgi* in monkeys. Dr. Tu and her colleagues bravely volunteered to be the first subjects to take the extract and had no ill effects. Initial clinical trials were carried out in the Hainan province in China on subjects infected with both *Plasmodium vivax* and *P. falciparum* (Tu 2011). It was an immediate success. The *A. annua* extract out-performed chloroquine, causing rapid disappearance of malarial symptoms and a reduction of parasites in the blood (Tu 2011). Dr. Tu shared the 2015 Nobel Prize in Medicine for her "discoveries concerning a novel therapy against Malaria."

Phytochemistry

Glandular trichomes on *Artemisia* leaves produce essential oils rich in volatile terpenes, such as camphor, 1,8-cineole, α,β-thujone, and borneol (Yang *et al.* 2012; Abad *et al.* 2012), that vary depending on species and chemotype (Brown 2010). High doses of α,β-thujone are neurotoxic, causing seizures (National Toxicology Program 2011). However, the levels of thujone present in wormwood wine, absinthe, vermouth, and so on, are too low to produce clinical effects, described as "absinthism." In addition, α,β-thujone has no known psychotropic effects. Therefore, any claims that the intoxicating effects of absinthe are attributable to thujone are conjecture (Lachenmeier 2008).

In addition to monoterpenes, *Artemisia* produces at least 20 known sesquiterpenes including artemisinin (arteannuin A) (Figure 6.20), arteannuin B, artemisitene, and artemisinin acid (Woerdenbag *et al.* 1994). In 1972, the sesquiterpene lactone artemisinin was identified as the chief antimalarial component in qinghaosu (Brown 2010; Tu 2011). In the genus *Artemisia*, only *A. annua* produces artemisinin, ranging from 0.01% to 1.4% in cultivated strains (Brown 2010). Artemisinin's short-acting antimalarial action involves the formation of free radicals via cleavage of the endoperoxide bond (Ho *et al.* 2014). It has been completely effective in treating chloroquine-resistant *Falciparum* malaria, capable of killing the parasite while it is infecting human blood

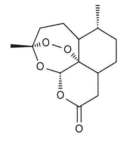

Figure 6.20 Structure of the sesquiterpene lactone artemisinin, determined by the Chinese Academy of Sciences, 1975.

cells (Sriram *et al.* 2004) and liver cells (Whirl-Carrillo *et al.* 2012). Since it has low solubility in both water and oil, semi-synthetic analogs were developed with improved potency and solubility. These analogs include artemether, arteether, and artesunate (Brown 2010).

Modern research has substantiated other ethnomedical uses of *Artemisia* spp. Most of these health benefits can be attributed to *Artemisia* sesquiterpenes, sesquiterpene lactones, and poly-phenols. Artemisinins have been reported to possess robust inhibitory effects against viruses, protozoa, helminthes, and fungi and have activity against autoimmune diseases, allergic disorders, and inflammation (Ho *et al.* 2014). Clinical trials showed that artemisinins are highly effective against the juvenile form of the blood fluke *Schistosomiasis japonica*, but only mod-erately effective against adults (Liu *et al.* 2014). This data suggests artemisinins may be more effective as a prophylactic for schistosomiasis (Panic *et al.* 2014). There have also been clinical trials showing the effectiveness of *A. absinthium* extracts against Crohn's disease (Omer *et al.* 2007; Krebs *et al.* 2010), *A. spicigera* against peptic ulcers (Park *et al.* 2008), and *A. princeps* against diabetes (Choi *et al.* 2011). Ethanolic extracts from *A. dracunculus* may be effective for diabetes and metabolic syndrome by restoring insulin sensitivity, stimulating insulin release, and improving carbohydrate metabolism (Aggarwal *et al.* 2015; Cefalu *et al.* 2008).

Modern Uses

Within the past decade, reports of *P. falciparum* resistance to artemisinin have begun to appear (Sriram *et al.* 2004), with the epicenter of resistance in western Cambodia (Leang *et al.* 2015). The World Health Organization (WHO) reports resistance to artemisinin in five countries as of February 2015, all in the Greater Mekong subregion: Cambodia, the Lao People's Democratic Republic, Myanmar, Thailand, and Vietnam. Currently, WHO recommends artemisinin-based combination therapy (ACT) for the treatment of uncomplicated malaria due to *P. falciparum* (WHO 2012). ACT combines a short-acting artemisinin with a long-acting drug from a different class such as lumefantrine or synthetic quinine analogs (mefloquine, piperaquine, etc.). Continued use of artemisinin monotherapies is considered to be a major contributing factor to the development of resistance to artemisinin derivatives (WHO 2012, 2015). Combination therapy is intended to reduce the rate of drug resistance development.

Beyond malaria, the Chinese Ministry of Health approved artemisinins for treating schistosomiasis in 1996 (Wang 2000). But, as *P. falciparum* resistance to artemisinin monotherapies becomes more widespread, there is a danger of potentiating this drug resistance if artemisinins are used for schistosomiasis in regions with a high incidence of malaria (Liu *et al.* 2014). Therefore, artemisinins may never receive approval in these regions for helminthic infections or other disorders. There are other bioactive mole-cules in *Artemisia* species that show promise for Crohn's disease, peptic ulcers, and diabetes. It would not be surprising to see new *Artemisia*-derived drugs on the market in the next decade.

References

Abad, M.J., Bedoya, L.M., Apaza, L., and Bermejo, P. (2012) The *Artemisia* L. genus: A review of bioactive essential oils. *Molecules*, **17**, 2542–2566.

Adams, J. (2004). *Hideous Absinthe: A History of the Devil in a Bottle*. University of Wisconsin Press, Madison.

Aggarwal, S., Shailendra, G., Ribnicky, D.M., Burk, D., Karki, N., and Qingxia Wang, M.S. (2015) An extract of *Artemisia dracunculus* L. stimulates insulin secretion from β cells, activates AMPK and suppresses inflammation. *Journal of Ethnopharmacology*, **170**, 98–105.

Brown, G. (2010) The biosynthesis of artemisinin (qinghaosu) and the phytochemistry of *Artemisia annua* L. (qinghao). *Molecules*, **15**, 7603–7698.

Brown, J., and Miller. A.R. (2011) *The Mixellany Guide to Vermouth & Other Aperitifs*, Mixellany Limited, London.

Cefalu, W.T., Ye, J., Zuberi, A. *et al.* (2008) Botanicals and the metabolic syndrome. *American Journal of Clinical Nutrition*, **87** (2) 481S–487S.

Choi, J.Y., Shin, S.K., Jeon, S.M. *et al.* (2011) Dose-response study of sajabalssuk ethanol extract from Artemisia princeps Pampanini on blood glucose in subjects with impaired fasting glucose or mild type 2 diabetes. *Journal of Medicinal Food*, **14** (1–2) 101–107.

Culpeper, N. (1995) *Culpeper's Complete Herbal: A Book of Natural Remedies of Ancient Ills*. NTC/Contemporary Publishing Company, New York.

Eisenman, S.W., Struwe, L., and Zaurov, D.E. (2013) *Medicinal Plants of Central Asia: Uzbekistan and Kyrgyzstan*. Springer, New York.

Hayat, M.Q., Khan, M.A., Ashraf, M., and Jabeen, S. (2009) Ethnobotany of the genus *Artemisia* L. (Asteraceae) in Pakistan. *Ethnobotany Research & Applications*, **7**, 147–162.

Ho, W.E., Peh, H.Y., Chan, T.K., and Wong, W.S. (2014) Artemisinins: pharmacological actions beyond anti-malarial. *Pharmacology & Therapeutics*, **142** (1), 126–139.

Krebs, S., Omer, T.N., and Omer, B. (2010) Wormwood (*Artemisia absinthium*) suppresses tumour necrosis factor alpha and accelerates healing in patients with Crohn's disease - A controlled clinical trial. *Phytomedicine*, **17** (5), 305–309.

Lachenmeier, D.W. (2008) Thujone-attributable effects of absinthe are only an urban legend – toxicology uncovers alcohol as real cause of absinthism. *Medizinische Monatsschrift für Pharmazeuten*, **31** (3), 101–106.

Leang, R., Taylor, W.R., and Bouth, D.M. (2015) Evidence of falciparum malaria multidrug resistance to artemisinin and piperaquine in western Cambodia: dihydroartemisinin-piperaquine open-label multicenter clinical assessment. *Antimicrobial Agents and Chemotherapy*, **59** (8), 4719–4726.

Liu, Y.X., Wu, W., Liang, Y.J. *et al.* (2014) New uses for old drugs: The tale of *Artemisinin* derivatives in the elimination of Schistosomiasis Japonica in China. *Molecules*, **19**, 15058–15074.

McGovern, P.E., Zhang, J., Tang, J. *et al.* (2004) Fermented beverages of pre- and proto-historic China. *Proceedings of the National Academy of Sciences of the United States of America*, **101** (51), 17593–17598.

Moreman, D.E. (1998) *Native American Ethnobotany*. Timber Press, Portland

National Toxicology Program. (2011) Toxicology and carcinogenesis studies of alpha,beta-thujone (CAS No. 76231-76-0) in F344/N rats and B6C3F1 mice (gavage studies). *National Toxicology Program Technical Report Series*, **2011** (570), 1–260.

Omer, B., Krebs, S., Omer, H., and Noor, T.O. (2007) Steroid-sparing effect of wormwood (*Artemisia absinthium*) in Crohn's disease: A double-blind placebo-controlled study. *Phytomedicine*, **14** (2–3), 87–95.

Panic, G., Duthaler, U., Speich, B., and Keiser, J. (2014) Repurposing drugs for the treatment and control of helminth infections. *International Journal for Parasitology – Drugs and Drug Resistance*, **4** (3), 185–200.

Park, S.W., Oh, T.Y., Kim, Y.S. *et al.* (2008) *Artemisia asiatica* extracts protect against ethanol-induced injury in gastric mucosa of rats. *Journal of Gastroenterology and Hepatology*, **23** (6), 976–984.

Shah, N.C. (2014) The economic and medicinal *Artemisia* species in India. *Scitech Journal*, **1** (1), 29–38.

Sriram, D., Rao, V.S., Chandrasekhar, K.V.G., and Yogeeswari, P. (2004) Progress in the research of artemisinin and its analogues as antimalarials: An update. *Natural Products Research*, **18**, 503–527.

Swanston-Flatt, S.K., Flatt, P.R., Day, C., and Bailey, C.J. (1991) Traditional dietary adjuncts for the treatment of diabetes mellitus. *Proceedings of the Nutrition Society*, **50**, 641–651.

Tu, Y. (2011) The discovery of artemisinin (qinghaosu) and gifts from Chinese medicine. *Nature Medicine*, **17**, 1217–1220.

Valles, J. and McArthur, E.D. (2001) *Artemisia* systematics and phylogeny: Cytogenetic and molecular insights. In *Shrubland Ecosystem Genetics and Biodiversity* (eds. E.D. McArthur & D.J. Fairbanks), US Department of Agriculture, Rocky Mountain Research Station, Ogden.

Wang, Z.H. (2000) Research progress of schistosomiasis japonica chemotherapy. *Hubei Journal of Preventive Medicine*, **11**, 1–3.

Whirl-Carrillo, M., McDonagh, E.M., Hebert, J.M. *et al.* (2012) Pharmacogenomics knowledge for personalized medicine. *Clinical Pharmacology & Therapeutics*, **92** (4), 414–417.

World Health Organization (2012) *Effectiveness of Non-Pharmaceutical Forms of Artemisia annua L. against malaria*. World Health Organization, Geneva.

World Health Organization (2015) *Withdrawal of Oral Artemisinin-Based Monotherapies*. World Health Organization, Geneva.

Willcox, M., Benoit-Vical, F., Fowler, D. *et al.* (2011) Do ethnobotanical and laboratory data predict clinical safety and efficacy of anti-malarial plants? *Malaria Journal*, **10** (Suppl. 1), S7.

Woerdenbag, H.J., Pras, N., and Chan, N.G. (1994) Artemisinin, related sesquiterpenes, and essential oil in *Artemisia annua* during a vegetation period in Vietnam. *Planta Medica*, **60** (3), 272–275.

Yang, Z.-N., Zhu, S.-Q., and Yu, Z.-W. (2012) Comparison of terpene components from flowers of *Artemisia annua*. *Bangladesh Journal of Pharmacology*, **7**, 114–119.

Traditional Treatment of Jaundice in Manipur, Northeast India

(S. S. Ningthoujam, D. S. Ningombam, A. D. Talukdar, M. D. Choudhury, K. S. Potsangbam)

Ethnomedicine

Northeast India is a melting pot of various medical systems, with influences of Ayurveda, traditional Chinese medicine (TCM), Burmese medicine, and indigenous folk medicine. Its diversity is reflected in the treatment of jaundice. Manipur is one of the states of Northeast India. It is inhabited by different communities, such as Meiteis, Meitei Pangals, Nagas, Kukis, and Nepalis, belonging to the Mongoloid and Indo-Aryan groups. Meiteis are one of the indigenous communities in Manipur following either Hinduism, Sanamahism, or Christianity. The Meiteis' traditional medicine system is an amalgamation of Ayurveda, indigenous folk medicine and some traces of Burmese medicine. A healer who practices the indigenous medical system is known as *Maiba*.

Jaundice is a manifestation of yellow pigmentation of the skin, conjunctiva membranes (Figure 6.21) of the sclera, and other mucous membranes caused by high bilirubin levels in the blood (Click *et al.* 2013; Roche and Kobos 2004). Dysfunction in any phase of bilirubin metabolism may lead to jaundice (Roche and Kobos 2004). Symptoms are often associated with liver and bile duct diseases but may also be caused by biochemical abnormalities, hematological abnormalities, autosomal recessive conditions, autoimmune conditions, bacterial infections, or the effects of drugs. Hyperbilirubinemia is also commonly observed in newborn babies because of liver immaturity, lack of breast milk, and/or incompatibility of blood groups. Neonatal jaundice usually disappears within 10–14 days. During this period, the liver becomes more efficient in removing bilirubin in the blood, so children may recover without undergoing phototherapy or exchange transfusion.

Various traditional medicine systems view the causes of jaundice differently and recommend different herbal medicines accordingly. In Ayurveda, jaundice is known as *Kamala* and is believed to be caused by an aggravated condition of *Pitta* (Ayurvedic humor representing fire). Its treatment is confined to improvement of the digestive system and liver functions. In TCM, jaundice is believed to be caused by external and internal factors that affect dampness in the stomach and spleen areas. Its treatment in TCM involves removal of cold dampness (*Yin* jaundice) and heat dampness (*Yang* jaundice). In spite of these varying interpretations, there are similarities in the diagnosis of the symptoms (Jang *et al.* 2010). Herbal recipes used in these treatments may serve as sources of lead compounds for development of new drugs for jaundice.

Interviews were conducted to discover common jaundice treatments used by traditional healers in Manipur. Despite the presence of many traditional healers in the state,

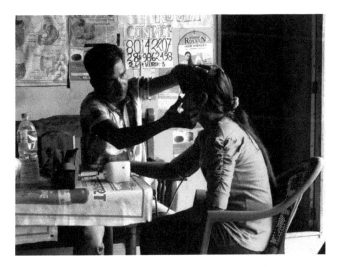

Figure 6.21 Diagnosis of jaundice by checking color of the eye.

only seven actively treat jaundice with specialized formulations. Four *Maibas* (two from Thoubal district, and one each from Imphal West and Imphal East District), one Muslim healer from Thoubal district, one Nepali healer from Imphal West district, and two Naga healers (one from the Tangkhul tribe in Ukhrul district, the other from the Tarao tribe in Chandel district) were interviewed during the course of this study. They believed jaundice (*Thongak Nungshit* or *Yaikabi* in Meithei or Manipuri language) is caused by improper dietary habits that affect the liver (MNI = *Phirak*). Meitei jaundice treatments can be divided into physical (Figure 6.22), mantras, and purely herbal treatments involving monoherbal (Table 6.8) and polyherbal recipes (Table 6.9).

In the monoherbal category, Meitei healers prescribed *Carica papaya* L. root decoction, whereas papaya seed juice and unripe fruit are used in the Ayurvedic system for jaundice and enlargement of the spleen and liver (Krishna *et al.* 2008). Lime water and banana were listed as a jaundice treatment in one Meitei healer's personal collection of ancient medicinal texts (Figure 6.23) and claimed to have a long tradition in Meitei medicine. The Kandha tribe in Orissa, India, also uses this treatment (Behera *et al.* 2006), possibly indicating cross-cultural transmission of traditional knowledge.

Phyllanthus fraternus G.L.Webster is used in the preparation of the Ayurvedic hepatoprotective drug *Bhumyamalki* (Singh 2008) and in a polyherbal combination (Table 6.9). Two of the polyherbal drugs were recommended to mothers for neonatal jaundice in the belief that bioactive components can be transferred during breastfeeding.

Some healers prescribed adjuvants to supplement the herbal medicine regimen such as drinking a lot of water (Meitei), ripe *C. papaya* juice (Muslim healers), a decoction of *Cucumis sativus* L. (MNI = Thabi) and *Allium ramosum* L (MNI = Maroi Nakuppi; Muslim healers), and sugarcane (*Saccharum officinarum* L.) juice with water (MNI = *Chu*; Meitei and Muslim healers). Sugarcane juice is also used for jaundice in other parts of the world such as Equatorial Guinea (Akendengué 1992).

Meitei healers restricted the use of yellow-colored food items during the treatment, perhaps as a safeguard against pseudojaundice (Roche and Kobos 2004). However, tribal and Muslim healers recommended use of *Cuscuta reflexa* Roxb., which is yellow in

Figure 6.22 Meitei healer preparing for ear candling.

Table 6.8 Monoherbal treatments for jaundice in Manipur, India.

Plant	Traditional name	Formulation	Traditional use
Carica papaya L.	MNI = *Awa thabi*	Decoction of roots	Meitei healers: Ayurveda treatment of jaundice and enlargement of spleen and liver
Musa × paradisiaca L.	MNI = *Laphoi*	Lime water with fruits	Meitei healers: Kandha tribe in Orissa, India
Phyllanthus fraternus G.L.Webster	MNI = *Chakpa Heikru*	Extracted juice	Ayurveda hepatoprotective drug *Bhumyamalki*
Iris domestica (L.) Goldblatt & Mabb.	NEP = *Tori bari, Tarware Phool* or *Tarbar lei*	Rhizomes chewed before eating; rhizomes in oil applied near scalp of children	Nepali healers: for jaundice, stomach disorders, and food poisoning
Mukia maderaspatana (L.) M.Roem.	TRO = *Nom Bil* NMF = *Nayong Karopthei*	Decoction	Siddha and Ayurvedic jaundice treatment
Pavetta indica L.	TRO = Chu-dar, NMF = *Sipchanghan Kathura*	Decoction	Naga healers: jaundice treatment
Cuscuta reflexa Roxb.	NMF = *Sangrei*	Decoction	Tribal and Muslim healers: jaundice treatment
Aegle marmelos (L.) Corrêa	MNI = *Heikhagok*	Fresh extracts with honey	Muslim healers: jaundice treatment
Cuscuta reflexa Roxb.	MNI = *Uri Sana manbi*	Decoction	Muslim healers: jaundice treatment
Mukia maderaspatana (L.) M.Roem	MNI = *Lam thabi*	Decoction	Muslim, Siddha, Ayurvedic, and Naga healers: jaundice treatment
Mimosa pudica L.	MNI = *Kangphal Ikaithabi*	Decoction with white rock sugar	Muslim healers: jaundice treatment

MNI: Meithei language; NEP: Nepali language; TRO: Tarao language; NMF: Tangkhul language.

color. Use of *C. reflexa* is prevalent in other parts of India and China as well (Patel *et al.* 2012). Some healers restricted food items that were thought to cause stomach inflammation such as alcohol, fermented foods (Nepali healers), and other food listed in Table 6.10.

Since the polyherbal medicines were complex mixtures mostly prescribed for neonatal jaundice (a disorder likely to heal on its own), the following section focuses on frequently prescribed monoherbal recipes. Most of the monoherbal recipes in this study

Figure 6.23 One healer reading from the transcribed ancient medicinal texts.

Table 6.9 Polyherbal treatments for jaundice by Meitei healers in Manipur, India.

Plants	Traditional name	Formulation	Traditional use[†]
Andrographis paniculata (Burm.f.) Nees	MNI = *Saban Maringkha*	Solution with or without honey	Given to mother for neonatal jaundice
Phlogacanthus thyrsiformis (Roxb. ex Hardw.) Mabb.	MNI = *Nongmangkha*		
Ocimum gratissimum L.	MNI = *Tulsi amuba*		
Centella asiatica (L.) Urb.	MNI = *Peruk*		
Ocimum tenuiflorum L.	MNI = *Tulsi*		
Momordica charantia L.	MNI = *Karot akhabi*		
Mimosa pudica L.	MNI = *Kangphal Ikaithabi*	Decoction	Given to mother for neonatal jaundice
Phlogacanthus thyrsiformis (Roxb. ex Hardw.) Mabb.	MNI = *Nongmangkha*		
Phyllanthus fraternus G.L.Webster	MNI = *Chakpa Heikru*	Juice	Neonatal jaundice
Allium ascalonicum L.	MNI = *Meitei Tilhou*		
Oroxylum indicum (L.) Kurz	MNI = *Shamba*	Bark juice of *O. indicum* mixed with *P. nigrum* juice and milk	Children with jaundice
Piper nigrum L.	MNI = *Gul Mirch*		
Terminalia chebula Retz.	MNI = *Manahi*	Fruit juice	Jaundice
Vitis vinifera L.	MNI = *Angoor*		
Pavetta indica L.	TRO = *Chu-dar*, NMF = *Sipchanghan Kathura*	Decoction	Jaundice
Cuscuta reflexa Roxb.	NMF = *Sangrei*		

MNI: Meithei language; TRO: Tarao language; NMF: Tangkhul language.
[†] All treatments were administered orally.

Table 6.10 Restricted food items for jaundice patients.

Plant	Traditional name	Traditional medicine system
Curcuma longa L.	MNI = *Yaingang*	Meitei, Nepali
Cucurbita maxima Duchesne	MNI = *Mairen*	Meitei
Tamarindus indica L.	MNI = *Mangge*	Meitei
Colocasia esculenta (L.) Schott	MNI = *Pan*	Meitei
Pyrus pyrifolia (Burm.f.), Nakai	MNI = *Nashpati*	Meitei
Artocarpus lacucha Buch.-Ham.	MNI = *Harikokthong*	Meitei
Colocasia esculenta (L.) Schott	NEP = *Karkalo*	Nepali
Abelmoschus esculentus (L.) Moench	NEP = *Bhindi*	Nepali
Vigna radiata (L.) R.Wilczek	NEP = Mugi dal	Nepali
Vigna mungo (L.) Hepper	NEP = *Kalo Dal*	Nepali

MNI: Meithei language; NEP: Nepali language.

were targeted towards non-obstructive jaundice in adult patients. According to traditional belief, development of hyperbilirubinemia is attributed to digestive-system-related disorders. Their treatments mainly focused on improving liver and digestive system functions. As such, curative properties in these recipes might be attributed to the presence of hepatoprotective compounds in these plants.

Mukia maderaspatana, C. papaya, and *P. fraternus* crude extracts have documented hepatoprotective activities (Adeneye *et al.* 2009; Gomathy *et al.* 2012; Khurnbongmayum *et al.* 2005; Petrus 2013; Singh 2008; Thabrew *et al.* 1995). For example, aqueous extracts of *M. maderaspatana* protected rat hepatocytes from damage induced by D-galactosamine and tert-butyl hydroperoxide (Thabrew *et al.* 1995). Three components of *M. maderaspatana*, ursolic acid, luteolin, and eugenol (Figure 6.24A–C) may be responsible for the observed activity. Ursolic acid protected against carbon tetrachloride– and galactosamine-induced liver injury in rats and acetaminophen-induced cholestasis (Liu 1995). The flavonoid luteolin was reported to ameliorate ethanol-induced hepatic steatosis and injury in mice (Liu *et al.* 2014). Likewise, eugenol was reported to ameliorate hepatic steatosis and fibrosis by downregulating SREBP1 gene expression (Jo *et al.* 2014). An aqueous seed extract of *C. papaya* was reported to prevent carbon tetrachloride–induced hepatotoxicity in rats (Adeneye *et al.* 2009). The possible role of zinc in treatment of jaundice needs to be considered as this element is reported in *C. papaya, Centella asiatica*, and *Piper nigrum*. Zinc has positive effects in animal models of jaundice but was ineffective in neonatal jaundice (Rana *et al.* 2011). In a clinical study, a fresh decoction of *P. fraternus* had hepatoprotective effects in patients with viral hepatitis (Singh 2008). The exact role of bioactive constituents of these medicinal plants in reducing hyperbilirubinemia is not known. However, the polyphenols phyllanthin and hypophyllanthin (Figures 6.24D–E) present in *P. fraternus* (Tripathi *et al.* 2006) were reported to protect against carbon tetrachloride– and galactosamine-induced cytotoxicity in primary cultured rat hepatocytes (Syamasundar *et al.* 1985). In addition to primary recipes, sugarcane juice recommended as a dietary regimen was also reported to protect against alcohol-induced liver injury in rats (Ri-ming 2011).

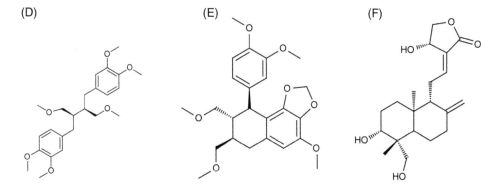

Figure 6.24 Putative active components of *M. maderaspatana*: (A) ursolic acid, (B) luteolin, (C) eugenol; *P. fraternus*, (D) phyllanthin, (E) hypophyllanthin; and *A. paniculata*, (F) andrographolide, (G) andrographiside, and (H) neoandrographolide.

Modern Applications

Some of the plants reported in the present study are used as constituents in other known marketed hepatoprotective polyherbals produced in India (Table 6.11).

Although there are several commercial formulations of *Andrographis paniculata*, it was only mentioned in one polyherbal recipe during the traditional healer interviews.

Table 6.11 Marketed Ayurvedic product with constituents similar to those reported in the present study.

Ayurvedic product	Constituent plant(s) reported in the present study
Himoliv	*Andrographis paniculata, Terminalia chebula, Carica papaya*
Bhuiamlki (Bhumyamalki)	*Phyllanthus fraternus*
Arogyavardhini Vati	*Terminalia chebula*
Yakrit Plihantak Churna	*Andrographis paniculata*
Stimuliv	*Andrographis paniculata*
Tefroliv	*Andrographis paniculata, Ocimum tenuiflorum, Piper longum, Terminalia chebula*
Livergen	*Andrographis paniculata*
Livokin	*Andrographis paniculata, Terminalia chebula*

There is some evidence of its effectiveness as a hepatoprotective agent. In mouse studies, pretreatment with the diterpenes present in *A. paniculata* (andrographolide, andrographiside, and neoandrographolide; Figures 6.24 F–H) protected against mouse liver damage (Kapil *et al.* 1993; Lee *et al.* 2014). None of the monoherbal and polyherbal recipes reported in the healer interviews were produced commercially as over-the-counter drugs, but were administered only by the healers during treatments. Plants common in many formulations used for similar type of diseases across cultures might indicate the presence of significant bioactive compounds. Considering their importance in traditional treatment of jaundice, further pharmacological studies are needed, especially in the case of *M. maderaspatana, C. papaya*, and *P. fraternus*. Herbal formulations encoded in traditional knowledge may make a significant contribution to the development of modern therapeutic drugs. The four healers interviewed for this study are directly coordinating with the Institute of Bioresources and Sustainable Development, Imphal, India, for such efforts.

References

Adeneye, A., Olagunju, J., Banjo, A.A.F., Abdul, S.F., Sanusi, O.A., and Sanni, O.O. (2009) The aqueous seed extract of *Carica papaya* Linn. prevents carbon tetrachloride induced hepatotoxicity in rats. *International Journal of Applied Research in Natural Products*, **2** (2), 19–32.

Akendengué, B. (1992) Medicinal plants used by the Fang traditional healers in Equatorial Guinea. *Journal of Ethnopharmacology*, **37**, 165–173.

Behera, S.K., Panda, A., Behera, S.K., and Misra, M.K. (2006) Medicinal plants used by the Kandhas of Kandhamal district of Orissa. *Indian Journal of Traditional Knowledge*, **5** (4), 519–528.

Click, R., Dahl-Smith, J., Fowler, L., DuBose, J., Deneau-Saxton, M., and Herbert, J. (2013) An osteopathic approach to reduction of readmissions for neonatal jaundice. *Osteopathic Family Physician*, **5**, 17–23.

Gomathy, G., Vijay, T., Gunasekaran, S., and Palani, S. (2012) Phytochemical screening and GC-MS analysis of *Mukia maderaspatana* (L.) leaves. *Journal of Applied Pharmaceutical Science*, **2** (12), 104–106.

Jang, H., Kim, J. Kim, S.K., Kim, C., Bae, S.H., Kim, A., Eom, D.M., and Song, M.Y. (2010) Ontology for medicinal materials based on traditional Korean medicine. *Bioinformatics*, **26**, 2359–2360.

Jo, H.K., Kim, G.W., Jeong, K.J., Kim do, Y., and Chung, S.H. (2014) Eugenol ameliorates hepatic steatosis and fibrosis by down-regulating SREBP1 gene expression via AMPK-mTOR-p70S6K signaling pathway. *Biological and Pharmaceutical Bulletin*, **37** (8), 1341–1351.

Kapil, A., Koul, I. B., Banerjee, S.K., and Gupta, B.D. (1993) Antihepatotoxic effects of major diterpenoid constituents of *Andrographis paniculata*. *Biochemical Pharmacology*, **46** (1), 182–185.

Khurnbongmayum, A. D., Khan, M., and Tripathi, R.S. (2005) Ethnomedicinal plants in the sacred groves of Manipur. *Indian Journal of Traditional Knowledge*, **4**, 21–32.

Krishna, K., Paridhavi, M., and Patel, J.A. (2008) Review on nutritional, medicinal and pharmacological properties of papaya (*Carica papaya* Linn.). *Natural Product Radiance*, **7** (4), 364–373.

Lee, T.Y., Chang, H.H., Wen, C.K., Huang, T.H., and Chang, Y.S. (2014) Modulation of thioacetamide-induced hepatic inflammations, angiogenesis and fibrosis by andrographolide in mice. *Journal of Ethnopharmacology* **158** (Pt. A), 423–430.

Liu, J. (1995) Pharmacology of oleanolic acid and ursolic acid. *Journal of Ethnopharmacology*, **49** (2), 57–68.

Liu, G., Zhang, Y., Liu, C. *et al.* (2014) Luteolin alleviates alcoholic liver disease induced by chronic and binge ethanol feeding in mice. *The Journal of Nutrition*, **144** (7), 1009–1015.

Petrus, A. (2013) Ethnobotanical and pharmacological profile with propagation strategies of Mukia maderaspatana (L.) M. Roem. – A concise overview. *Indian Journal of Natural Products and Resources*, **4**, 9–26.

Patel, S., Sharma, V., Chauhan, N.S., and Dixit, V.K. (2012) An updated review on the parasitic herb of *Cuscuta reflexa* Roxb. *Journal of Chinese Integrative Medicine*, **10** (3), 249–255.

Rana, N., Mishra, S., Bhatnagar, S., Paul, V., Deorari, A.K., and Agarwal, R. (2011) Efficacy of zinc in reducing hyperbilirubinemia among at-risk neonates: a randomized, double-blind, placebo-controlled trial. *The Indian Journal of Pediatrics*, **78** (9), 1073–1078.

Ri-ming, W. (2011) Research of the hepatoprotective effects of sugar cane juice against alcohol-induced liver injury in rats. *Journal of Anhui Agricultural Sciences*, **6**, 173.

Roche, S.P., and Kobos, R. (2004) Jaundice in the adult patient. *American Family Physician*, **69**, 299–308.

Singh, H. (2008) Hepatoprotective effect of Bhumyamalki (*Phyllanthus fraternus* Webster) and Phaltrikadi decoction in patients of acute viral hepatitis. *Indian Journal of Traditional Knowledge*, **7**, 560–565.

Syamasundar, K.V., Singh, B., Thakur, R.S., Husain, A., Kiso, Y., and Hikino, H. (1985) Antihepatotoxic principles of *Phyllanthus niruri* herbs. *Journal of Ethnopharmacology*, **14** (1), 41–44.

Thabrew, M.I., Gove, C. D., Hughes, R.D., McFarlane, I.G., and Williams, R. (1995) Protection against galactosamine and tert-butyl hydroperoxide induced hepatocyte damage by *Melothria maderaspatana* extract. *Phytotherapy Research*, **9** (7), 513–517.

Tripathi, A.K., Verma, R.K. Gupta, A.K., Gupta, M.M., and Khanuja, S.P. (2006) Quantitative determination of phyllanthin and hypophyllanthin in *Phyllanthus* species by high-performance thin layer chromatography. *Phytochemical Analysis*, **17** (6), 394–397.

Ethnobotany and Phytochemistry of Sacred Plant Species
Betula utilis (bhojpatra) and *Quercus oblongata* (banj) from Uttarakhand Himalaya, India

(H. Singh, S. Khatoon)

Ethnobotany and Ethnopharmacology

Tree worship is an ancient practice in India; their parts are often used in worshipping gods and deities or treated as the abode of gods. Sacred plant species have social, cultural, and religious values. Many of the sacred species such as basil (*Ocimum tenuiflorum* L.) and neem (*Azadirachta indica* A. Juss.) are multipurpose medicinal plants. Apart from their medicinal uses, sacred species are often ecologically important, playing a role in nutrient cycling and conservation, as well as in ensuring water balance within the soil.

Uttarakhand Himalaya lies on the southern slope of the Himalaya range, where climate and vegetation vary greatly with elevation, from glaciers at the highest elevations to subtropical forests at the lower elevations. It has a total area of 53,484 km^2, of which 93% is mountainous and 65% is covered by forest. The highest elevations are covered by ice and bare rock. This state has a multiethnic population spread across two geo-cultural regions: the Gahrwal and the Kumaon. A large portion of the population is Rajput including members of the native Garhwali, Kumaoni, and Gujjar communities, as well as a number of immigrants. Approximately one-fifth of the population belongs to the scheduled castes and tribes such as the Raji (living near the border with Nepal and constituting less than 5% of the population), Jaunsari, Tharu, Boxa, and Bhotia. Local communities conserve their forest by dedicating them to local deities or ancestors. The deities differ depending upon the local communities, like Golu, Haat kali, Hokra Devi, Gangnath, Kotgari, Harju, Shyamju, and so on. People of this region are closely associated with nature and the environment and have devoted their entire forest to local deities and ancestors. Plant species often seen in the sacred groves of Uttarakhand include Himalayan cedar [*Cedrus deodara* (Roxb. ex D.Don) G.Don], fragrant olive (*Osmanthus fragrans* Lour.), Chir pine (*Pinus roxburghii* Sarg), figs (*Ficus* spp.), oaks (*Quercus* spp.), Himalayan birch (*Betula utilis* D. Don), junipers (*Juniperus* spp.), and tree rhododendron (*Rhododendron arboreum* Sm.). Local communities of the sacred groves have several ethnobotanical uses. Folk music, dance, literature, and poetry in the Kumaonies and Garhwali languages are based on these sacred plants (Singh *et al.* 2014).

The tribes of Uttarakhand have widely used the ethnobotanically important trees *Betula utilis* and *Quercus oblongata* D. Don for various ailments. They are described here with respect to their importance and utilization in sacred groves.

The Himalayan birch or *bhojpatra* is considered a holy tree and was used centuries ago in India as paper for writing Sanskrit scripts and texts. The bark is still used for writing sacred mantras that are placed in an amulet for protection from evil and enemies. It is believed that keeping *bhoj* sticks and *bhojpatra* in the home wards off evil spirits and black magic; sticks are valued as charms and used in religious ceremonies (Pusalkar and Singh 2012; Tiwari *et al.* 2010).

Bark paste is used for the treatment of broken bones, gout, piles, and rheumatism as well as a spermicide, stomachic, and blood purifier. Bark paste is mixed with cow urine and applied externally on sores. Gum or bark exudate is also applied externally to get relief from fever and body aches. In one formula, 50 g *bhojpatra* resin and 20 g *Prunus*

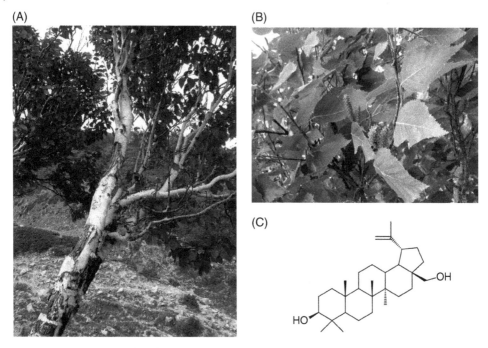

(A)

(B)

(C)

Figure 6.25 *Betula utilis* D. Don; local name *Bhojpatra*; synonyms, *B. bhojpatra* Wall. and *B. jacquemontii* Spach; family, Betulaceae; leaves alternate, ovate-elliptic, or rhomboid, with irregularly serrated margins; flowering and fruiting: June-September; grows along with small shrubs and herbs in alpine regions with an altitude of 3000–3600 m. (A) Tree and (B) flowering twig; (C) chemical structure of betulin.

persica (L.) Batsch seed kernels are ground into a paste, mixed with milk, and taken as a drink to aid conception and enhance internal strength. *P. persica* bark is also used as a psychomedicine (Phondani *et al.* 2010; Rawat and Jalal 2011; Singh *et al.* 2014; Tiwari *et al.* 2010), an antiseptic, for ear complaints, hysteria, jaundice (Malhotra and Balodi 1984), and wounds (Negi *et al.* 1985).

The Himalayan oak or *banj* is a sacred tree; its wood is often used in *Yagna* and other *poojas* (prayers). In *Yagna*, wood of *Quercus* spp. is used for making fire in the center of the temples or in sacred groves. The gum of the tree is traditionally used for gonorrheal and digestive disorders. Stem bark and resin are used as an energetic and for the treatment of bronchial problems and stomachache. Seed paste is prepared and applied to snakebites and scorpion stings (Gaur 1999; Kumar *et al.* 2011; Pande *et al.* 2006; Phondani *et al.* 2010; Singh *et al.* 2014). The leaves, seeds, and bark are also used in livestock healthcare.

Chemistry and Bioactivity

While the trees have no chemical properties that explain their religious significance, both species contain a range of phytochemicals that may explain their use in ethnomedicine. *B. utilis* bark is rich in triterpenoids including karachic acid, β-sitosterol, betulin, betulinic acid, oleanolic acid, acetyloleanolic acid, lupeol, lupenone, methyl betulonate, and methyl betulate (Khan *et al.* 2012; Rastogi *et al.* 2015).

Figure 6.26 *Quercus oblongata* D. Don; local name *Banj*; syn. *Q. leucotrichophora*, *Q. incana*; Fagaceae. Evergreen tree, 20–30 m high, leaves sharply toothed, 6–15 cm long, dense white-wooly hairs on the underside. Flowers arranged in catkins; male catkins are woolly-haired. Nuts are ovoid, approximately 1.5 cm, half covered by the involucral cup at maturity; flowering and fruiting April–June. Found in alpine regions with altitude 1700–2400 m. (A) Tree and (B) fruiting twig; chemical structure of (C) catechin and (D) gallic acid.

Various *B. utilis* extracts and their phytoconstituents (i.e., betulin and betulinic acid) have documented biological activities. Betulin has anticancer, antimicrobial, anti-inflammatory, and differentiation-promoting effects and wound-healing properties, while betulinic acid has anticancer, antibacterial, antimalarial, anti-inflammatory, anti-helminthic, antinociceptive, and antiviral activity against HSV-1 and HIV (Rastogi *et al.* 2015; Singh *et al.* 2012). Possible modes of action include reduction of pro-inflammatory cytokines, interaction with the opioid system, and multiple anticancer targets (Lucetti *et al.* 2010). The aforesaid pharmacological activities could explain many of its ethnomedicinal uses such as stomachic, blood purifier, gout, piles, and rheumatism. Betulin binds to γ-aminobutyric acid (GABA) receptors, possibly explaining its use in traditional psychomedicine (Muceniece *et al.* 2008).

Quercus oblongata ethanolic bark extract contains up to 23% tannins by weight (Watt 1908) and flavonoids such as 3-O-[{α-L rhamno- pyranosyl-(1‴ → 4″)} {α-L rhamnopyranosyl-(1‴′ → 6″)}] -β-D-glucopyranosyl quercetin, and 7-methoxy kaempferol (Sati *et al.* 2011). The volatile bark extract contains 86.4% monoterpenoids, 6.5% sesquiterpenoids, and 0.11% aliphatic aldehydes. 1,8-Cineol (40.4%) and γ-terpinene (16.4%) are the major monoterpene constituents (Sati *et al.* 2012). Tannins (e.g., gallic acid), condensed tannins (e.g., catechins), and flavonoids (e.g., quercetin and quercetin 3-O-disaccharide) have all been isolated from leaf extracts (Heuzé and Tran 2014; Sati *et al.* 2011). Oak galls contain large amounts of gallic and tannic acid (Patni *et al.* 2012).

The volatile and ethanolic bark extracts showed antimicrobial activity (Sati *et al.* 2011, 2012). Gallic acid is a polyphenol reported to have anticancer, antidiabetic, anti-inflammatory, antimicrobial, antioxidant, antiviral, cytoprotective, hepatoprotective, and neuroprotective effects (Kasture *et al.* 2009). Catechin, a natural antioxidant, plays a protective role in neurodegeneration (Ruan *et al.* 2009). Gallic acid and catechin also inhibited *Helicobacter pylori* cultures (Díaz-Gómez *et al.* 2013). These studies provide some support for the ethnomedicinal claims of *Q. oblongata*, especially digestive disorders and stomach ailments that could be caused by microbial or viral infections.

Modern Uses

Today, *B. utilis* is used for packaging material, construction, livestock fodder, and as traditional medicine. Its triterpenoid constituents have a range of biological activities, but are especially potent anti-inflammatory and anticancer agents. Given the biological activities and its history of human use, clinical studies are warranted to better characterize the pharmacological potential of *B. utilis*.

Preliminary research shows that *Q. oblongata* extracts have potent antimicrobial activity, especially against *H. pylori*. This activity may explain the traditional use of gum/resins to treat digestive disorders and stomachache. Further research is required to see if *Q. oblongata* may be developed into a botanical drug for *H. pylori*–related gastrointestinal disorders.

References

Díaz-Gómez, R., López-Solís, R., Obreque-Slier, E., and Toledo-Araya, H. (2013) Comparative antibacterial effect of gallic acid and catechin against *Helicobacter pylori*. *LWT – Food Science and Technology*, **54**, 331–335.

Gaur, R.D. (1999) *Flora of the District Garhwal North West Himalaya*, 1st edition. Trans Media, Srinagar.

Heuzé, V., and Tran, G. (2014) Banj oak (*Quercus leucotrichophora*), http://www.feedipedia.org/node/108 (accessed 6 January 2016).

Kasture, V.S., Katti, S.A., Mahajan, D., Wagh, R., Mohan, M., and Kasture, S.B. (2009) Antioxidant and antiparkinson activity of gallic acid derivatives. *Pharmacologyonline*, **1**, 385–395.

Khan, I., Sangwan, P.L., Dhar, J.K., and Koul, S. (2012) Simultaneous quantification of five marker compounds of *Betula utilis* stem bark using a validated high-performance thin-layer chromatography method. *Journal of Separation Science*, **35** (3), 392–399.

Kumar, M., Sheikh, M.A., and Bussmann. R.W. (2011) Ethnomedicinal and ecological status of plants in Garhwal Himalaya, India. *Journal of Ethnobiology and Ethnomedicine*, **7**, 32.

Lucetti, D.L., Lucetti, E.C., Bandeira, M.A. *et al.* (2010) Anti-inflammatory effects and possible mechanism of action of lupeol acetate isolated from *Himatanthus drasticus* (Mart.) Plumel. *Journal of Inflammation (London, England)*, **7**, 60.

Malhotra, C.L., and Balodi, B. (1984) Wild medicinal plants in the use of Johari tribals. *Journal of Economic and Taxonomic Botany*, **5**, 841–843.

Muceniece, R., Saleniece, K., Rumaks, J. *et al.* (2008) Betulin binds to gamma-aminobutyric acid receptors and exerts anticonvulsant action in mice. *Pharmacology Biochemistry and Behavior*, **90** (4), 712–716.

Negi, K.S., Tiwari, J.K., and Gaur, R.D. (1985) Economic importance of some common trees in Garhwal Himalaya: an ethnobotanical study. *Indian Journal of Forestry*, **8**, 276–289.

Pande, P.C., Tiwari, L., and Pande, H.C. (2006) *Folk-medicine and aromatic plants of Uttaranchal*. Bishen Singh Mahendra Pal Singh, Dehradun.

Patni, V., Sharma, N., and Mishra, P. (2012) Oak (*Quercus leucotrichophora*) gall, as an intense source of natural gallic acid. *International Journal of Life Sciences*, **1** (3), 186–191.

Phondani, P.C., Maikhuri, R.K., Rawat, L.S. *et al.* (2010) Ethnobotanical uses of plants among the Bhotiya tribal communities of Niti Valley in Central Himalaya, India. *Ethnobotany Research and Applications*, **8**, 233–244.

Pusalkar, P.K., and Singh, D.K. (2012) *Flora of Gangotri National Park*. Botanical Survey of India, Kolkata.

Rastogi, S., Pandey, M.M., and Rawat, A.K.S. (2015) Medicinal plants of the genus *Betula* – traditional uses and a phytochemical–pharmacological review. *Journal of Ethnopharmacology*, **159**, 62–83.

Rawat, V.S., and Jalal, J.S. (2011) Sustainable utilization of medicinal plants by local community of Uttarkashi District of Garhwal, Himalaya, India. *European Journal of Medicinal Plants*, **1** (2), 18–25.

Ruan, H., Yang, Yi., Zhu, X., Wang, X., and Chen, R. (2009) Neuroprotective effects of (±)-catechin against 1-methyl-4-phenyl-1,2,3,6-tetrahydropyridine (MPTP)-induced dopaminergic neurotoxicity in mice. *Neuroscience Letters*, **450** (2), 152–157.

Sati, S.C., Sati, N., and Sati, O.P. (2011) Chemical investigation and screening of antimicrobial activity of stem bark of *Quercus leucotrichophora*. *International Journal of Pharmacy and Pharmaceutical Science*, **3** (3), 89–91.

Sati, S.C., Sati, N., Sati, O.P., Biswas, D., Chauhan, B.S. (2012) Analysis and antimicrobial activity of volatile constituents from *Quercus leucotrichophora* (Fagaceae) bark. *Natural Products Research*, **26** (9), 869–872.

Singh, H., Husain, T., Agnihotri, P., Pande, P.C., and Khatoon, S. (2014) An ethnobotanical study of medicinal plants used in sacred groves of Kumaon Himalaya, Uttarakhand. *Journal of Ethnopharmacology*, **154** (1), 98–108.

Singh, S., Yadav, S., Sharma, P., and Thapliyal, A. (2012) *Betula utilis*: A potential herbal medicine. *International Journal of Pharmaceutical & Biological Archives*, **3** (3), 493–498.

Tiwari, J.K., Dangwal, L.R., Rana, C.S., Tiwari, P., and Ballabha, R. (2010) Indigenous uses of plant species in Nanda Devi Biosphere Reserve, Uttarakhand, India. *Report and Opinion*, **2** (2), 67–70.

Watt, G. (1908) *The Commercial Products of India: Being an Abridgment of the Dictionary of the Economic Products of India*. J Murray, London.

Neem-Based Insecticides

(M. Isman)

History and Ethnobotany

Human use of parts and derivatives of the Indian neem tree, *Azadirachta indica* A. Juss. (Figure 6.27A–B), in Ayurvedic medicine dates back at least 2000 years. Its name in Sanskrit, *aristha,* translates as "reliever of sicknesses" or "health and bestower," and even today the tree is revered throughout rural India as "the village pharmacy" (National Research Council 1992; Schmutterer 2002). Various preparations from its foliage, twigs, and seeds have been used externally for dermatological disorders and internally for a wide range of maladies as well. The twigs are commonly used as a dentifrice in rural India. As a result of the last-mentioned use, neem toothpastes incorporating extracts of neem bark are sold in Europe and North America as well as in India. Neem seeds are rich in oil with demonstrated antibiotic activity that likely accounts for its reputed therapeutic effects. It has also been shown to be a potent natural spermicide for intervaginal use, with the added benefit of protection against cervical cancer (Vasenwala *et al.* 2012). Young leaves are eaten as a vegetable in parts of India and Southeast Asia. Another traditional use of neem is to repel pests by layering dried leaves with clothing and rice stores. More recently the seed oil has been used to produce soaps, lubricants, and cosmetics.

Neem's uses in traditional medicine as well as the fast growth and wide-ranging climate suitability had a role in its introduction to sub-Saharan Africa, Central America, the Caribbean, and the South Pacific early in the twentieth century by Indian immigrants. *A. indica* has two sibling species: the Thai neem tree, *Azadirachta siamensis* (Val.), indigenous to Thailand, and the sentang or marrango tree, *Azadirachta excelsa* (Jack), indigenous to Thailand, Malaysia, Myanmar, and the Philippines (Schmutterer 2002). Both of the latter species are restricted to lowland monsoon forests, unlike *A. indica*, which can thrive with limited rainfall.

Indian scientists investigated the pest control properties of neem as far back as the 1920s, but Western scientific interest in neem was spurred by the German entomologist Heinrich Schmutterer in 1959. Witness to a passing plague of desert locusts while in the Sudan, Schmutterer observed a denuded landscape – except for a stand of neem trees

(A)

(B)

Figure 6.27 Neem tree (*Azadirachta indica*) and foliage.

that remained almost completely undamaged by the voracious insects. This serendipi-tous observation culminated (more than a decade later) in the isolation and identifica-tion of azadirachtin – a fascinating natural insecticide and the most potent natural insect "antifeedant" discovered to date (Schmutterer 1990).

The most potent neem insecticides are prepared via alcohol extraction of "neem cake," the solid residue from neem seeds following cold-pressing of the seeds to expel the seed oil. The insecticidal principles are limonoid-type triterpenoids, foremost of which is azadirachtin (Figure 6.28). Its complex chemical structure, including more than a dozen chiral centers and a natural epoxide group, has defied de novo synthesis except on a strictly lab-bench scale. Detailed structure-activity studies indicated that the full chemical skeleton and a number of the oxygenated substituent groups are required for bioactivity in insects (Simmonds *et al.* 1990). Altogether, about a dozen natural analogs of azadirachtin have been isolated, although the only one of consequence in neem insecticides is 3-tigloylazadirachtol, sometimes referred to as azadirachtin B (Figure 6.28). It is about five times less abundant than azadirachtin; none of the other azadirachtins make up more than 1% of the total weight of azadirachtin in neem seeds. Other major limonoids in neem seeds are salannin (and analogs) and nimbin (and analogs). While often more abundant than azadirachtin in neem seeds, they are 30–300 times less bioactive in insects when tested in isolation (Isman *et al.* 1996). Therefore, their contribution to the overall bioactivity of neem seed extracts or neem insecticides is likely minimal (Isman *et al.* 1990).

Azadirachtin has unique biological actions in insects. The primary mechanism of action is inhibition of the synthesis and release of α-ecdysone, a hormone triggering molting in insects and other arthropods (Mordue 2004). Insects consuming neem-treated foliage succumb from incomplete molts, often becoming "trapped in their old skin," or at higher doses, failing to molt entirely. In adult insects, azadirachtin can cause sterility through endocrine disruption. A more immediate effect on insects is their cessation of feeding. While a direct antifeedant effect has been observed, namely, stimulation of a deterrent receptor in the insect's gustatory sensilla on the mouthparts, an anorexic effect, likely mediated by the neuroendocrine system, has also been reported (Nisbet *et al.* 1993). Dissimilarity between the neuroendocrine system of arthropods and vertebrates probably accounts for the lack of toxicity, and hence safety, of neem insecticides to mammals (including humans; Boeke *et al.* 2004) as well as to birds, fish (Wan *et al.* 1996), and other wildlife. The chinaberry tree, *Melia azedarach* L., another member of the mahogany family (Meliaceae) is closely related but more cosmopolitan than the Indian neem tree. *M. azedarach* also produces natural insecticides in its seeds, but they are less potent than

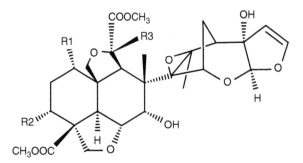

Figure 6.28 Chemical structure of azadirachtin and 3-tigloylazadirachtol, the primary insecticidal principles in neem seeds (Tig = tigloyl = (E)-2-methyl-2-butenoyl; Ac = acetate).

R_1	R_2	R_3	
OTig	OAc	OH	Azadirachtin
OH	OTig	H	3-Tigloylazadirachtol ("Aza B")

neem and have documented toxicity to mammals. Another rare and valuable property of azadirachtin is its systemic action (translocation) in many plants, making it especially effective against phloem-feeding pests such as aphids (Lowery and Isman 1996).

Present Uses

The first "modern" neem insecticides were based on patented alcohol extracts of neem seeds. Margosan-O, an extract containing 0.3% azadirachtin developed by Vikwood Botanicals (USA), was approved for use on nonfood crops (ornamentals) by the US Environmental Protection Agency in 1985. In the same year, ITC Ltd. (India) developed RD-9 Repelin, also containing 0.3% azadirachtin, which was subsequently registered with India's Central Insecticides Board. In 1993, neem insecticides were granted approval for use on food crops in the United States. Rapid improvements in the extraction and refinement of azadirachtin in the early 1990s led to technical-grade materials containing 30–50% azadirachtins, from which emulsifiable concentrated insecticides were formulated containing 3.0–4.5% azadirachtins (Walter 1999). Even today, most of the technical-grade neem for production of insecticides originates in India. Contemporary neem insecticides sold in the United States, EU, and other industrialized countries typically contain between 1.0% and 4.5% azadirachtins, whereas in developing countries (e.g., Kenya, Cuba) crude neem seed extracts formulated as insecticides typically contain 0.1% azadirachtin. In North America and Europe, neem insecticides have enjoyed modest commercial success, in part because of their relatively high cost, but an important exception is in organic agriculture, where they are preferred products owing to their perceived safety to users and the environment (Isman 2004).

References

Boeke, S.J., Boersma, M.G., Alink, G.M. *et al.* (2004) Safety evaluation of neem (*Azadirachta indica*) derived pesticides. *Journal of Ethnopharmacology*, **94**, 25–41.

Isman, M.B. (2004) Factors limiting commercial success of neem insecticides in North America and Western Europe, in, *Neem: Today and in the New Millennium*, (eds. O. Koul and S. Wahab). Kluwer Academic Press, Dordrecht, pp. 33–41.

Isman, M.B., Koul, O., Luczynski, A., and Kaminski, J. (1990) Insecticidal and antifeedant bioactivities of neem oils and their relationship to azadirachtin content. *Journal of Agricultural and Food Chemistry*, **38**, 1406–1411.

Isman, M.B., Matsuura, H., MacKinnon, S., Durst, T., Towers, G.H.N., and Arnason, J.T. (1996) Phytochemistry of the Meliaceae: So many terpenoids, so few insecticides. *Recent Advances in Phytochemistry*, **30**, 155–178.

Lowery, D.T., and Isman, M.B. (1996) Inhibition of aphid (Homoptera: Aphididae) reproduction by neem seed oil and azadirachtin. *Journal of Economic Entomology*, **89**, 602–607.

Mordue (Luntz), A.J. (2004) Present concepts of the mode of action of azadirachtin from neem, in, *Neem: Today and in the New Millennium* (eds. O. Koul and S. Wahab). Kluwer Academic Press, Dordrecht, pp. 229–242.

National Research Council. (1992) *Neem: A Tree for Solving Global Problems*. National Academy Press, Washington, DC.

Nisbet, A.J., Woodford, J.A.T., Strang, R.H.C., and Connolly, J.D. (1993) Systemic antifeedant effects of azadirachtin on the peach-potato aphid *Myzus persicae. Entomologia Experimentalis Appllicata*, **68** (1), 87–98.

Schmutterer, H. (1990) Properties and potential of natural pesticides from the neem tree. *Azadirachta indica. Annual Review of Entomology*, **35**, 271–297.

Schmutterer, H. (ed) (2002) *The Neem Tree*, 2nd edition. Neem Foundation, Mumbai.

Simmonds, M.S.J., Blaney, W.M., Ley, S.V., Anderson, J.C., and Toogood, P.L. (1990). Azadirachtin: Structural requirements for reducing growth and increasing mortality in lepidopterous larvae. *Entomologia Experimentalis et Applicata*, **55**, 169–181.

Vasenwala, S.M., Seth, R., Haider, N. *et al.* (2012) A study on antioxidant and apoptotic effect of *Azadirachta indica* (neem) in cases of cervical cancer. *Archives of Gynecology and Obstetrics*, **286** (5), 1255–1259.

Walter, J.F. (1999) Commercial experience with neem products. In *Methods in Biotechnology, Vol. 5: Biopesticides: Use and Delivery* (eds. F.R. Hall, J.J. Menn, and N.J. Totawa). Humana Press, New York, pp. 155–170.

Wan, M.T., Watts, R.G., Isman, M.B., and Strub, R. (1996) Evaluation of the acute toxicity to juvenile Pacific northwest salmon of azadirachtin, neem extract and neem-based products. *Bulletin of Environmental Contamination and Toxicology*, **56**, 432–439.

7

Europe

T. B. Tumer, B. M. Schmidt, and M. Isman

© 2014 Christian Fraaß

Ethnobotany: A Phytochemical Perspective, First Edition. Edited by B. M. Schmidt and D. M. Klaser Cheng.
© 2017 John Wiley & Sons Ltd. Published 2017 by John Wiley & Sons Ltd.

Introduction

Figure 7.1 Political map of Europe (courtesy of United States Central Intelligence Agency).

In many ways, Europeans were the founders and authors of ethnobotany as a scientific discipline. With some notable exceptions from other regions, it was the Greek, Roman, and Islamic botanists who recorded and built upon botanical knowledge. Their curiosity about the natural world led to plant-finding expeditions in neighboring regions, bringing home new medicines, perfumes, dyes, crops, fibers, and building materials. Starting in the 1400s, British, Dutch, Spanish, and Portuguese explorers set out on global expeditions with botanists or naturalists in tow in order to catalog new plant species and bring back specimens for study. In most cases, native peoples were already familiar with the plants and their uses and shared this information with the Europeans. In turn, European botanists communicated this information to the scientific community.

European civilization began in the Levant, an area in the eastern Mediterranean (Figure 7.1), during the Neolithic period. Evidence of European agriculture can be traced as far back as the Neolithic village of Makri, Greece. Phytoliths show that villagers cultivated wheat and barley and raised free-range animals (Tsartsidoua *et al.* 2009). Lentils (*Lens culinaris* Medik.), peas (*Pisum sativum* L.), bitter vetch [*Vicia ervilia* (L.) Willd.], barley (*Hordeum vulgare* L.), emmer wheat [*Triticum timopheevii* (Zhuk.) Zhuk.], and einkorn wheat (*Triticum boeoticum* Boiss.) were domesticated in southeastern Turkey and the Fertile Crescent (Western Asia) c. 9000 BCE and spread to the Mediterranean (see Abbo *et al.* 2013 for a review). The cooler, wetter climate of Northern Europe and the dry climate of the Sahara limited the expansion of the cereal grains. *Brassica* crops like rapeseed (*Brassica napus* L.) were better suited for cooler northern climates. Turnip (*Brassica rapa* L.) seeds were uncovered at neolithic sites in Switzerland, and wild species are common throughout the Alps. The humble turnip was an important crop in Ancient Greece, and oilseed *Brassica* spp. were cultivated north of the Alps for vegetable oil (Reiner *et al.* 1995). The domestication and origin of grapes remains uncertain, but recent evidence points to Western Asia (Myles *et al.* 2011). Grape (*Vitis vinifera* L.) cultivation (Figure 7.2) was widespread throughout southern Europe and the Fertile Crescent since the Neolithic age. Neolithic hunter-gatherers were likely the first to make wine from wild grapes. The oldest winery, discovered in present-day Armenia, dates to the Chalcolithic period, c. 4100 BCE (Barnard *et al.* 2011). Wild hops (*Humulus lupulus* L.) (Figure 7.3) grow throughout the Northern Hemisphere, including Europe. It is not known how they were domesticated or when they were first used for brewing beer in Europe, but fossilized pollen evidence suggests that hops were already used in the early Roman era (Murakami *et al.* 2006). Table 7.1 lists several other crops that originated in Europe.

During the Bronze Age (c. 3200–600 BCE), everything north of the Mediterranean region was considered "Northern Europe." Various Indo-European cultural groups inhabited Northern Europe including Balto-Slavic and Dacian peoples, and the Hallstatt Culture of Central Europe that branched off to form the Italic, Nordic/Germanic, and Celts. The Bronze Age Minoan civilization arose on the island of Crete c. 2700–1100 BCE. Minoan pottery and frescoes featured abundant floral imagery, giving them the reputation of being a peaceful flower-loving society. Crocuses (*Crocus* spp.) and lilies (*Lilium* spp.) were the most frequently depicted flowers, suggesting they held special religious or cultural significance (Day 2013). Paleobotany can often substantiate plant uses that were not recorded in literary texts written by and for the upper class. For example, Linear B, a tablet from the Late Bronze Age Greece, focuses on high-value crops like saffron (*Crocus sativus* L.), wheat, and barley but never mentions lower-value

Figure 7.2 Vineyard on the banks of the Rhine River, Germany.

pulses. Capers (*Capparis spinosa* L.) were depicted in Minoan art, and paleobotanists have recovered fossilized caper seeds, suggesting that capers were part of the Bronze Age diet. Late Bronze Age artifacts of wine and olive oil (*Olea europaea* L.) production and cereal grain processing have been recovered in Crete, giving further insight into the relationships of Minoans and plants.

In addition to providing insights into Mycenean crop production, Linear B depicts Mycenaean-era perfume production at the palace at Pylos. Olive oil and essential oil plants like sage (*Salvia officinalis* L.) and coriander (*Coriandrum sativum* L.), as well as henna (*Lawsonia inermis* L.) were delivered to the palace for compounding. Hellenistic-period perfumeries with plant extraction equipment have been uncovered on the island of Delos and the region around Corinth. Theophrastus had some understanding of plant preservatives in his book *On Odors*, as he described adding an astringent to olive oil before using it as a perfume base. The book provided instruction for adding various plant extracts from flowers, bark, resins, and so on, to olive oil. Rose-scented perfume

(A)

(B)

Figure 7.3 (A) Field of cultivated hops near Munich, Germany; (B) wild hops growing near Rust, Germany.

and aromatic resins were particularly popular during classical antiquity. Canaanite jars containing terebinth resin (*Pistacia terebinthus* L.) recovered from a late Bronze Age shipwreck provide further support for the use of perfumes during the Bronze Age.

Natural dyes were also widely used from the Bronze Age to modern times. Saffron is native to Greece and Southwest Asia and was used as a yellow fabric dye during the Aegean Bronze Age. Safflower (*Carthamus tinctorius* L.) and chamomile [*Cota tinctoria* (L.) J.Gay] were also used, but considered inferior yellow dyes. Madder (*Rubia tinctorum* L.) was used as a red dye, requiring a mordant to fix the color onto fabric. Woad (*Isatis tinctoria* L.) was the most popular blue dye throughout Europe in prehistoric times. The Mycenaeans used indigo (*Indigofera tinctoria* L.) as a paint pigment in frescos (see Indigo case study in the section titled "Indigo: The Devil's Dye and the American Revolution" in Chapter 7 for further reading).

Fossilized remains from Roman latrines (c. fifth–seventh century CE) suggest that Romans ate a diet of fig (*Ficus carica* L.), plum (*Prunus* spp.), grape, coriander, and dill (*Anethum graveolens* L.); however, olives were conspicuously absent (Day 2013). Roman elite had the luxury of cooking with exotic spices from Southeast Asia, made available through African trade routes. Renowned Greek scholars such as Hippocrates, Theophrastus, Galen, and Dioscorides published formulas for plant-based medicines in treatises and *materica medicas*. Table 7.2 lists some of the more commonly used

Table 7.1 Crops that originated in Europe.

Fruits and vegetables		Nuts and oils	
Asparagus	*Asparagus officinalis* L., Asparagaceae	Chestnut	*Castanea sativa* Mill., Fagaceae
Black currant	*Ribes nigrum* L., Grossulariaceae	Mustard	*Brassica nigra* (L.) K.Koch, Brassicaceae
Cabbage	*Brassica oleracea* L., Brassicaceae	Rapeseed	*Brassica napus* L., Brassicaceae
Celery	*Apium graveolens* L., Apiaceae	**Dyes**	
Chicory	*Cichorium intybus* L., Asteraceae	Chamomile	*Cota tinctoria* (L.) J.Gay, Asteraceae
Chives	*Allium schoenoprasum* L., Amaryllidaceae	Indigo	*Indigofera tinctoria* L., Fabaceae
Grapes	*Vitis vinifera* L., Vitaceae	Madder	*Rubia tinctorum* L., Rubiaceae
Hops	*Humulus lupulus* L., Cannabaceae	Saffron	*Crocus sativus* L., Iridaceae
Plums	*Prunus domestica* L., Rosaceae	Safflower	*Carthamus tinctorius* L., Asteraceae
Radish	*Raphanus raphanistrum* subsp. *sativus* (L.) Domin, Brassicaceae	Woad	*Isatis tinctoria* L., Brassicaceae
Raspberry	*Rubus* spp., Rosaceae		
Rhubarb	*Rheum officinale* Baill., Polygonaceae		
Strawberry	*Fragaria* spp., Rosaceae		
Turnips	*Brassica rapa* L., Brassicaceae		

medicinal plants. Opium, poison hemlock (*Conium maculatum* L.), henbane (*Hyoscyamus niger* L.), and aconite (*Aconitum napellus* L.) (Figure 7.4A–C) were used as sedatives and analgesics, as well as poisons. Socrates notoriously used poison hemlock to commit suicide. Poisoning was rampant in ancient Greece, so a universal antidote was developed called mithridatium containing storax (*Styrax officinalis* L.), myrrh (*Commiphora* spp.), spikenard [*Nardostachys jatamansi* (D.Don) DC.], and around 60 other herbs. To enter a trance, the Oracle of Apollo at Delphi likely consumed or inhaled psychotropic plants such as laurel leaves (*Laurus nobilis* L.), henbane, *Datura*, or "mad honey" (see mad honey case study in the section titled "Mad Honey" in Chapter 7).

The early botanical scholars had a significant influence on ethnobotany worldwide. Muslim scholars such as Al-Dinawari, Ibn Juljul, and Ibn al-Baytar based their work on Dioscorides' *Materia Medica*. They studied and collected plants throughout the Islamic world, describing their anatomy, reproduction, habitat, diseases, and cultivation. Ibn

Table 7.2 Medicinal plants native to Europe.

Angelica	*Angelica archangelica* L., Apiaceae
Apothecary rose	*Rosa gallica* L., Rosaceae
Autumn crocus	*Colchicum autumnale* L., Colchicaceae
Bugloss	*Anchusa officinalis* L., Boraginaceae
Catnip	*Nepeta cataria* L., Lamiaceae
Comfrey	*Symphytum officinale* L., Boraginaceae
Elecampane	*Inula helenium* L., Asteraceae
Feverfew	*Tanacetum parthenium* (L.) Sch.Bip., Asteraceae
Foxglove	*Digitalis purpurea* L., Plantaginaceae
German chamomile	*Matricaria chamomilla* L. Asteraceae
Hollyhock	*Alcea rosea* L., Malvaceae
Horehound	*Marrubium vulgare* L., Lamiaceae
Hyssop	*Hyssopus officinalis* L., Lamiaceae
Lady's Mantle	*Alchemilla xanthochlora* Rothm. Rosaceae
Lavender	*Lavandula angustifolia* Mill.
Lemon balm	*Melissa officinalis* L., Lamiaceae
Lily-of-the-valley	*Convallaria majalis* L., Asparagaceae
Mallow	*Malva sylvestris* L., Malvaceae
Mint	*Mentha* spp., Lamiaceae
Mullein	*Verbascum thapsus* L., Scrophulariaceae
Roman chamomile	*Chamaemelum nobile* (L.) All., Asteraceae
Rosemary	*Rosmarinus officinalis* L., Lamiaceae
Saint John's Wort	*Hypericum perforatum* L., Hypericaceae
Sage	*Salvia officinalis* L., Lamiaceae
Snapdragon	*Antirrhinum majus* L., Scrophulariaceae
Squill	*Drimia maritima* (L.) Stearn
Sweet woodruff	*Galium odoratum* (L.)Scop., Rubiaceae
Tansy	*Tanacetum vulgare* L., Asteraceae
Thyme	*Thymus vulgaris* L. Lamiaceae
Valerian	*Valeriana officinalis* L., Caprifoliaceae
Willow	*Salix* spp., Salicaceae
Yarrow	*Achillea millefolium* L., Asteraceae

al-Baytar described more than 1400 plant-based medicines [see the section titled "The Middle Ages (500–1500 CE)" in Chapter 1 for further reading]. Likewise, the work of early European explorers, botanists, chemists, and physicians introduced important medicines such as quinine, digitalis, colchicine, reserpine, pilocarpine, salicin (aspirin), and tubocurarine. The British established the use of *Cinchona* bark as a malaria treatment in the mid 1800s when they began worldwide cultivation of the plant. Digitalis

(A)

(C)

(B)

Figure 7.4 (A) Poison hemlock (*Conium maculatum*) flowers, Lincolnshire, England. © 2009 Mick Talbot; (B) henbane (*Hyoscyamus niger*) growing in Suomenlinna Fortress near Helsinki, Finland. © 2011 Anneli Salo; (C) aconite or monk's hood (*Aconitum napellus*) © 2003 Joan Simon.

Figure 7.5 Foxglove, *Digitalis purpurea.*

(Figure 7.5) had been known as a poison and herbal remedy since the Dark Ages, but Scottish physician William Withering is credited with discovering its usefulness for treating heart conditions. The indole alkaloid reserpine was first isolated from Indian *Rauvolfia* spp. in 1952 by Swiss scientists Schlittler, Müller, and Bein. Since then, it has been used as an antipsychotic for schizophrenia and to treat hypertension. Brazilian physician Symphrondio Coutinhou brought samples of *Pilocarpus jaborandi* Holmes leaves to Europe in 1873. By the following year, A.W. Gerrard from England and E. Hardy from France had independently isolated pilocarpine, an imidazole alkaloid that induces salivation and decreases intraocular pressure in glaucoma. Assyrians and Sumerians were aware of the analgesic and anti-inflammatory properties of willow (*Salix* spp.), but the active compound salicin was not extracted and purified until 1828 by German chemist Johann Buchner. British scientist Harold King isolated the bis-benzyltetrahydroisoquinoline alkaloid tubocurarine from a sample of South American *Chondrodendron tomentosum* Ruiz & Pav. arrow poison in 1935. It was used as a muscle relaxant during surgical anesthesia until safer derivatives were developed.

References and Additional Reading

Abbo, S., Lev-Yadun, S., Heun, M., and Gopher, A. (2013) On the "lost" crops of the neolithic Near East. *Journal of Experimental Botany*, **64** (4), 815–822.

Barnard, H., Dooley, A.N., Areshian, G., Gasparyan, B., and Faull, K.F. (2011) Chemical evidence for wine production around 4000 BCE in the Late Chalcolithic Near Eastern highlands. *Journal of Archaeological Science*, **38** (5), 977–984.

Day, J. (2013) Botany meets archaeology: People and plants in the past. *Journal of Experimental Botany*, **64** (18), 5805–5816.

Murakami, A., Darby, P., Javornik, B. *et al.* (2006) Molecular phylogeny of wild hops. *Humulus lupulus* L. *Heredity (Edinb)*, **97** (1), 66–74.

Myles, S., Boyko, A.R., Owens, C.L. *et al.* (2011) Genetic structure and domestication history of the grape. *Proceedings of the National Academy of Sciences*, **108** (9), 3530–3535.

Reiner, H., Holzner, W., and Ebermann, R. (1995) The development of turnip-type and oilseed-type *Brassica rapa* crops from the wild type in Europe – An overview of botanical, historical and linguistic facts. In *Rapeseed Today and Tomorrow*, Vol. 4, 9th International Rapeseed Congress, Cambridge, pp 1066–1069.

Tsartsidoua, G., Lev-Yadunb, S., Efstratiouc, N., and Weiner, S. (2009) Use of space in a Neolithic village in Greece (Makri): Phytolith analysis and comparison of phytolith assemblages from an ethnographic setting in the same area. *Journal of Archaeological Science*, **36** (10), 2342–2352.

Differential Use of *Lavandula stoechas* L. among Anatolian People against Metabolic Disorders

(T. B. Tumer)

Ethnobotany and Ethnopharmacology

Anatolia (Asia Minor), surrounded by the Black Sea on the north, Aegean Sea on the west, and Mediterranean Sea on the south, is a large peninsula that covers the Asian part of Turkey (Türkiye). Located at the crossroads of Asia and Europe, it has great phytogeographical and cultural diversity contributing to a rich ethnobotanical and ethno-medical heritage. The flora of Anatolia contains over 12,000 species and subspecies including nearly 3,300 endemics (Davis *et al.* 1988; Güner *et al.* 2000). The origin and diversity of many plant genera (spanning crops, fruit trees, and ornamentals) in Europe center around Anatolia. Among this remarkable variation, nearly 800 plant species have been used in traditional remedies based on a deep ethnobotanical knowledge of various historical Anatolian civilizations including Sumerian, Assyrians, Akkadians, Hittites, Anatolian Seljuks, and Ottomans (Yeşilada and Sezik 2003).

Figure 7.6 *Lavandula stoechas.*

Table 7.3 The classification of *L. stoechas.*

Kingdom	Plantae
Subkingdom	Tracheobionta
Superdivison	Spermatophyta
Division	Magnoliophyta
Class	Magnoliopsida
Subclass	Asteridae
Order	Lamiales
Family	Lamiaceae
Genus	*Lavandula*
Species	*stoechas*

One of the most widely used species in traditional folk medicine of Anatolia is *Lavandula stoechas,* commonly known as topped or butterfly lavender. It is also known as French or Spanish lavender in Europe. However, there is confusion over the common names for different lavender species. *L. stoechas* is native to countries bordering the Mediterranean Sea, and is also found in northern and southern Africa, and the Arabian Peninsula (Staicov *et al.* 1969). Currently, it is cultivated all over the world including the United States and Australia. It is an evergreen shrub growing 30–100 cm tall with fragrant gray-green leaves and unusual purple spikes of flowers at the top (Figure 7.6). It is one of the 28 species in the very famous genus of lavender within the large mint family of Lamiaceae (Table 7.3).

The genus *Lavandula* has worldwide popularity and commercial value in aromatherapy and in the soap and fragrance industries due its intense and pleasant aroma. The traditional use of *Lavandula* species as therapeutic agents has a very long history, and the earliest documents on medical uses were found in the records of ancient Romans and Byzantium residents of Anatolia (Cavanagh and Wilkinson 2002). *Lavandula* species have been most commonly used for their essential oils to treat a wide range of stress- and anxiety-related ailments including depression, fatigue, colic, ulcer, headache, motion sickness, and spasms (reviewed in Chu and Kemper 2001). The most common use is aromatherapy, especially massage with lavender oils, to relieve tension occurring throughout the body. Antimicrobial and antifungal activities have also been reported for the essential oils of different varieties (Cavanagh and Wilkinson 2002). Lavenders that have medicinal values fall into four main species: *L. dentata* L., *L. angustifolia* Mill., *L. latifolia* Medik., and *L. stoechas.* Although they all share similar ethnobotanical properties and common chemical constituents, among them *L. stoechas* has very distinct therapeutic uses among Anatolian people compared to Europeans. This species grows wild and is found abundantly in the Mediterranean and Aegean region of Turkey. The numerous ethnobotanical studies conducted in these regions demonstrated that *L. stoechas* has been traditionally used for the treatment of a group of metabolic diseases like circulation system disorders: heart disease, high cholesterol and atherosclerosis (Abay and Kılıç 2001; Bulut 2006; Tuzlacı and Eryasar-Aymaz 2001; Ugulu *et al.* 2009); as well as diabetes, hypertension, and obesity (Everest and Ozturk 2005; Gürdal and Kültür 2013; Polat and Satıl 2012; Uysal 2008). In some regions of Anatolia, *L. stoechas* is named "head sweeper" among local people owing to a common traditional belief for its clearing effect on clogged arteries around the areas of the head and neck. Typical preparations of *L. stoechas* in Anatolia involves decoction or infusion of the flowered tips together with leaves and stems (all the above ground parts), which is drunk hot two or three times a day. This is especially important when considering its differential traditional use in Anatolia, which shows that the "aqueous" components, rather than essential oil part, might be "bioactive" and likely to play significant role against the metabolic disorders listed earlier.

Either fresh or dried plants can be used in the preparations, depending on the season. In late spring (mid-May), plants are collected by local people from the hills and woodlands, tied in bunches, and sold in the district bazaar (Figure 7.7A). In the

(A)

(B)

Figure 7.7 (A) A countrywoman is selling fresh bunches of *Lavandula stoechas* in the district bazaar. (B) The color of infusion prepared from fresh plant.

fresh preparation, flowers release indigo blue color into hot water (see Figure 7.7B) if the plant is collected when the buds just burst into blossom. In addition to its common use, in some regions, indigenous people also believed that the chewing of *L. stoechas* (stem part with leaves) is highly effective against addiction to cigarettes.

Unlike other countries, ethnobotanical use of *L. stoechas* in Pakistan, Tunisia, and Morocco show some similarities with Anatolia (Gilani *et al.* 2000; Leporatti and Ghedira 2009; Tahraoui *et al.* 2007). The rest of the *L. stoechas* ethnobotanical literature predominantly emphasizes the use of its essential oils in aromatherapy and psychological distress conditions (Cavanagh and Wilkinson 2002; Chu and Kemper 2001). Different uses of this specific lavender species between Anatolian people and other countries may be the natural result of the peculiar Turkish chemotype, which we discuss below.

Chemistry and Bioactivity

The essential oil compositions of lavender species have been extensively studied. Accordingly, the main constituents of oil are derivatives of monoterpenes (linalool, linalyl acetate, 1,8-cineole (eucalyptol), β-ocimene, fenchone, and camphor), benzenoids (coumarin and rosmarinic acid), and sesquiterpenes. The relative concentration of these compounds can vary in different *Lavandula* species. Aroma analysis of four main therapeutically important *Lavandula* species showed that oils derived from *L. latifolia* and *L. stoechas* have high camphor levels. On the other hand, *L. dentata*, and *L. angustifolia* are low (<2%) in camphor (Cavanagh and Wilkinson 2002). High-camphor *Lavandula* varieties are generally used in insecticides because of its insect repellent and insecticidal properties. The antibacterial effect of lavender oil obtained from various *Lavandula* species has been partially attributed to eugenol, rosmarinic acid, α-terpineol, terpinen-4-ol, and camphor (Chu and Kemper 2001; Pattnaik *et al.* 1997). Cineole, which is found in almost all *Lavandula* species, has been known for its antispasmodic and local anesthetic activities (Pattnaik and Subramanyam 1997).

Considering *L. stoechas* specifically, studies from different Mediterranean regions including Italy, Greece, Corsica, Tunisia, and Morocco reported that bicyclic monoterpenes, namely, fenchone, camphor (Figure 7.8), and 1,8-cineole, are the main components of the essential oil obtained from this species (Messaoud *et al.* 2012; Skoula *et al.* 1996). Among them, either fenchone/camphor or fenchone/1,8-cineole can be dominant depending on the geographical origin of the plant (Skoula *et al.* 1996). However, the essential oil composition of *L. stoechas* from Turkey (Gören *et al.* 2002) showed a completely different profile, which is dominated by pulegone (40.4%), menthol (18.1%), and menthone (12.1%). Similar to camphor, pulegone has very potent insecticidal activity (Figure 7.8). Menthone also shows strong activity as an insecticide; however, it is less effective compared to pulegone (Franzios *et al.* 1997).

Although there is voluminous literature on the essential oil components of lavender species,

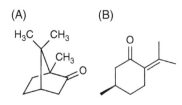

(A) camphor, (B) pulegone.

Figure 7.8 (A) camphor, (B) pulegone.

information on aqueous components is extremely limited. The only available literature on the polyphenol content of *L. stoechas* did not provide a complete profile; however, it reported the presence of flavone 7-glycosides, apigenin, and luteolin 7-*O* glucosides (Upson *et al.* 2000). Apigenin has been found to promote adult neurogenesis in vivo and in vitro by stimulating neuronal differentiation (Taupin 2009). The proposed promising role of apigenin for the treatment of neurological diseases, disorders, and injuries may also provide evidence for the antidepressant properties of *Lavandula* species. Recently, it was found that luteolin improves hepatic insulin sensitivity and ameliorates the deleterious effects of diet-induced obesity (Kwon *et al.* 2015) via the interplay between liver and adipose tissue, which may support the traditional use of *L. stoechas* among Anatolians. More studies are required to identify the aqueous components of *L. stoechas* and its bioactive compounds that are effective against metabolic disorders.

Along with the essential oils extracted from other *Lavandula* species, pure oil obtained from *L. stoechas* is still used in aromatherapy, incorporated into soaps, shampoos, skin care products, deodorants, air fresheners, disinfectants, and many other products of cosmetic and detergent industries due its pleasant fragrance and antimicrobial/antifungal activity. In the pharmaceutical industry, lavender oil is used for its soothing, carminative, and painkilling properties (Cavanagh and Wilkinson 2002; Chu and Kemper 2001). It is also used in the formulation of insecticides due to insecticidal and insect repellent constituents. In addition, lavender honey is also in great demand in the food industry. When the aqueous bioactive components of *L. stoechas* are completely identified, it will probably gain popularity in the dietary supplement and pharmaceutical industries.

References

Abay, G., and Kılıç, A. (2001) Pürenbeleni ve yanıktepe (Mersin) yörelerindeki bazı bitkilerin yöresel adları ve etnobotanik özellikleri. *Ot Sistematik Botanik Dergisi*, **8** (2), 97–104.

Bulut, Y. (2006) *Manavgat (Antalya) Yöresinin Faydalı Bitkileri*. M.Sc. thesis, Süleyman Demirel Üniversitesi, Fen Bilimleri Enstitüsü, Biyoloji Anabilim Dalı, Çünür.

Cavanagh, H.M., and Wilkinson, J.M. (2002) Biological activities of lavender essential oil. *Phytotherapy Research*, **16** (4), 301–308.

Chu, C.J., and Kemper, K.J. (2001). Lavender (*Lavandula* spp.), http://www.longwoodherbal.org/lavender/lavender.pdf.

Davis, P.H., Mill, R.R., and Tan, K. (1988) *Flora of Turkey and the East Aegean Islands*, Vol. 10. Edinburgh University Press, Edinburgh.

Everest, A., and Ozturk, E. (2005) Focusing on the ethnobotanical uses of plants in Mersin and Adana provinces (Turkey). *Journal of Ethnobiology and Ethnomedicine*, **1**, 6.

Franzios, G., Mirotsou, M., Hatziapostolou, E., Kral, J., Scouras, Z.G., and Mavragani-Tsipidou, P. (1997) Insecticidal and genotoxic activities of mint essential oils. *Journal of Agricultural and Food Chemistry*, **45** (7), 2690–2694.

Gilani, A.H., Aziz, N., Khan, M.A. *et al.* (2000) Ethnopharmacological evaluation of the anticonvulsant, sedative and antispasmodic activities of *Lavandula stoechas* L. *Journal of Ethnopharmacology*, **71** (1–2), 161–167.

Gören, A.C., Topçu, G., Bilsel, G., Bilsel, M., Aydoğmus, Z., and Pezzuto, J.M. (2002) The chemical constituents and biological activity of essential oil of *Lavandula stoechas ssp. Stoechas. Zeitschrift für Naturforschung*, **57** (9–10), 797–800.

Güner, A., Özhatay, N., Ekim, T., and Başer, K.H.C. (eds) (2000) *Flora of Turkey and the East Aegean Islands*, Vol. 11. Edinburgh University Press, Edinburgh.

Gürdal, B., and Kültür, S. (2013) An ethnobotanical study of medicinal plants in Marmaris (Muğla, Turkey). *Journal of Ethnopharmacology*, **146** (1), 113–126.

Kwon, E.Y., Jung, U.J., Park, T., Yun, J.W., and Choi, M.S. (2015) Luteolin attenuates hepatic steatosis and insulin resistance through the interplay between the liver and adipose tissue in mice with diet-induced obesity. *Diabetes*, **64** (5), 1658–1669.

Leporatti, M.L., and Ghedira, K. (2009) Comparative analysis of medicinal plants used in traditional medicine in Italy and Tunisia. *Journal of Ethnobiology and Ethnomedicine*, **5**, 31.

Messaoud, C., Chograni, H., and Boussaid, M. (2012) Chemical composition and antioxidant activities of essential oils and methanol extracts of three wild *Lavandula* L. species. *Natural Product Research*, **26** (21), 1976–1984.

Pattnaik, S., Subramanyam, V.R., Bapaji, M., and Kole, C.R. (1997) Antibacterial and antifungal activity of aromatic constituents of essential oils. *Microbios*, **89**, 39–46.

Polat, R., and Satıl, F. (2012) An ethnobotanical survey of medicinal plants in Edremit Gulf (Balıkesir-Turkey). *Journal of Ethnopharmacology*, **139**(2), 626–641.

Skoula, M., Abidi, C., and Kokkalou, E. (1996) Essential oil variation of *Lavandula stoechas* L. *ssp. stoechas* growing wild in Crete (Greece). *Biochemical Systematics and Ecology*, **24**, 255–260.

Staicov, V., Chingova, B., and Kalaidjiev, I. (1969) Studies on several lavender varieties. *Soap, Perfumery & Cosmetics*, **42**, 883–887.

Tahraoui, A., El-Hilaly, J., Israili, Z.H., and Lyoussi, B. (2007) Ethnopharmacological survey of plants used in the traditional treatment of hypertension and diabetes in south-eastern Morocco (Errachidia province) *Journal of Ethnopharmacology*, **110** (1), 105–117.

Taupin, P. (2009). Apigenin and related compounds stimulate adult neurogenesis. Mars, Inc., the Salk Institute for Biological Studies: WO2008147483. *Expert Opinion on Therapeutic Patents*, **19** (4), 523–527.

Tuzlacı, E., and Eryasar-Aymaz, P. (2001) Turkish folk medicinal plants. Part IV: Gonen (Balıkesir). *Fitoterapia*, **72**, 323–343.

Ugulu, I., Baslar, S., Yorek, N., and Dogan, Y. (2009). The investigation and quantitative ethnobotanical evaluation of medicinal plants used around Izmir province, Turkey. *Journal of Medicinal Plants Research*, **3** (5), 345–367.

Upson, T.M., Grayer, R., Greenham, J.R., Williams, C., Al-Ghamdi, F., and Chen, F. (2000) Leaf flavonoids as systematic characters in the genera *Lavandula* and *Sabaudia*. *Biochemical Systematics and Ecology*, **28** (10), 991–1007.

Uysal, G. (2008) *Etnobotany of Koycegiz (Mugla)*. M.Sc. thesis, Mugla University Institute of Science and Technology, Mugla.

Yeşilada, E., and Sezik, E. (2003) A survey on the traditional medicine in Turkey: semi-quantitative evaluation of the results, in, *Recent Progress in Medicinal Plants, Ethnomedicine and Pharmacognosy II*, Vol. 7 (eds. V.K. Singh, J.N. Govil, S. Hashmi, and G. Singh). Stadium Press, LLC, Houston, pp. 389–412.

Mad Honey

(B. M. Schmidt)

Honey has an ancient history, but the story of mad honey begins in the time of the Oracle of Delphi, the Pythia, in the eighth century BCE. The Homeric Hymn to Hermes mentions "bee oracles" from Delphi's Mt. Parnassos that could only prophesize after ingesting *meli chloron* or "green honey." Some speculate that these bee oracles were the Pythia, who required intoxicating honey to utter her prophesies (Mayor 1995; Ott 1998). In 401 BCE, Xenophon mentions intoxicating honey in his account of the Retreat of the Ten Thousand, a military campaign near Colchis on the eastern coast of the Black Sea:

> ... the swarms of bees in the neighborhood were numerous, and the soldiers who ate of the honey all went off their heads, and suffered from vomiting and diarrhea, and not one of them could stand up, but those who had eaten a little were like people exceedingly drunk, while those who had eaten a great deal seemed like crazy, or even, in some cases, dying men ... On the next day, however, no one had died, and at approximately the same hour as they had eaten the honey they began to come to their senses; and on the third or fourth day they got up, as if from a drugging. *(Anabasis, 4.8.18–21)*

Xenophon may have inspired a sort of mad honey biological warfare nearly 500 years later in the same region. In 67 BCE, Pompeii the Great (Pompeidus) led a Roman military campaign against the Heptakometes in what is presently known as İkizdere Valley near the Black Sea in eastern Turkey. The Heptakometes were barbarians living in the valley. Familiar with the effects of mad honey, they set out honeycombs (some sources say honey pots) for the invading army to find. The Roman soldiers consumed the honey, became intoxicated, and were killed by the Heptakometes. In 77 CE, Pliny the Elder coined the term *meli maenomenon*, "mad honey" (*Naturalis Historia* XXl Chapter 45). He attributed the affects to Rhododendrons growing in the area, a fact that was well known to the inhabitants of the Black Sea region. In the sixteenth century, mad honey was exported to Europe as an ingredient to create beverages with greater intoxicating effects than alcohol alone. It was known as *miel fou*, or "crazy honey" in France.

There are further accounts of mad or toxic honey from around the world. Other members of Ericaceae have been implicated in grayanotoxin honey poisonings in Japan (*Elliottia paniculata*) (Tsuchiya 1977) and the Southeastern United States (*Kalmia latifolia* L., mountain laurel) (Jones 1947). *K. latifolia* honey sold under the name "Metheglin" was intentionally added to liquor in New Jersey, the United States, during the eighteenth century to produce a very intoxicating drink (Kebler 1896; Ott 1998).

Outside of Ericaceae, mad honey has been traced back to various toxic plants. For example, yellow jasmine (*Gelsemium sempervirens*, Gelsemiaceae) was the source of fatal honey poisonings in Branchville, South Carolina, the United States, in the nineteenth century (Kebler 1896; King and Aspinwall 1885). The Mayans had a mad honey called xtabentún, derived from *Turbina corymbosa* (L.) Raf. (Convolvulaceae). It was highly valued for its uterotonic and psychoactive properties. In New Zealand, toxic honey is unintentionally produced when bees collect honeydew from vine hopper insects (*Scolypopa australis* Walker) feeding on the tutu plant (*Coriaria arborea* Lindsay, Coriariaceae). Poisoning symptoms include nausea, vomiting, seizures, and coma. There are reports of toxic honey from *Atropa belladonna* L. or *Datura* spp. from Poland and Hungary (Ott, 1998) and toxic honeys from various plant sources throughout South America.

Chemistry and Bioactivity

Mad honey from the Eastern Black Sea region, also called Pontic Honey or *deli bal*, contains intoxicating grayanotoxins. Bees collecting nectar from Rhododendron flowers, especially *Rhododendron ponticum* L., inadvertently concentrate grayanotoxins in the honey. Plants produce nectar composed primarily of mono- and disaccharides and amino acids to attract pollinators (González-Teuber and Heil 2009). Other compounds like proteins, lipids, phenols, alkaloids, and volatile organic compounds have also been found in nectar. Some plants produce "toxic" compounds in their nectar to ward off nectar robbers that consume nectar but do not pollinate (González-Teuber and Heil 2009). Bees are attracted to the odor of nectar and its nutritional rewards. After a bee drinks nectar, it is stored in the crop or "honey stomach" where invertase and digestive acids convert sucrose into glucose and fructose. The nectar is regurgitated and ingested several times before being deposited in the honeycomb cells. Once in the cells, honeybees fan their wings to dehydrate the honey as a way to preserve it against undesired fermentation. This dehydration process also serves to concentrate any toxic compounds that were present in the nectar.

   Grayanotoxins (also called andromedotoxin, acetylandromedol, rhodotoxin, and asebotoxin; Figure 7.9) are a group of diterpenes found in ericaceous plants. They are synthesized via the mevalonic acid pathway from the precursor (-)-kaurene. Although more than 25 grayanotoxins have been identified, grayanotoxin 3 is considered the principal toxic isomer, and grayanotoxin 1 is the main toxin responsible for cardiac manifestations (Aliyev *et al.* 2009). Grayanotoxins 1 and 2 were identified in the honey, leaves, and flowers of *R. ponticum* and *R. flavum* (Hoffmanns.) G.Don in the eastern Black Sea area (Jansen *et al.* 2012). Grayanotoxins are neurotoxins that bind to

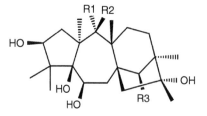

Figure 7.9 **Grayanotoxin structure.**

	R1	R2	R3
Grayanotoxin 1	OH	CH₃	COCH₃
Grayanotoxin 2	CH₂	CH₂	OH
Grayanotoxin 3	OH	CH₃	OH
Grayanotoxin 4	CH₂	CH₂	COCH₃

(A)   (B)   (C)

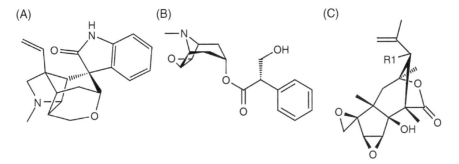

Figure 7.10 Structures of the indole alkaloid gelsemine (A), the tropane alkaloid scopolamine (B), and the sesquiterpenes tutin (C, R1=H) and hyenanchin (C, R1=OH).

group II receptor sites of sodium ion channels located in the cell membrane, creating a maintained state of depolarization (Onat *et al.* 1991). Skeletal and heart muscles, nerves, and the central nervous system are primarily affected. Shortly after ingestion, dizziness, weakness, excessive perspiration, nausea, and vomiting may occur (Dilber *et al.* 2002). More severe cardiac symptoms may develop, including hypotension, bradyarrhythima, sinus bradycardia, nodal rhythm, anomalous atrioventricular excitation, and atrioventricular block (Akinci *et al.* 2008; Dursunoglu *et al.* 2007; Gündüz *et al.* 2007; Özkan *et al.* 2004). Grayanotoxin poisoning is not typically fatal, with symptoms disappearing after 24 hours.

Mad honey poisonings can also be caused by alkaloids and sesquiterpenes. The toxic agent in *G. sempervirens* is the indole alkaloid gelsemine (Figure 7.10A). In xtabentún, psychoactivity is attributed to ergoline alkaloids. The tropane alkaloid scopolamine (Figure 7.10B) was responsible for the toxic honey poisonings traced back

Figure 7.11 Jars of "mad honey" are still available for purchase in areas along the Black Sea.

to *Datura* spp. in Poland (Lutomski *et al.* 1972) and was suspected in the poisonings in Hungary (Islam *et al.* 2014). In New Zealand mad honey, the coriamyrtin-type sesquiterpene neurotoxin tutin (Figure 7.10C) from the tutu plant (*Coriaria arborea*) is transformed into hyenanchin (Figure 7.10C) in the vine hopper insect's gut. Since bees collect both tutu nectar and vine hopper honeydew, both tutin and hyenanchin are found in the toxic honey.

Pontic mad honey is still produced in the same region along the Black Sea. Tourists travel to the Turkish province of Trabzon looking for mad honey may find it sold under the names *deli bal* or *orman komar bali* (Figure 7.11). Apiarists take their beehives to mountain slopes with large swaths of *Rhododendron ponticum* and *R. luteum* early in the spring. The concentration of grayanotoxin in mad honey varies year to year, but since nearly all of the nectar and pollen that the bees collect are from Rhododendrons, the intoxicating properties are certain. Local people use *deli bal* for medicinal purposes, consuming small amounts to treat hypertension, diabetes, and stomach ailments. It is also known as a sexual performance enhancer, leading to several reports of "mad honey sex" ending with a trip to the hospital (Demircan *et al.* 2009). Thanks to the global economy, tourists may skip the trip to Turkey altogether, purchasing *deli bal* via the Internet or from a Turkish friend instead.

References

Akinci, S., Arslan, U., Karakurt, K., and Cengel, A. (2008) An unusual presentation of mad honey poisoning: Acute myocardial infarction. *International Journal of Cardiology*, **129**, e56–58.

Aliyev, F., Türkoğlu, C., Çeliker, C., Firatli, I., Alici, G., and Uzunhasan, I. (2009) Chronic mad honey intoxication syndrome: A new form of an old disease? *Oxford Journals*, **11** (7), 954–956.

Demircan, A., Keleş, A., Bildik, F., Aygencel, G., Doğan, N.O., and Gómez, H.F. (2009) Mad honey sex: Therapeutic misadventures from an ancient biological weapon. *Annals of Emergency Medicine*, **54** (6), 824–829.

Dilber, E., Kalyoncu, M., Yaris, N., and Ökten, A. (2002) A case of mad honey poisoning presenting with convulsion: Intoxication instead of alternative therapy. *Turkish Journal of Medical Sciences*, **32**, 361–362.

Dursunoglu, D., Gur, S., and Semiz, E. (2007) A case of complete atrioventricular block related to mad honey intoxication. *Annals of Emergency Medicine*, **50**, 484–485.

González-Teuber, M., and Heil, M. (2009) Nectar chemistry is tailored for both attraction of mutualists and protection from exploiters. *Plant Signaling & Behavior*, **4** (9), 809–813.

Gündüz, A., Durmus, I., Turedi, S., Nuhoglu, I., and Ozturk, S. (2007) Mad honey poisoning-related asystole. *Emergency Medicine Journal*, **24**, 592–593.

Islam, M.N., Khalil, M., Islam, M.A., and Gan, S.H. (2014) Toxic compounds in honey. *Journal of Applied Toxicology*, **34** (7), 733–742.

Jansen, S.A., Kleerekooper, I., Hofman, Z.L., Kappen, I.F., Stary-Weinzinger, A., and Van der Heyden, M.A. (2012) Grayanotoxin poisoning: 'Mad honey disease' and beyond. *Cardiovascular Toxicology*, **12** (3), 208–215.

Jones, W.R. (1947) Honey poisoning. *Gleanings in Bee Culture*, **75**, 76–77.

Kebler, L.F. (1896) Poisonous Honey. *Proceedings of the American Pharmaceutical Association*, **44**, 167–174.

King and Aspinwall (1885) Poisonous Honey. *The Bee Keepers' Magazine*, **13**, 248–249.

Lutomski, J., Debska, W., and Gorecka, M. (1972) Przypadek skazenia miodu skopolamina [Poisoning honey with scopolamine]. *Pszczelarstwo* 23 (10), 5–6. [Chemical Abstracts 83:109560x].

Mayor, A. (1995) Bees and the baneful Rhododendron: Mad honey! *Archaeology*, **48** (6), 32–40.

Ott, J. (1998) Bees and toxic honeys as pointers to psychoactive and other medicinal plants. *Economic Botany*, **52** (3), 260–266.

Onat, F.Y., Yegen, B.C., and Lawrence, R. (1991) Mad honey poisoning in man and rat. *Reviews on Environmental Health* **9**, 3–9.

Özkan, H., Akdemir, R., Yazici, M., Gündüz, H., Duran, S., and Uyan, C. (2004) Cardiac emergencies caused by honey ingestion: A single center experience. *Emergency Medicine Journal*, **21**, 742–744.

Tsuchiya, H. (1977). Studies on a poisonous honey originated from azalea, *Tripetaleia paniculata*. *Kanagawa-ken Eisei Kenkyusho Kenkyu Hokoku*, 7, 19–28. [Chemical Abstracts 91:84673m].

Indigo: The Devil's Dye and the American Revolution

(B. M. Schmidt)

Ethnobotany

Indigo is a blue dye produced by several different species in the Brassicaceae and Fabaceae families. Knowledge of indigo is widespread wherever these plants were found and dates back at least 3000 years, as an indigo-dyed kerchief was found in King Tutankhamun's tomb. The type of species used for indigo extraction largely depended on the climate. The plant commonly known as indigo, *Indigofera tinctoria* L. (Fabaceae) (Figure 7.12), has a tropical distribution; its native habitat is uncertain. Natal indigo (*Indigofera arrecta* A.Rich.) originated in tropical Africa and was widely cultivated in India and Southeast Asia. Gambian indigo [*Philenoptera cyanescens* (Schum. & Thonn.) Roberty] is widely distributed in tropical Africa and has the same range of colors as Natal indigo. It is used to make *bara siti* cloth in Gambia and is used as food and medicine in several African countries (Jansen and Cardon 2005). In Europe, where a tropical climate is lacking, woad (*Isatis tinctoria* L., Brassicaceae) (Figure 7.13) filled the demand

Figure 7.12 *Indigofera tinctoria*. © 2004 Kurt Stüber.

Figure 7.13 *Isatis tinctoria.* © 2011 Matt Lavin.

for blue indigo dye. The name "indigo" comes from the Greek *indikon* meaning a substance from India, reflecting India's historical role as a supplier of indigo.

Although India had been a center of indigo production for centuries, the European story of indigo actually begins with woad. Many regions in Germany specialized in the production of woad, the only blue dye available in Europe until the introduction of indigo from Asia. Woad is produced from the leaves of *Isatis tinctoria*, a biennial or short-lived perennial native to southeastern Europe. It produces a rosette of leaves in the first season, followed by a single stem of yellow flowers in the second season. Woad was extensively used as a pigment for paints and textile dyes in Western Europe during Roman times. It was used as body paint in the British Isles to scare away enemies. Roman soldiers called these people "picts," Celtic for "painted." The Saxon green garments worn by Robin Hood and his men were made by first dyeing the cloth in woad, then in deep yellow weld (*Reseda luteola* L.). Woad production became so economically important during the 1200s that the government controlled its production to protect their profits.

Woad production was an elaborate process. The Papyrus Graecus Holmiensis (Stockholm Papyrus, c. 300–400 CE) provides the following protocol:

> Cut off the woad and put together in a basket in the shade. Crush and pulverize, and leave it a whole day. Air thoroughly on the following day and trample about in it so that by the motion of the feet it is turned up and uniformly dried. Put together in baskets lay it aside. ... Put about a talent of woad in a tube, which stands in the sun and contains not less than 15 metretes [37.4 L], and pack it in well. Then pour urine in until the liquid rises over the woad and let it be warmed by the sun, but on the following day get the woad ready in a way so that you (can) tread around in it in the sun until it becomes well moistened. One must do this, however for 3 days together. (*Caley 1927*)

Figure 7.14 Illustration of German woad mill in Thuringia from Daniel Gottfried Schreber's book on woad.

In a somewhat more modern process (c. 1200), leaves were picked, crushed with wooden rollers (Figure 7.14), and formed into 3-inch diameter balls. The balls were dried and stored until use. To use the dye, the woad balls were ground into powder by rollers and piled into layers in a "couching house." Water was added to the layers, and they were left to ferment for two weeks, producing a foul odor. When finished, the dye is a yellowish green color, but turns blue with exposure to oxygen. Queen Elizabeth I banned woad production within five miles of her residence on account of the odor. The Skirlbeck Mill in Lincolnshire, England, produced the last commercial batch of Woad in 1932.

Woad production in Europe ended largely in response to the introduction of indigo from the subtropical shrub *Indigofera tinctoria*. It is native to the Malay archipelago and was used as a blue fabric dye and paint in Asia since 3000 BCE. It is produced in a similar manner as woad: leaves are crushed and fermented to produce a yellowish-green dye that turns blue with oxidation. The wet dyestuff was often dried in slabs and cut into "cakes" for ease of commerce (Figure 7.15). Among all the natural dyes, indigo had the greatest influence on global society. India was the center of indigo production for

Figure 7.15 Indigo dye cake.
© 2006 Evan Izer.

centuries; they controlled indigo production just as Europe controlled woad. Europeans feared that introduction of indigo would reduce woad profits, and so many countries passed laws banning its import. In France and Germany, dyers had to take an oath never to use indigo under punishment of death. An international woad producers union was formed called the Woadites. Their main purpose was to fight importation of indigo, the "devil's dye." But by the 1500s, Armenian and Portuguese traders brought large quantities of indigo dyestuff to Europe from trading posts in India, followed by the Dutch and English. With a sustainable supply of affordable indigo, Europe dropped their opposition to the "devil's dye." Gradually, indigo replaced woad as the preferred blue fabric dye across Europe.

The Spanish decided to produce indigo in their newly acquired territories in the New World. Starting in the mid 1500s, plantations were established in Central America, Mexico, Florida, and eventually Hispaniola and Louisiana. Other colonial powers established indigo plantations in the West Indies. Guatemalan and French West Indies indigo were considered the highest quality for New World indigo. However, New World indigo was still considered inferior to indigo from India. This may have been a result economically motivated adulteration on the part of the colonists, adding sand to the cakes or using the whole plant instead of just the leaves. The earliest plantation workers were enslaved Native Americans, but they quickly became ill from working in the indigo processing plants. The Spanish and other colonial powers imported slaves from Africa, perhaps because they were considered less susceptible to the ills of the Native Americans or perhaps because African slaves already had extensive knowledge of indigo cultivation and dye production from Africa. In Western Africa (Figure 7.16), women had attained great power and wealth through their expertise in indigo production.

Thanks to African slaves, indigo profits in the New World outpaced sugar and cotton by the 1700s. Eliza Lucas Pinckney brought indigo production to South Carolina in 1739. When she was 16, her father, a British army officer, was called to be the governor of Antigua. He left Eliza in charge of his South Carolina plantations while he was gone. She experimented with growing several crops without success until her father sent back some indigo seeds. Her slaves were already skilled at indigo cultivation and dye production, leading to great success. She encouraged other plantation owners in South Carolina to grow indigo.

During the time leading up to the American Revolution, the American dollar had no value, and so indigo cakes were used as currency. Indigo became the de facto color of the "blue coat" uniforms since it was the only commercial dye produced by the Colonies. Indigo was used to dye the star field on the original American flag. Before the war, South Carolina was exporting 1.1 million pounds of indigo to Europe. At that time, the American colonies, West Indies, and South and Central American indigo plantations were the major suppliers of indigo to Europe. They were also outpacing India in the development of modern indigo production methods.

Nevertheless, New World indigo production could not keep pace with European demand. The English East India Company began establishing new large-scale

Figure 7.16 A girl in Tanzania wearing a traditional indigo-dyed garment over top of indigo-dyed denim © 2008 Ilya Raskin.

plantations in Bengal, India, based in part on the methods developed in the New World. By the beginning of the nineteenth century, India was the largest exporter of indigo in the world. European demand for indigo put great pressure on India to maintain a constant supply. Farmers were required to produce indigo for England under threat of torture, displacing subsistence food crops required for their survival. In response to continued oppression, farmers lead a two-year movement called the Indigo Revolt of 1859.

All the controversy and competition surrounding indigo would soon enter a different realm when Alfred Von Baeyer began efforts to synthesize indigo in 1870. He was unsuccessful at creating an economically feasible reaction, but by 1897, two German companies, Hoechst and BASF, finally developed a cheaper synthetic substitute to natural indigo. This was a major threat to the natural indigo industry, especially in British-controlled India. To remain competitive on the global market, Indian researchers attempted to increase the yield of indigo in *Indigofera tinctoria*. They began trials using *Indigofera arrecta* cultivars from around the world, but were largely unsuccessful. The industry experienced a major routing by synthetic indigo from 1909 to 1914. The demand for natural indigo rebounded during World War I when the supply of synthetic indigo was cut off from Germany. But most farmers had given up on indigo at that point and were reluctant to return to a dyeing industry. The final nail in the coffin was when Britain obtained the secret chemical process for synthetic indigo after the war and annihilated any remaining interest in natural indigo. Today, nearly all indigo comes from the synthetic process.

Phytochemistry

In the process of drying and fermenting both indigo and woad leaves, oxidation and β-glucosidase activity of anaerobic bacteria convert indole alkaloids isatan A and isatan B from the plant material to indican (Oberthür *et al.* 2004) (Figure 7.17A–C). Indican is converted to indoxyl, which is then converted to indigotine, otherwise known as indigo. Indigo is not water soluble, but it can be reduced to leucoindigo, or "white indigo," a water-soluble molecule (Figure 7.17D–E). Traditionally, this was achieved through the addition of urine to the dye extraction process. After leucoindigo is deposited on the fibers, it can be oxidized back to indigo during the drying process, turning blue and becoming water insoluble. Indigo is a substantive or direct vat dye; the molecules are attracted to fabric fibers by hydrogen bonding and do not require a mordant to set the dye. The traditional dyeing process is simple; fabric is soaked in a dye bath containing woad or indigo fermentation liquid with an ideal concentration around 0.02g/L solids (the plant material is typically removed before dyeing). The fabric is then air-dried, which serves as a second oxidation process to deposit indigo pigment on the surface of the fibers.

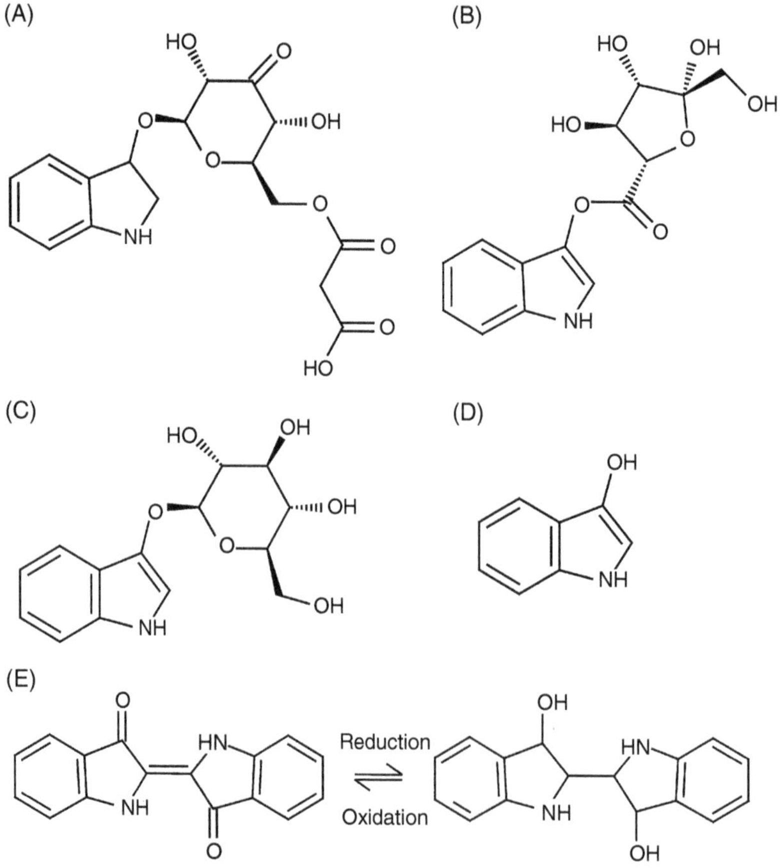

Figure 7.17 Chemical constituents of indigo dyes: (A) isatan A, (B) isatan B, (C) indican, (D) indoxyl, and (E) the reduction of indigo to leucoindigo and oxidation back to indigo.

Native Americans and African slaves both experienced morbidity and mortality related to their working conditions in the indigo industry. According to James Roberts (1858), a soldier under General Washington:

> From fifty to sixty hands work in the indigo factory; and such is the effect of the indigo upon the lungs of the laborers, that they never live over seven years.

Workers came into direct contact with the dye vats, often standing inside them to stir the mix with poles. They hand-spooned the indigo into bags to drain over night. The next day, they packed the blue sludge into square molds with drainage holes and cut the indigo into cakes once it was dry (Figure 7.15). In dye houses where textiles were dyed with indigo, the indigo cake was mixed with urine, tannic acid, or wood ash to produce water-soluble leucoindigo. Anecdotal reports claim indigo workers worldwide had an increased susceptibility to cancer and tuberculosis. There may be some scientific evidence for these claims. Indigo and other related bis-indole alkaloid indigoids have demonstrated mutagenic activity (Calvo *et al.* 2011), may elicit toxic effects of other compounds, or may be non-genotoxic carcinogens (Rannug *et al.* 1992). Indospicine and 3-nitropropionic acid, present in several *Indigofera* spp., have neurotoxic properties, occasionally causing death in livestock and humans (Gracie *et al.* 2010; Labib *et al.* 2012; Lima *et al.* 2012). Most acute and chronic toxicity studies in rodents use purified indigo (as opposed to a fermented indigo crude extract) and conclude that indigo is only mildly toxic, even when ingested at high doses (Ferber 1987). Unfortunately, there are no comprehensive studies identifying the large range of compounds that would have been produced in the indigo fermentation vats. One study characterizing the bacterial community in indigo fermentation (Aino *et al.* 2010) identified numerous species including *Bacillus*, *Clostridium*, and *Halomonas*. So, although it is plausible that compounds present in the fermentation vats were toxic, especially after prolonged exposure, scientific evidence is lacking.

Indigo synthesis is a far cry from natural indigo vat fermentation, using petroleum-derived starting materials instead of plant material. Adolf Von Baeyer's first indigo synthesis was a seven-step process starting with phosphorous pentachloride and other reactants to yield isatin chloride. Isatin chloride was then converted to indigo using zinc. His second attempt started from *o*-nitrobenzaldehyde and acetone and became known as the Baeyer-Drewden indigo synthesis reaction. But *o*-nitrobenzaldehyde was not readily available on a commercial scale. Zurich Professor Karl Heumann (1850–1894) was able to obtain indigo from phenyl-glycine (from aniline and chloroacetic acid) in an alkali melt (Schaefer 2014). Where Baeyer was attempting to close the indole ring with a C-N bond, Heumann closed the ring with a C-C bond, which happens to correspond to the biosynthetic pathway (Schaefer 2014). The reaction gave only a 10% yield, so Heumann came up with a higher-yielding second indigo synthesis using anthranilic acid instead of aniline. Meanwhile, Hoechst and BASF were developing their own efficient process. During their experiments, a careless BASF employee broke a mercury thermometer, serendipitously creating an experiment that provided the highest yields yet. In 1897, after 17 years of research, BASF brought the first synthetic indigo to market (Figure 7.18). To remain competitive, Hoechst joined forces with Degusa, forming the company "Indigo GmbH." Five years after BASF, they brought their own synthetic indigo to market with an improved production process.

Figure 7.18 BASF synthetic indigo production plant, 1890.

Modern Uses

Indigo dye is no longer commercially produced from plants using traditional methods. But synthetic versions still play a powerful role in modern society and the global economy. Indigo is the currently the most important textile dye by volume (Clark 2011). Dystar (formerly part of BASF), the largest producer of synthetic indigo, manufactures more than 30,000 tons per year, with global indigo production around 80,000 tons in 2010 (Ghaly *et al.* 2014). Many people are familiar with indigo-dyed jeans or denim fabric. Both terms originate in Europe; jeans from fabric produced in Genoa, Italy (in French, *Gênes*), and denim likely from a fabric produced in Nimes, France (*de Nimes*). The term "dunga-ree" came from work clothes worn in the Indian town of Dongri, in Hindi *dungri*. Denim jeans became popular in the United States after Levi Strauss and Jacob Davis produced riveted jeans for miners during the San Francisco Gold Rush in 1873. One pair of denim blue jeans requires about 10 grams of indigo. Only the warp threads of denim fabric are dyed, with the weft threads remaining white. The warp threads make up the surface of denim fabric and eventually fade to lighter blue with washing and abrasion. Jean fabric was traditionally made with colored warp and weft threads. The first pairs of "jeans" (orig-inally called "waist overalls") made by Levi Strauss were constructed of denim fabric man-ufactured by the Amoskeag Manufacturing Company in Manchester, New Hampshire (Downey 2014). The term "jeans" eventually replaced the term "overalls" after denim pants became widely popular in the United States after WWII (Downey 2014).

References

Aino, K., Narihiro, T., Minamida, K., Kamagata, Y., Yoshimune, K., and Yumoto, I. (2010) Bacterial community characterization and dynamics of indigo fermentation. *FEMS Microbiology Ecology*, **74** (1) 174–183.

Caley, E.R. (1927) The Stockholm Papyrus. An English translation with brief notes. *Journal of Chemical Education*, **4** (8), 979–1002.

Calvo, T.R., Cardoso, C.R., da Silva Moura, A.C. *et al.* (2011) Mutagenic activity of *Indigofera truxillensis* and *I. suffruticosa* aerial parts. *Evidence Based Complement Alternative Medicine*. Available from: http://www.hindawi.com/journals/ecam/2011/323276/ (accessed 8 January 2016).

Clark, M. (2011) *Handbook of Textile and Industrial Dyeing: Principles, Processes and Types of Dyes*. Elsevier, Philadelphia.

Downey, L. (2014) *A Short History of Denim*. Levi Straus & Co., San Francisco.

Ferber, K.H. (1987) Toxicology of indigo: A review. *Journal of Environmental Pathology, Toxicology and Oncology*, **7** (4), 73–83.

Ghaly, A.E., Ananthashankar, R., Alhattab, M., and Ramakrishnan, V.V. (2014) Production, characterization and treatment of textile effluents: A critical review. *Journal of Chemical Engineering & Process Technology*, **5**, 182.

Gracie, A., Brown, A., and Saville, P. (2010) *Birdsville disease (Birdsville indigo – Indigofera linnaei)*, serial no. 633, Agdex No. 657, Australia Northern Territory Government, Darwin.

Jansen, P.C.M., and Cardon, D. (eds) (2005) *Plant Resources of Tropical Africa 3. Dyes and Tannins*. PROTA Foundation, Wageningen, Backhuys Publishers, Leiden, CTA, Wageningen.

Labib, S., Berdai, M.A., Bendadi, A., Achour, S., and Harandou, M. (2012) Fatal poisoning due to *Indigofera*. *Archives de pédiatrie*, **19** (1), 59–61.

Lima, E.F., Riet-Correa, F., Gardner, D.R. *et al.* (2012) Poisoning by *Indigofera lespedezioides* in horses. *Toxicon*, **60** (3), 324–328.

Oberthür, C., Schneider, B., Graf, H., and Hamburger, M. (2004) The elusive indigo precursors in woad (*Isatis tinctoria* L.) – identification of the major indigo precursor, isatan A, and a structure revision of isatan B. *Chemistry & Biodiversity*, **1** (1), 174–182.

Roberts, J. (1858) *The Narrative of James Roberts, a Soldier Under Gen. Washington in the Revolutionary War, and Under Gen. Jackson at the Battle of New Orleans, in the War of 1812: "A Battle Which Cost Me a Limb, Some Blood, and Almost My Life"* James Roberts, Chicago, 1858. Available at: http://docsouth.unc.edu/neh/roberts/roberts.html (8 January, 2016).

Rannug, U., Bramstedt, H., and Nilsson, U. (1992) The presence of genotoxic and bioactive components in indigo dyed fabrics – a possible health risk? *Mutation Research*, **282** (3), 219–225.

Schaefer, B. (2014) *Natural Products in the Chemical Industry*. Springer, New York.

Insecticides Based on Plant Essential Oils

(M. Isman)

History and Ethnobotany

The use of aromatic oils from plants by humans to ward off pests including blood-feeding ectoparasites can be traced back more than 6000 years to the ancient Egyptians. Distillation of plant essential oils (as practiced today) has been attributed to the Persian physician Avicenna, who lived approximately 1000 years ago. While there are many anecdotal reports of the "traditional" uses of plant oils, well-documented use of essential oils as insect repellents appears to date back to 1901, with the accidental discovery of citronella as a repellent from its intended use as a fragrance in hairdressing. However, the use of citronella to repel insects fell out of favor following the introduction of the synthetic repellent DEET (*N,N*-diethyl-*m*-toluamide) in 1956 (Peterson and Coats 2001).

The discovery of the insecticidal action of plant essential oils and their constituents is considerably more recent. Some of the earliest reports include those on the insecticidal action to the housefly (*Musca domestica*) of 1,8-cineole from rosemary oil (Jacobson and Haller 1947), *d*-limonene from orange oil (Sharma and Saxena 1974), thymol from thyme oil (Miki 1978), and eugenol from clove oil (Marcus and Lichtenstein 1979). Many of the earlier reports dealing with the effects on agricultural crop and stored product pests are reviewed by Regnault-Roger (1997). A recent bibliometric study revealed that more than 2200 papers documenting the effects of plant essential oils on insects and related pests were published between 1980 and 2011 (Isman and Grieneisen 2014). The majority of these papers describe plant essential oils widely used in commerce as flavorings and fragrances and/or describe aromatic plants grown extensively in many geographic regions. Among the better-known commodity essential oils investigated and used as insecticides and repellents are those from rosemary (*Rosmarinus officinalis* L., Lamiaceae), peppermint (*Mentha* × *piperita* L., Lamiaceae), cinnamon (*Cinnamomum verum* J. Presi, Lauraceae), clove (*Syzygium aromaticum* L., Myrtaceae), lemongrass (*Cymbopogon* spp., Poaceae), thyme (*Thymus vulgaris* L., Lamiaceae), orange [*Citrus sinensis* (L.) Osbeck; Rutaceae], and eucalyptus (*Eucalyptus* spp., Myrtaceae) (Figure 7.19A–D).

Chemistry and Bioactivity

Plant essential oils obtained through hydrodistillation are often complex mixtures of monoterpenoid alcohols, aldehydes and esters, sesquiterpenes, and biogenically related phenols and phenylpropenes (Figure 7.20). It is not uncommon for a plant essential oil to comprise 50 or more such compounds; in some cases one or a small number of constituents dominate by weight, while in others no single constituent will make up more than 10–20% by weight. There are, however, certain essential oils in which one major compound constitutes more than 90% of the natural composition (e.g., eugenol in clove bud oil). Beyond the complexity of plant essential oils is their natural chemical variation arising from genetic, geographic, seasonal, and climatic influences (Hussain *et al.* 2008; Razafimamonjison *et al.* 2014; Salido *et al.* 2003). Chemical composition can even vary diurnally (Kpoviessi *et al.* 2012).

Plant essential oil terpenoids can have acute insecticidal activities and sublethal behavioral effects; the latter especially as repellents and deterrents. As insecticides, their rapid action points to the nervous system as the proximate site of action, although there appear to be multiple sites and mechanisms of action including antagonism of the invertebrate neuromodulator, octopamine (Enan 2001; Gross *et al.* 2014), and agonism of γ-aminobutyric acid-gated chloride channels in nerve axons (Priestley *et al.* 2003). One particularly intriguing aspect of the insecticidal nature of plant essential oils is that for some, toxicity cannot be readily attributed to a single or major constituent, but instead results from synergistic interactions between constituents, some of which individually appear inactive (Miresmailli *et al.* 2006; Tak *et al.* 2016). Another

Figure 7.19 Members of Lamiaceae used as essential oil insecticides include (A) peppermint; (B) rosemary; (C) sage; and (D) thyme.

interesting aspect is the idiosyncratic differences in toxicity of specific essential oil terpenoids between species of insects, making predictions of bioactivity or efficacy on pests risky in the absence of empirical data.

Apart from the long-established use of citronella oil as an insect repellent, the use of essential oils as insecticides in the United States dates back to the late 1990s when several products based on rosemary, peppermint, clove, cinnamon, and thyme oils were introduced for professional (urban) pest control. Introduction of these products into the marketplace was facilitated by their exemption from registration by the Environmental Protection Agency (EPA). This relieved the manufacturer from the need for the expensive and time-consuming toxicological and environmental data normally

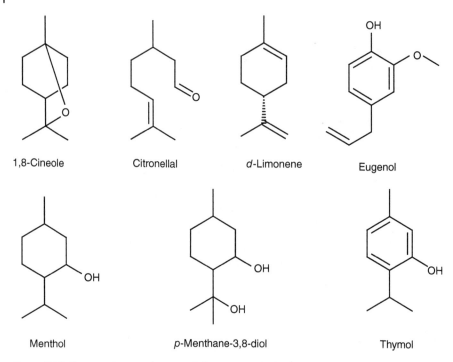

1,8-Cineole Citronellal *d*-Limonene Eugenol

Menthol *p*-Menthane-3,8-diol Thymol

Figure 7.20 Some major constituents of plant essential oils with insecticidal activity.

needed to support registration of a pesticide. The exemption was based on the long-standing use of these plants as culinary herbs and spices and their oils as flavorings and fragrances in foods, beverages, and cosmetics. Plant essential-oil-based insecticides are currently used in agriculture, especially on protected (greenhouse) crops and in organic food production, in animal health (primarily for flea and tick control on companion animals), for professional pest control (e.g., for cockroaches, ants, flies) in restaurants, schools, and warehouses, and as consumer products for home and garden use (Isman *et al.* 2011). Closely related products based on common essential oils are also used as fungicides (for control of plant disease) and as herbicides. An insecticide based on orange oil (containing mostly *d*-limonene) is sold in Europe, whereas insecticides containing 1,8-cineole (= eucalyptol) from *Eucalyptus* are sold in Australia, China, and India.

References

Enan, E. (2001) Insecticidal activity of essential oils: Octopaminergic sites of action. *Comparative Biochemistry and Physiology Part C*, **130** (3), 325–337.

Gross, A., Kimber, M.J., Day, T.A., Ribeiro, P., and Coats, J.R. (2014) Investigating the effect of plant essential oils against the American cockroach octopamine receptor (Pa oa1) expressed in yeast. *American Chemical Society Symposium Series*, **1172**, 113–130.

Hussain, A.I., Anwar, F., Hussain Sherazi, S.T., and Przybylski, R. (2008) Chemical composition, antioxidant and antimicrobial activities of basil (*Ocimum basilicum*) essential oil depends on seasonal variations. *Food Chemistry*, **108**, 986–995.

Isman, M.B., and Grieneisen, M.L. (2014) Botanical insecticide research: Many publications, limited useful data. *Trends in Plant Science*, **19** (3), 140–145.

Isman, M.B., Miresmailli, S., and Machial, C. (2011) Commercial opportunities for pesticides based on plant essential oils in agriculture, industry and consumer products. *Phytochemistry Reviews*, **10**, 197–204.

Jacobson, M., and Haller, H.L. (1947) The insecticidal component of *Eugenia haitiensis* identified as 1,8-cineol. *Journal of the American Chemical Society*, **69** (3), 709–710.

Kpoviessi, B.G.H.K., Ladekan, E.Y., Kpoviessi, D.S.S. *et al.* (2012) Chemical variation of essential oil constituents of *Ocimum gratissimum* L. from Benin, and impact on antimicrobial properties and toxicity against *Artemia salina* Leach. *Chemistry and Biodiversity*, **9**, 139–150.

Marcus, C., and Lichtenstein, E.P. (1979) Biologically active components of anise: Toxicity and interactions with insecticides in insects. *Journal of Agricultural and Food Chemistry*, **27**(6), 1217–1223.

Miki, K. (1978) *Insecticide against Maggots.* Japan patent 53–066420.

Miresmailli, S., Bradbury, R., and Isman, M.B. (2006) Comparative toxicity of *Rosmarinus officinalis* L. essential oil and blends of its major constituents against *Tetranychus urticae* Koch (Acari: Tetranychidae) on two different host plants. *Pest Management Science*, **62**, 366–371.

Peterson, C., and Coats, J. (2001) Insect repellents – past, present and future. *Pesticide Outlook*, **12** (4), 154–158.

Priestley, C.M., Williamson, E.M., Wafford, K.A., and Sattelle, D.B. (2003) Thymol, a constituent of thyme essential oil, is a positive allosteric modulator of human $GABA_A$ receptors and a homo-oligomeric GABA receptor from *Drosophila melanogaster*. *British Journal of Pharmacology*, **140** (8), 1363–1372.

Razafimamonjison, G., Jahiel, M., Duclos, T., Ramanoelina, P., Fawbush, F., and Danthu, P. (2014) Bud, leaf and stem essential oil composition of *Syzygium aromaticum* from Madagascar, Indonesia and Zanzibar. *International Journal of Basic and Applied Sciences*, **3**, 224–233.

Regnault-Roger, C. (1997) The potential of botanical essential oils for insect pest control. *Integrated Pest Management Reviews*, **2**, 1–10.

Salido, S., Altarejos, J., Nogueras, M., Saanchez, A., and Luque, P. (2003) Chemical composition and seasonal variations of rosemary oil from southern Spain. *Journal of Essential Oil Research*, **15**, 10–14.

Sharma, R.N., and Saxena, K.N. (1974) Orientation and developmental inhibition in the housefly by certain terpenoids. *Journal of Medical Entomology*, **11** (5), 617–621.

Tak, J.-H., Jovel, E., and Isman, M.B. (2016) Comparative and synergistic activity of *Rosmarinus officinalis* L. essential oil constituents against the larvae and an ovarian cell line of the cabbage looper, *Trichoplusia ni* (Lepidoptera: Noctuidae). *Pest Management Science*, **72**, 474–480.

8

Oceania

B. M. Schmidt

© 2012 Daniel Ramirez

Ethnobotany: A Phytochemical Perspective, First Edition. Edited by B. M. Schmidt and D. M. Klaser Cheng.
© 2017 John Wiley & Sons Ltd. Published 2017 by John Wiley & Sons Ltd.

Introduction

Oceania is a series of islands in the tropical Pacific Ocean including Melanesia, Micronesia, and Polynesia (Figure 8.1). This chapter will also include the ethnobotany of Australasia (Australia, New Zealand, and New Guinea). There is little documentation of ethnobotany until the arrival of European explorers. Humans likely have inhabited Australasia and western Melanesia for the past 40,000 years. New DNA evidence suggests people in remote eastern Oceania (including the Solomon Islands, Vanuatu, New Caledonia, Fiji, Micronesia, and Polynesia) are descended from the Lapita people who left Taiwan c. 1400 BCE, traveling by boat to Australasia, the Malay Archipelago, and eventually northward to Polynesia c. 900 BCE. The Lapitas and Melanesian peoples may both be common ancestors of the Polynesians, based on biological, archaeological, and linguistic evidence. Easter Island and the Hawaiian Islands were the last to be settled, but the time period is uncertain. Some archeological evidence suggests Polynesians arrived between 300–400 CE; however, recent radiocarbon dating of artifacts seems to suggest Easter Island was not inhabited until 1200 CE (Hunt and Lipo 2006). There is a possibility that Polynesians reached the coast of South America, bringing back New World crops such as coconut (*Cocos nucifera* L.) and sweet potato [*Ipomoea batatas* (L.) Lam.] before the arrival of European explorers.

As a result of Oceania's remote location, native inhabitants did not encounter Europeans until the sixteenth century or later. Ferñao de Magalhães (Ferdinand Magellan) and Juan Sebastian Elcano encountered the Chamorro people in the Marianas during their trans-Pacific voyage in 1520–1521. During this same voyage, native Filipino Lapu-Lapu and his tribe from the Philippine island of Mactan killed Magellan during a battle. Dutchman Willem Janszoon was the first European to visit Australia and meet with the Aboriginal people in 1606. As an employee of Dutch East India Company, his intention was to explore the west coast of New Guinea for economic opportunities, but he sailed farther south to Cape York Peninsula and Cape Keerweer in Queensland. First contact with the east coast of Australia happened much later, when Captain James Cook landed near Bawley Point and explored Botany Bay in 1770. Cook was also the first European to circumnavigate New Zealand and made first European contact with New Caledonia. In 1778, Cook landed in Waimea, Kaua'i, making first contact with indigenous Hawaiian tribes. Reports of cannibalism[1] and warfare between tribes deterred European contact and colonization in Fiji and several other islands, for good reason. On Cook's second Pacific voyage, ten crew members were attacked, killed, and reportedly cannibalized by Maori tribesmen in New Zealand. When Captain Cook returned to Hawaii in 1779, he was killed in battle and by some accounts, cannibalized, since only parts of his body were returned to the crew.

The islands of Oceania contain many different types of ecosystems from saltwater sea grass beds, coastal strands, and mangrove forests to grasslands, freshwater wetlands, and upland rainforests. Plants that arrived on the islands without human intervention are considered native. Table 8.1 lists several traditional native crops of Oceania.

Macadamia nuts are one of the few crops native to Australia to gain commercial importance; Eucalyptus trees (Figure 8.2) are another. *Eucalyptus* spp. produce cineole-rich essential oils that are used for relief of cold and flu symptoms. The origin of coconut (*Cocos nucifera* L.) is uncertain. While it may have originated in Oceania, others speculate it originated in South America and was brought to Oceania by humans or

---

[1] It is possible some accounts of cannibalism in Oceania were exaggerated by Europeans to make the native people seem more savage, thereby justifying their exploitation.

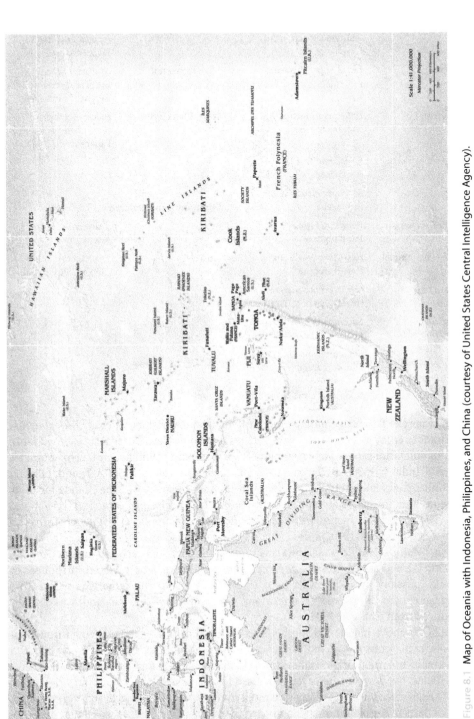

Figure 8.1 Map of Oceania with Indonesia, Philippines, and China (courtesy of United States Central Intelligence Agency).

Table 8.1 Crops that originated in Oceania.

Fruits and vegetables		Grains, legumes, and nuts	
Arrowroot	*Tacca leontopetaloides* (L.) Kuntze, Dioscoreaceae	Wingbean	*Psophocarpus tetragonolobus* (L.) DC., Fabaceae
Banana	*Musa* spp., Musaceae	Macadamia nut	*Macadamia integrifolia* Maiden & Betche, Proteaceae
Breadfruit	*Artocarpus altilis*, Moraceae	Polynesian chestnut	*Inocarpus fagifer* (Parkinson) Fosberg, Fabaceae
Coconut	*Cocos nucifera* L., Arecaceae		
Noni	*Morinda citrifolia* L., Rubiaceae	**Stimulants and sedatives**	
Pandan/screwpine	*Pandanus* spp., Pandanaceae	Betel	*Areca catechu* L., Arecaceae
Purple passionfruit	*Passiflora edulis* Sims, Passifloraceae	Kava	*Piper methysticum* G.Forst., Piperaceae
Swamp taro	*Cyrtosperma merkusii* (Hassk.) Schott, Araceae		
Sugarcane	*Saccharum officinarum*, Poaceae		
Yams	*Dioscorea* spp. Dioscoreaceae		

natural causes. The progenitor to modern sugarcane varieties, *Saccharum robustum* E.W.Brandes & Jeswiet ex Grassl, probably originated in New Guinea, spreading east to the Solomon Islands, New Hebrides, and New Caledonia around 8000 BCE; to Southeast Asia and India by 6000 BCE; and to Fiji and Hawaii between 500 CE and 110 CE. Early inhabitants undoubtedly introduced a number of species to the islands for food, medicine, building materials, and so on, including taro [*Colocasia esculenta* (L.) Schott], yams (*Dioscorea* spp.), mango (*Mangifera indica* L.), sweet potato [*Ipomoea batatas* (L.) Lam.], rose apple [*Syzygium jambos* (L.) Alston], sea hibiscus (*Hibiscus tilliaceus* L.), and cassava (*Manihot esculenta* Crantz). Guava (*Psidium guajava* L.), soursop (*Annona muricata* L.), and papaya (*Carica papaya* L.) are New World tropical fruits introduced to Oceania, possibly by indigenous people making trips to the Americas well before European contact. Kiwifruit, an important, iconic crop in New Zealand, was actually introduced from China in 1904.

Five of the above-mentioned species have been essential to everyday living throughout Oceania: noni, screwpine, breadfruit, coconut, and sea hibiscus. Noni (*Morinda citrifolia* L.) is a multi-use plant for medicine, famine food, dye, building materials, and animal feed. In medicine, the flowers and leaves were used to relieve stomach pain. Fruits (Figure 8.3) were used to clear sinuses, and in more modern times, to treat diabetes and hypertension. Various parts of the plant have been used to treat gonorrhea, blisters, swellings, diarrhea,

Figure 8.2 Rainbow eucalyptus (*Eucalyptus deglupta* Blume), native to New Guinea.

and colic. The inner bark produces a yellow dye that can be made reddish by adding lime. The roots can be ground to make a substitute for turmeric in cooking.

Pandanus spp. are some of the most useful plants on the island atolls. Leaves were used as breadfruit wrappers, sails, roof thatching stitching, and for making mats, baskets, brace-lets, and hats. The fruits (Figure 8.4) are edible (some species taste better than others); prop roots were used to make lobster and fish traps; tree trunks were split and used for roof beams; root fibers were used for cordage; and fruit fibers were used for dental floss.

Breadfruit (*Artocarpus altilis*, Figure 8.5) is a staple food steamed, fried, or baked like a potato before eating. It can also be crushed into a stiff pulp that is formed into loaves and covered with leaf wrappers. Whole or mashed fruit can be stored in leaf-lined pits where it ferments, imparting a distinctive aroma and flavor. White breadfruit tree sap (latex) was used for glue and to seal canoes. *Calophyllum inophyllum* L. sap was used in a similar fashion.

Coconut palms (Figure 8.6) grow wild and as cultivated varieties. Fresh or dried coco-nut meat (copra) is consumed as a staple food. Coconut water is consumed fresh or

Figure 8.3 Noni fruit (*Morinda citrifolia*).

Figure 8.4 Pandan or screwpine fruit (*Pandanus* sp.) growing in Moreton Bay, Australia. © 2005 J. Brew.

fermented into palm toddy or palm wine. The leaves are used as fish sweeps and floor coverings, and woven into mats, baskets, and fans. The husk fibers are used as cordage and nets. Coconut shells have multiple uses as vessels, jewelry, handicrafts, and adornments. The wood is used to make various types of spears and building materials. Coconut oil was a basic ingredient in cosmetics and personal care products, much as it is in modern times. Given the aforementioned uses, some people believed coconut to be the perfect plant, a gift from the gods providing complete nutrition and other essentials for living. German August Engelhardt famously took this belief to an extreme. He relocated to the island of Kabakon in 1902 and began a coconut cult called *Sonnenorden*. Several cult members caught malaria and died. Engelhardt became emaciated (presumably from malnutrition and disease) and was found dead on a Kabakon beach in 1919 (Martyris 2015).

Hibiscus tilliaceus L. has light wood used to build canoe parts (paddles, bailers, etc.), fishing equipment, and breadfruit harvesting poles. Wood bast fibers were used to make traditional clothing and can be woven into mosquito netting. Leaves are mixed with coconut juice and drunk out of a coconut cup during pregnancy to facilitate childbirth. Flowers were rubbed on skin rashes and for treating eye disorders.

Fishing is important to the survival of costal communities, and so it is not surprising that people discovered plants which could kill or stun fish. *Derris elliptica* (Wall.) Benth.

Figure 8.5 Breadfruit (*Artocarpus altilis*).

may be the best known of all fish poisons. The crushed roots contain rotenone, a compound that effectively kills both fish and insects. When the crushed roots are thrown into the water, fish become stunned and float to the top, making them easy to catch. The poison apparently does not pose a hazard to humans consuming the fish, although it is highly toxic when consumed directly. Crushed seeds and other parts of the mangrove tree *Barringtonia asiatica* (L.) Kurz work in much the same way. The active compounds are thought to be saponins. *Cerbera manghas* L. seeds contain the cardiac glycoside cerberin. They are used to stun fish as well as to treat cancer.

Native inhabitants of Oceania use hundreds of medicinal plants for a range of disorders. Mental illness, birth defects, and neurodevelopmental disorders were common in isolated island communities. Historically, malevolent spirits or curses were believed to be the cause, but herbs were still the preferred treatment. For example, *Vigna marina* (Burm.) Merr. leaves were applied externally to calm patients with acute mental symptoms (e.g., shouting, talking nonstop, and insomnia) in Samoa. In the Marshall Islands and Pohnpei, *Cassytha filiformis* was mixed with water and used as a shower for mental illness. Skin disorders also occurred regularly, as the number of herbal remedies for skin disorders in Oceania pharmacopeias often outnumbers remedies for other medical problems. In Pohnpei, people would put akelel (*Rhizophora mucronata* Lam.) leaves and water in a tree cavity to ferment, and then drink the water to rejuvenate the skin. Longrunner [*Rorippa sarmentosa* (Soland. ex Forst.) J.F. Macbr.] and malbau (*Premna serratifolia* L.) bark and/or leaves were used to heal wounds and carbuncles in Samoa. Malbau leaves were also used to treat ringworm and baby heat rash. Native to Australia, *Melastoma malabathricum* L. fruits and flowers were used to treat shingles in Pohnpei and skin wounds across the region. People frequently experienced a scaly rash from

Figure 8.6 Coconut palm (*Cocos nucifera*) on the island of Hawaii.

overconsumption of kava [see kava case study in the section titled "Pharmacological Effects of Kavalactones from Kava (*Piper methysticum*) Root" in this chapter], for which many herbal remedies were developed.

Several native plants produce dyes used for textiles, body paint, and tattoos. Many of the same plants were also used for personal care (i.e., dental care, skincare, and perfumes). Candlenut [*Aleurites moluccanus* (L.) Willd.] nut oil was used for skincare, and wood ash was used to make tattoos. Soot from boiled breadfruit sap was also used to make tattoos. Tamanu oil from *Calophyllum inophyllum* nuts is another oil commonly used for skincare. *Pittosporum ferrugineum* W.T.Aiton bark was used for dye, and the sap was used for perfume. Leaves from the weedy primrose-willow *Ludwigia octovalvis* (Jacq.) P.H.Raven were used as a black textile dye fixative. *Curcuma australasica* Hook.f. roots were used to make a yellow dye for face and body paint. Velvetleaf (*Tournefortia argentea* L. f.) leaves were used in steam baths and to cover bruises. Beach cabbage [*Scaevola taccada* (Gaertn.) Roxb.] flowers have been used as a perfume, and the leaves produce a greenish yellow fabric dye. Woody stems from *Triumfetta procumbens* G. Forst. were used as toothbrushes and *Pandanus tectorius* Parkinson ex Du Roi fruit was used as dental floss.

References and Additional Reading

Balick, M.J. (2009) *Ethnobotany of Pohnpei: Plants, People, and Island Culture.* University of Hawaii Press, Honolulu.

Denham, T.P., Haberle, S.G., Lentfer, C. *et al.* (2003) Origins of agriculture at Kuk Swamp in the highlands of New Guinea. *Science,* **301** (5630), 189–193.

Gibbons, A. (2001) The peopling of the Pacific. *Science,* **291** (5509), 1735–1737.

Gurib-Fakim, A. (2006) Medicinal plants: Traditions of yesterday and drugs of tomorrow. *Molecular Aspects of Medicine,* **27** (1), 1–93.

Hunt, T.L., and Lipo, C.P. (2006) Late colonization of Easter Island. *Science,* **311** (5767), 1603–1606.

Kirch, P.V., Hartshorn, A.S., Chadwick, O.A. *et al.* (2004) Environment, agriculture, and settlement patterns in a marginal Polynesian landscape. *Proceedings of the National Academy of Sciences,* **101** (26), 9936–9941.

Martyris, N. (2015) Death by coconut: A story of food obsession gone too far. *National Public Radio,* 4 December, http://www.npr.org/sections/thesalt/2015/12/03/457124796/death-by-coconut-a-story-of-food-obsession-gone-too-far (accessed 5 December, 2015).

Sauer, J.D. (1993) *Historical Geography of Crop Plants: A Select Roster,* 1st edition. CRC Press, Boca Raton.

University of Hawaii (2015) University of Hawaii Plants and People of Micronesia Database, http://manoa.hawaii.edu/botany/plants_of_micronesia/index.php/plants-and-their-uses (accessed July 17, 2015).

Banana (*Musa* spp.) as a Traditional Treatment for Diarrhea

Ethnobotany

Musa spp. (Musaceae) are herbaceous perennials native to Southeast Asia and Melanesia. No wild *Musa* spp. have been found in the Americas or Africa, but they are extensively cultivated throughout the tropics as bananas (sweet fruit varieties) and plantains (starchy fruit varieties). The succulent pseudostem is a cylinder of leaf-petiole sheaths rising up from a fleshy rhizome or corm. The glossy ovate leaves can reach 3.5 m in length. *Musa* spp. are monocarpic; they produce only one drooping inflorescence of green-whitish flowers. The plant dies after fruit are produced. The fruit come from an inferior ovary and are classified as berries. The exocarp is green and leathery, turning yellow when fully ripe. The mesocarp consists of white to yellowish orange pulp, with an inner epithelium covering the ovarian cavity making up the endocarp. Cultivated bananas are seedless due to their triploid state, but wild bananas have numerous small black seeds covered in a sticky sap (Figure 8.7A–C). A single banana is called a finger while a cluster of bananas is called a hand. A bunch of bananas consists of about 8 hands with 15 fingers each (Figure 8.8).

(A)

(B)

(C)

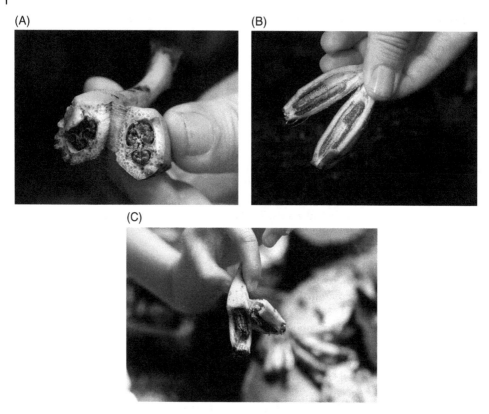

Figure 8.7 (A) Banana in cross section; and (B & C) longitudinal section showing the ovarian cavity with abundance of seeds and reduced mesocarp compared to cultivated varieties.

Nearly all edible bananas and plantains are diploid or triploid, derived from *Musa acuminata* Colla and/or *Musa balbisiana* Colla with parthenocarpic fruit (De Langhe *et al.* 2009). Parthenocarpy in bananas means they develop fleshy fruit without pollination, but they are not necessarily seedless (Kennedy 2009). The terms "banana" and "plantain" are often used interchangeably and do not represent different species. Earliest banana domestication likely occurred in New Guinea and the surrounding islands of Malesia (Kennedy 2009). Anthropogenic introduction of vegetatively propagated land races west to Africa and east to Oceania thousands of years ago led to a wide variety of diverse cultivars endemic to these regions. Complex hybridization produced many new varieties with seedless fruit. The introduction of bananas to the Americas remains vague. Some speculate they arrived with the African slave trade, but others suspect Polynesian explorers introduced their varieties to South America (Koeppel 2008; Langdon 1993). Today *Musa × paradisiaca* L. cvs. 'Gros Michel' and 'Cavendish' are two of the most popular dessert banana cultivars for worldwide consumption. 'Gros Michel', also called "Big Mike," was the dominant export banana to North America and Europe during the nineteenth century. It was grown in vast monocultures across Central and South America that easily succumbed to Panama disease (fusarium wilt, *Fusarium oxysporum*) in 1899, leading to a major banana shortage. By the 1950s, 'Gros

Figure 8.8 Bunch of wild bananas (*Musa* sp.)

Michel' was all but annihilated in the Western Hemisphere. In 1958, it was replaced by the smaller, less flavorful 'Cavendish.' However, 'Gros Michel' continued to be popular in Southeast Asia.

Almost every part of the banana plant has been used ethnobotanically. The leaves are used to line pit ovens and to wrap food for cooking, storage, and transport. The pseudostem core can be cooked and eaten, but it is also harvested for fibers such as abaca, machi, and basho, generating significant income for women living in rural areas. The terminal inflorescence is eaten cooked or raw. The fruit is a valuable food source, as it is high in vitamins and produces food reliably throughout the year. It is eaten cooked or raw and is often made into baby food. There are many different ethnomedical uses for *Musa* fruit (see Kennedy 2009 for a review). Both ripe and immature fruits, seed mucilage, flower, stem juice, and sap have been used to treat diarrhea by people in various regions including Nigeria, India, Bangladesh, Indonesia, Vietnam, Philippines, Papua New Guinea, and Minanao (Kennedy 2009; Abe and Ohtani 2013; Saha *et al.* 2013) (Figure 8.9). In countries where diarrhea causes significant morbidity and mortality, especially in children, effective management is critical. In many rural areas, medical facilities are limited or unavailable. Therefore, home treatment of diarrhea is common.

Figure 8.9 Wild banana is an important treatment for acute diarrhea in children throughout the tropics.

Phytochemistry and Bioactivity

There are numerous studies reporting the positive effects of consuming bananas for management of acute diarrhea. In a Bangladesh field trial, 2968 children ages 6–36 months old were randomly assigned to receive standard diet or standard diet plus green banana pulp when the children had diarrhea (Rabbani *et al.* 2010). Children with prolonged diarrhea receiving green banana had faster recovery rates. In 62 Bangladeshi boys ages 5–12 months with acute diarrhea, cooked green banana added to a rice diet significantly reduced amounts of stool, oral rehydration solution, intravenous fluid, and numbers of vomiting, and diarrheal duration compared to rice diet alone (Rabbani *et al.* 2001). In a hospital trial, 80 children ages 1–28 months with diarrhea persisting at least 14 days received either a yogurt-based or green-banana-supplemented diet (Alvarez-Acosta *et al.* 2009). The green banana group fared better with diminished stool output, weight, and diarrhea duration. They also gained more weight than with the yogurt-based diet.

Seed extract from *Musa × paradisiaca* delayed the onset of diarrhea in an experimental mouse model, decreased the volume of feces, and reduced gastrointestinal transit time (Hossain *et al.* 2011). It also had activity against several organisms known to cause infectious diarrhea including *Escherichia coli* and *Shigella dysenteriae* (Hossain *et al.* 2011). Venkatesh *et al.* (2013) found that ethanolic fruit extracts from *M. × paradisiaca* cv. 'Puttable' and *M. acuminata* cv. 'Grand Naine' inhibited a range of bacteria known to cause infectious diarrhea. This antibacterial activity may be especially important since antibiotic resistance is becoming more common. Shigellosis caused by *Shigella* spp. is characterized by clinical dysentery and is a significant cause of morbidity and mortality in impoverished regions. Kosek *et al.* (2010) found that green banana is an effective adjunctive therapy for shigellosis, reducing the duration of dysentery and persistent diarrhea and improving weight gain.

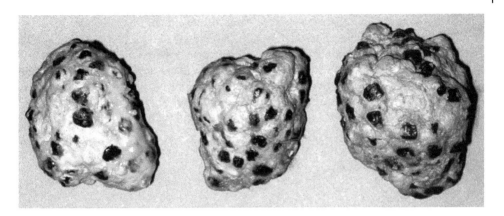

Figure 8.10 Wild banana phytobezoars (up to 5 cm diameter) extracted from the intestine of a Laotian man. Slesak *et al*. 2011. Reproduced with permission of John Wiley & Son, Ltd.

One important side effect worth noting is the possibility for wild bananas to form bowel obstructions called phytobezoars (Figure 8.10). In Laos, wild bananas are often eaten as famine food. One report found phytobezoars are most prevalent in young male Laotian laborers working in the fields before the rice harvest, a time of high food insecurity (Slesak *et al*. 2011). Eating the seeds and eating the wild bananas on an empty stomach increases the risk of forming phytobezoars (Schoeffl *et al*. 2004; Slesak *et al*. 2011). A study of neonates in Indonesia found that feeding banana only as a first solid food significantly increased the risk of small intestine obstruction compared to babies that received other solid food or liquid diet (Wiryo *et al*. 2003).

There appear to be several different compounds and mechanisms that play a role in reducing diarrhea. Green banana fruit has a high amylase-resistant starch content that is fermented into short-chain fatty acids in the colon, stimulating colonic salt and water absorption (Binder 2010; Rabbani *et al*. 2004). Soluble non-starch polysaccharides may prevent or improve diarrhea by inhibiting epithelial adhesion and translocation by pathogenic bacteria (Roberts *et al*. 2013). Hossain *et al*. (2011) speculated that flavonoids present in the seeds produce an antisecretory effect, perhaps by reducing the production of prostaglandins and nitric oxide, molecules that provoke net secretion of electrolytes. Flavonoids may also inhibit the growth and activity of pathogenic bacteria that cause diarrhea. Compounds in *Musa* × *paradisiaca* stem sap may act synergistically by binding to opioid receptors in the gastrointestinal mucosa; inhibiting intestinal motility and hydroelectrolytic secretion (flavonoids); denaturing proteins (tannins); and inhibiting histamine release (saponins) to reduce diarrhea (Yakubu *et al*. 2015).

Today, banana is widely used throughout the tropics as a cheap, effective way to reduce diarrhea in home and hospital settings. There are several dried banana flake products on the market for diarrhea control, some in combination with pectin, oligosaccharides, or other prebiotics. However, raw bananas are a widely available, effective, and economical choice. The World Health Organization (WHO) suggests eating bananas as a way to replenish lost potassium during acute diarrhea, especially in cases where oral rehydration salts are unavailable. Yet they do not formally recommend bananas to treat diarrhea. Moreover, bananas are not officially recommended for diarrhea reduction by the Center for Disease Control (CDC) or other North American health institutions, despite the numerous clinical trials showing positive effects on diarrhea duration and severity. Some point out that banana only reduced the duration and severity of diarrhea in clinical trials by a modest amount. Nevertheless, bananas are a nutritious and economical food, with the ability to replenish lost potassium and potentially reduce the overall morbidity of acute diarrhea.

References

Abe, R., and Ohtani, K. (2013) An ethnobotanical study of medicinal plants and traditional therapies on Batan Island, the Philippines. *Journal of Ethnopharmacology*, **145** (2), 554–565.

Alvarez-Acosta, T., León, C., Acosta-González, S. *et al.* (2009) Beneficial role of green plantain [*Musa paradisiaca*] in the management of persistent diarrhea: A prospective randomized trial. *Journal of the American College of Nutrition*, **28** (2), 169–176.

Binder, H.J. (2010) Role of colonic short-chain fatty acid transport in diarrhea. *Annual Review of Physiology*, **72**, 297–313.

De Langhe, E., Vrydaghs, L., Maret, P., Perrier, X., and Denham, T. (2009) Why bananas matter: an introduction to the history of banana domestication. *Ethnobotany Research & Applications*, **7**, 165–177.

Hossain, M.S., Alam, M.B., Asadujjaman, M. *et al.* (2011) Antidiarrheal, antioxidant and antimicrobial activities of the *Musa sapientum* seed. *Avicenna Journal of Medical Biotechnology*, **3** (2), 95–105.

Kennedy, J. (2009) Bananas and people in the homeland of genus *Musa*: Not just pretty fruit. *Ethnobotany Research & Applications* **7**, 179–197.

Koeppel, D. (2008) *Banana. The Fate of the Fruit that Changed the World*. Hudson Street (Penguin), New York.

Kosek, M., Yori, P.P., and Olortegui, M.P. (2010) Shigellosis update: Advancing antibiotic resistance, investment empowered vaccine development, and green bananas. *Current Opinion in Infectious Diseases*, **23** (5), 475–480.

Langdon, R. (1993) The banana as a key to early American and Polynesian history. *Journal of Pacific History*, **28**, 15–35.

Rabbani, G.H., Larson, C.P., Islam, R., Saha, U.R., and Kabir, A. (2010) Green banana-supplemented diet in the home management of acute and prolonged diarrhoea in children: A community-based trial in rural Bangladesh. *Tropical Medicine and International Health*, **15** (10), 1132–1139.

Rabbani, G.H., Teka, T., Zaman, B., Majid, N., Khatun, M., and Fuchs, G.J. (2001) Clinical studies in persistent diarrhea: Dietary management with green banana or pectin in Bangladeshi children. *Gastroenterology*, **121** (3), 554–560.

Rabbani, G.H., Teka, T., and Saha, S.K. (2004) Green banana and pectin improve small intestinal permeability and reduce fluid loss in Bangladeshi children with persistent diarrhea. *Digestive Diseases and Sciences*, **49** (3), 475–484.

Roberts, C.L., Keita, A.V., Parsons, B.N. *et al.* (2013) Soluble plantain fibre blocks adhesion and M-cell translocation of intestinal pathogens. *Journal of Nutritional Biochemistry*, **24** (1), 97–103.

Saha, S., Shilpi, J.A., and Mondal, H. (2013) Bioactivity studies on *Musa seminifera* Lour. *Pharmacognosy Magazine*, **9** (36), 315–322.

Schoeffl, V., Varatorn, R., Blinnikov, O., and Vidamaly, V. (2004) Intestinal obstruction due to phytobezoars of banana seeds: A case report. *Asian Journal of Surgery*, **27** (4), 348–351.

Slesak, G., Mounlaphome, K., Inthalad, S., Phoutsavath, O., Mayxay, M., and Newton, P.N. (2011) Bowel obstruction from wild bananas: A neglected health problem in Laos. *Tropical Doctor*, **41** (2), 85–90.

Venkatesh, R., Krishna, V., Girish Kumar, K., and Santosh Kumar, S.R. (2013) Antibacterial activity of ethanol extract of *Musa paradisiaca* cv. Puttabale and *Musa acuminate* cv. Grand naine. *Asian Journal of Pharmaceutical and Clinical Research*, **6**, 167–170.

Wiryo, H., Hakimi, M., Wahab, A.S., and Soeparto, P. (2003) Vomiting, abdominal distention and early feeding of banana (*Musa paradisiaca*) in neonates. *Southeast Asian Journal of Tropical Medicine and Public Health*, **34** (3), 608–614.

Yakubu, M.T., Nurudeen, Q.O., Salimon, S.S. *et al.* (2015) Antidiarrhoeal activity of *Musa paradisiaca* sap in Wistar rats. *Evidence Based Complementary Alternative Medicine*, Available from: http://dx.doi.org/10.1155/2015/683726 (8 January, 2016).

Pharmacological Effects of Kavalactones from Kava (*Piper methysticum*) Root

Ethnobotany

Kava (*Piper methysticum* G.Forst., Piperaceae) is a tropical shrub native to Oceania. Plants are sterile and must be propagated by cuttings. Its exact origins are uncertain; possibly all cultivars were originally cloned from the ancestral plant *P. methysticum* var. wichmannii (C. DC.) Lebot, found growing in Vanuatu (Applequist and Lebot 2006). Vanuatu folk classification system recognizes 247 cultivars, a number that may indicate its importance in ancient Vanuatu medicine (Lebot *et al.* 1997). From Vanuatu, kava was dispersed throughout Oceania following the migratory routes of the Pacific Islanders from Melanesia through Polynesia north to Hawaii (c. 500 CE).

The underground organs (rhizomes and roots) from this prized plant were used to make a relaxing drink, called kava or 'awa in Hawaiian. The custom of preparing and drinking kava is so ubiquitous in Oceania that some say it is the one cultural item that links together all the people of Oceania (Singh 1992). Kava was a sacred drink in old Hawaii, consumed by chiefs, offered to the gods, and prescribed by medical kahunas. In one native Hawaiian narrative, 'awa was brought to the islands by the gods Kāne and Kanaloa. There is even a chant (mele) asking a goddess to make 'awa grow abundantly. Commoners were allowed to drink common varieties of 'awa after a hard day's work in the fields or fishing boats. However, highly valued varieties such as moi (royal), hiwa (black), and papa (recumbent) were reserved for ceremonial use as offerings to the gods and goddesses (Beckwith 1970). Beckwith also notes that the most highly prized 'awa were plants that sprouted on other trees, growing with exposed roots, known *as kau laau* (resting on trees) or a *ka manu* (planted by the birds).

Johann George Forster, a naturalist on Captain Cook's second voyage (1772–1775) named kava *Piper methysticum* or "intoxicating pepper." In September 1773, Cook's ship *Resolution* was moored just southwest of Tahiti. Two Polynesian youths prepared and drank kava in Cook's cabin while Forster observed:

> [Kava] is made in the most disgustful manner that can be imagined, from the juice contained in the roots of a species of pepper-tree. This root is cut small, and the pieces chewed by several people, who spit the macerated mass into a bowl, where some water (milk) of coconuts is poured upon it. They then strain it through a quantity of fibers of coconuts, squeezing the chips, till all their juices mix with the coconut-milk; and the whole liquor is decanted into another bowl. They swallow this nauseous stuff as fast as possible; and some old topers value themselves on being able to empty a great number of bowls … The pepper-plant is in high esteem with all the natives of these islands as a sign of peace; perhaps, because getting drunk together, naturally implies good fellowship. (Forster 1777 in Lebot *et al.* 1997)

Cook had concerns about kava's excess consumption, as he witnessed side effects such as "kava dermopathy" (scaly skin), trembling, red eyes, and chronic pain. Ethnobotanist Margaret Titcomb (1948) noted Hawaiians induced kava dermopathy as a skin peel to cure skin conditions. After the scales peeled away, the skin was left smooth and soft. Kava is an important medicine in the pharmacopoeia of many Pacific Islands. Polynesians used kava to treat a range of conditions including women's health (e.g., menstrual problems), urogenital conditions, respiratory problems, gastrointestinal upset, headaches, weight gain, general weakness, chills, sleep difficulty, to prevent infections, and to soothe crying babies (Johnston and Rogers 2006; Lebot *et al.* 1997). This broad range of use in traditional pharmacopeias suggests a diverse array of active components in kava that likely vary with cultivar and preparation methods. The most typical preparation was a drink made by macerating the peeled rhizome or stump in a bowl (Figure 8.11), combining it with water or coconut milk, and squeezing it through a cloth to remove solids (Norton and Ruze 1994). Sometimes the macerated stump was used externally as a poultice for skin disorders including leprosy (Lebot *et al.* 1997).

As missionaries arrived in Hawaii in the 1820s, they persuaded Queen Ka'ahumanu to prohibit kava plantings and kava consumption. They associated kava with alcohol and other vices like opium. Despite the law, kava consumption continued, as it was growing all over the island and easily obtained. Mark Twain reported kava being sold in the market when he visited Hawaii in 1866. He wrote a letter to the *New York Times*, calling kava "… so terrific that mere whisky is foolishness to it. It turns a man's skin to white fish-scales that are so tough a dog might bite him, and he would not know it till he read about it in the papers" (Twain 1976). As alcohol consumption became more common in Hawaii, 'awa began to lose importance. In fact, some erroneously think that alcohol like brandy or gin was the traditional offering to Pele (Johnston and Rogers 2006). According to Titcomb (1948), 'awa, not alcohol, was a required offering for every god from Pele down to lesser gods.

Phytochemistry

Kavalactones (sometimes called kavapyrones) concentrated in the rhizomes and roots are thought to be the active principles in *P. methysticum*. Although 19 kavalactones have been identified to date, only six are believed to be responsible for most of the bioactivity

Figure 8.11 Young women preparing kava in a bowl, Samoa c. 1899 (photo by Malcolm Ross).

(Figure 8.12). Three dihydrochalcones (flavokavin A, B, and C) and eight minor compounds are also present in the underground organs (Lebot *et al.* 2014). Aerial parts of the plant contain toxic alkaloids such as pipermethystine, so they are rarely used in traditional medicine (Sarris *et al.* 2011). Chemotype and kavalactone content are responsible for kava's physiological effect. Total kavalactone content determines the intensity, while the chemotype determines the overall quality (Lebot *et al.* 2014). Cultivar, plant part (rhizomes, roots, stumps, basal stems), age of the plant, and growing conditions determine the chemotype. There are hundreds of *P. methysticum* cultivars and chemotypes, some with very high levels of kavalactones. For example, noble cultivars are consumed for everyday use in the Pacific Islands. They are rich in kavain; its fast absorption produces a rapid relaxing effect (Lebot *et al.* 2014). Two-day (*tudei*) cultivars are considered too strong for everyday consumption. They are rich in dihydromethysticin and dihydrokavain, often causing nausea and reportedly "two-days" worth of undesirable effects.

A Cochrane review (Pittler and Ernst 2003) uncovered 11 clinical trials using kava rhizome ethanolic or acetone extracts to treat anxiety. A meta-analysis of six of those studies showed kava was an effective symptomatic treatment for anxiety and was relatively safe for short-term use (1–24 weeks). Multiple mechanisms have been suggested to explain kava's anxiolytic (anti-anxiety) and sedating properties. These include kavalactones' ability to block voltage-gated sodium ion channels, enhance ligand binding to γ-aminobutyric acid (GABA) type A receptors, reduce norepinephrine reuptake, and inhibit monoamine oxidase B (Hannam *et al.* 2014; Singh and Singh 2002). Kava can also block calcium ion channels, thereby diminishing excitatory neurotransmitter release, and suppress the synthesis of thromboxane A2, which antagonizes γ-aminobutyric acid type A receptor function.

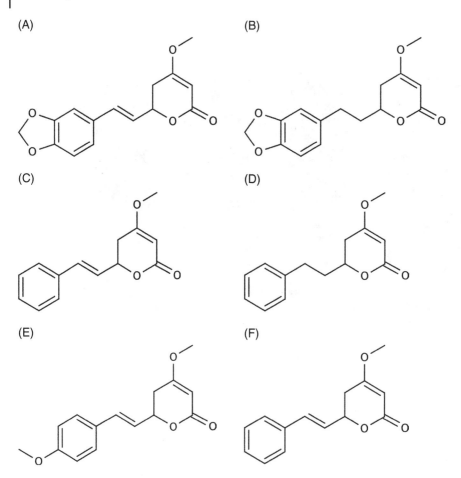

Figure 8.12 Structure six bioactive kavalactones: (A) methysticin; (B) dihydromethysticin; (C) kavain; (D) dihydrokavain; (E) yangonin; and (F) desmethoxyyangonin.

Kava supplements were withdrawn from EU and UK markets over hepatotoxicity concerns in 2002 and 2003, respectively (Ernst 2007). There were over 100 case reports of hepatotoxicity where kava was possibly responsible. In many of the reports, extenuating circumstances could have been to blame including concomitant ingestion with other hepatotoxic compounds (e.g., alcohol and drugs), preexisting hepatic insufficiency of cytochrome P450s (CYP450) required to metabolize kavalactones, taking higher than recommended doses of kava, and unknowingly taking adulterated kava (Sarris *et al.* 2011). Potential adulterations include using incorrect plant part for the extract (e.g., rhizome peels or aerial parts that contain pipermethystine), or incorrect species/variety. Most commercial kava formulations are produced by ethanol or acetone extraction, not the traditional water extraction method. Acetone and ethanol extracts would have contained flavokavins, which may be responsible for the observed hepatotoxicity. In one study, flavokavin B from two-day cultivar 'Isa' was toxic to cultured HepG2 liver cells (Jhoo *et al.* 2006). Flavokavin B was found to be ten times higher in two-day cultivar 'Palisi' from Vanuatu than the levels found in noble cultivar 'Ava

La'au' (DiSilvestro *et al.* 2007). Another study found that flavokavin B was five times higher in acetone extracts than in water extracts (Xuan *et al.* 2008).

Kava dermopathy is another side effect of kava consumption experienced across geographic regions and different populations. Known as kanikani in Fiji, heavy kava users experience a temporary scaly ichthyosiform skin eruption. Scientists speculate that kavalactones cause the condition, either by an allergic response (Schmidt and Boehncke 2000), phototoxicity (Xia *et al.* 2012), or a functional defect in one of the CYP450 enzymes required for kavalactone metabolism (Hannam *et al.* 2014). The CYP deficiency is plausible since symptoms mimic a genetic condition caused by CYP4F22 gene mutations, lamellar ichthyosis type 3. Kavalactones inhibit, and their metabolites form tight-binding complexes with, a number of CYP450 enzymes that are structurally similar to CYP4F22.

Modern Uses

People still drink kava in the Pacific Islands. It is still prepared by the traditional water extract methods, straining it by hand, and mixing it with water for social consumption. The doses of kavalactones in this setting are typically high, and nonstandardized. In contrast, dietary supplements are still typically ethanol or acetone extracts encapsulated for non-social consumption.

The legal status of kava outside Oceania is constantly in flux due to safety concerns over kava dietary supplements. In 2014, Germany and the EU lifted the kava ban. Kava remains legal in the United States and many other countries for personal use. In 2015, Australia took steps to ban kava at the national level in response to widespread abuse. It was already restricted in Western Australia and some Northern Territory Aboriginal communities. Kava imports are still not allowed in the United Kingdom; however, some sources state it is allowed for personal use. Compared to the 1990s, modern use of kava as a dietary supplement is negligible.

Vanuatu remains a major producer of kava. Their Parliament passed the Kava Act of 2002 to regulate the sale of kava cultivars (Lebot *et al.* 2014; Republic of Vanuatu 2002). The act prohibits sale or export of two-day, wichmannii, or medicinal kava, but farmers can grow them for personal use. Noble cultivars are the only type of kava approved for export. Two-day cultivars were widely planted in Vanuatu during the "kava boom" of the 1990s partly because the high yield of biomass and kavalactones was profitable for commercial extract producers. When the kava market crashed, farmers were left with fields of two-day types unsuitable for traditional consumption and prohibited for export. Sometimes two-day types were mixed in with or passed off as noble types for the international markets, reducing the overall quality, safety, and reliability of global kava products. As a result, the kava market has never rebounded to the heyday of the 1990s. In spite of safety concerns, kava research continues, especially in the areas of anxiety, insomnia, and other central nervous system (CNS) applications. In conclusion, kava prepared and consumed in the traditions of the Pacific Islands seems to provide relaxing benefits without widespread cases of hepatotoxicity. Although international kava quality standards have been proposed to reduce the safety concerns of hepatotoxic supplements reaching the market, they have not been universally implemented. As a result, kava dietary supplements still carry the potential for liver toxicity.

References

Applequist, W.L., and Lebot, V. (2006) Validation of *Piper methysticum* var. wichmannii (Piperaceae). *A Journal for Botanical Nomenclature* **16** (1), 3–4.

Beckwith, M. (1970) *Hawaiian Mythology*. University of Hawai'i Press, Honolulu.

DiSilvestro, R.A., Zhang, W., and DiSilvestro, D.J. (2007) Kava feeding in rats does not cause liver injury nor enhance galactosamine-induced hepatitis. *Food and Chemical Toxicology*, **45**, 1293–1300.

Ernst, E. (2007) A re-evaluation of kava (*Piper methysticum*). *British Journal of Clinical Pharmacology*, **64** (4), 415–417.

Hannam, S., Murray, M., Romani, L., Tuicakau, M., and Whitfeld, M.J. (2014) Kava dermopathy in Fiji: An acquired ichthyosis? *International Journal of Dermatology*, **53**, (12), 1490–1494.

Jhoo, J.W., Freeman, J.P., Heinze, T.M. *et al.* (2006) In vitro cytotoxicity of nonpolar constituents from different parts of kava plant (*Piper methysticum*). *Journal of Agricultural and Food Chemistry*, **54**, 3157–3162.

Johnston, E., and Rogers, H. (2006) Introduction, Hawaiian Cultivars, and Non-Hawaiian Cultivars Grown in Hawai'i Today. In *Hawaiian 'Awa: Views of an ethnobotanical treasure* (eds. E. Johnston and H. Rogers). Association for Hawaiian 'Awa, Hilo, pp. 30–63.

Lebot, V., Merlin, M., and Lindstrom, L. (1997) *Kava: The Pacific Elixir: The Definitive Guide to Its Ethnobotany, History, and Chemistry*. Healing Arts Press, Rochester.

Lebot, V., Do, T.K., and Legendre, L. (2014) Detection of flavokavins (A, B, C) in cultivars of kava (*Piper methysticum*) using high performance thin layer chromatography (HPTLC). *Food Chemistry*, **151**, 554–560.

Norton, S.A., and Ruze, P. (1994) Kava dermopathy. *Journal of the American Academy of Dermatology*, **31**, 89–97.

Pittler, M.H., Ernst, E. Kava extract for treating anxiety. *Cochrane Database of Systematic Reviews*, **2003**, CD003383.

Republic of Vanuatu Kava Act No. 7 of 2002, http://www.paclii.org/vu/legis/num_act/ka200252/(accessed 30 July, 2015).

Sarris, J., LaPorte, E., and Schweitzer, I. (2011) Kava: A comprehensive review of efficacy, safety, and psychopharmacology. *Australian and New Zealand Journal of Psychiatry*, **45**, 27–35.

Schmidt, P., and Boehncke, W.H. (2000) Delayed-type hypersensitivity reaction to kava-kava extract. *Contact Dermatitis*, **42**, 364.

Singh, Y.N. (1992) Kava: An overview. *Journal of Ethnopharmacology*, **37** (1), 13–45.

Singh, Y.N., and Singh, N.N. (2002) Therapeutic potential of kava in the treatment of anxiety disorders. *CNS Drugs*, **16**, 731–743.

Titcomb, M. (1948) Kava in Hawaii. *The Journal of the Polynesian Society*, **57** (2), 105–171.

Twain, M. (1976) *Mark Twain's Notebooks & Journals*, Vol. 1 (1855–1873). University of California Press, Berkeley.

Xia, Q., Chiang, H.M., Zhou, Y.T. *et al.* (2012) Phototoxicity of kava – formation of reactive oxygen species leading to lipid peroxidation and DNA damage. *American Journal of Chinese Medicine*, **40** (6), 1271–1288.

Xuan, T.D., Fukuta, A.A., Wie, A.C., Elzaawely, A.A., Khanh, T.D., and Tawata, S. (2008) Efficacy of extracting solvents to chemical compounds of kava (*Piper methysticum*) root. *Journal of Natural Medicine*, **62**, 188–194.

Botanical Index

a

Abelmoschus esculentus   66, 147, 156, 158, 263
Acacia   50, 51, 83, 147, 150, 155, 156
Acanthosicyos naudiniana   60
Acanthus   6
Acca sellowiana   60
Acer   63, 103, 186, 187
Aceraceae   63
Achillea   78, 88, 160–166, 283
Aconitum   7, 44, 243, 282
   A. ferox   45
   A. napellus   171, 284
   A. primum   11–12
Adansonia   65–66
   A. digitata   66
Aegle marmelos   261
Aesculus   63
   A. hippocastanum   63
Aframomum melegueta   40, 41, 91, 151, 156
Agavaceae, *see* Asparagaceae
Agave   35, 203–208
   A. americana   36, 204
   A. fourcroydes   204, 208
   A. parryi   203
   A. salmiana   204
   A. sisalana   204–205, 208
   A. tequilana   36, 203
Aibes   23
Ajuga turkestanica   210, 213
Alcea rosea   283
Alchemilla xanthochlora   283

Aleurites moluccanus   239, 316
Alfaroa   106
Allium   34, 149
   A. ampeloprasum   34, 149
   A. ascalonicum   262
   A. cepa   34, 98, 230
   A. ramosum   260
   A. sativum   34, 149, 230, 236
   A. schoenoprasum   34, 282
   A. tuberosum   34
Alnus rubra   200
Alocasia macrorrhizos   34
Aloe   6, 35–36, 155, 171
   A. vera   36
Alpinia galanga   40, 41
Alstroemeriaceae   34
Amaranthaceae   209–213, 222–226
Amaranthus   103, 226
Amaryllidaceae   34
Ambrosia   87
Ammi majus   149, 150
Amomum   40, 41, 234, 235
Amorphophallus paeoniifolius   33
Anacardiaceae   61
Anacardium occidentale   61, 62, 193, 194, 195
Ananas comosus   38, 105, 195
Anchusa officinalis   283
Andrographis paniculata   262, 264, 265, 266
Anenome   45
Anethum graveolens   88, 229, 232, 281
Angelica archangelica   283

Ethnobotany: A Phytochemical Perspective, First Edition. Edited by B. M. Schmidt and D. M. Klaser Cheng.
© 2017 John Wiley & Sons Ltd. Published 2017 by John Wiley & Sons Ltd.

Annona   28–29
  A. cherimola   28, 195
  A. diversifolia   29
  A. muricata   29, 193, 312
  A. purpurea   29
  A. reticulata   28
  A. squamosa   29
Annonaceae   28–29
Antirrhinum majus   283
Apiaceae   88–89
Apium graveolens   88, 282
Apocynaceae   78–79
Apocynum   78
Aquifoliaceae   86–87
Aquilegia   45
Araceae   33–34
Arachis hypogaea   50, 51, 195, 244
Araliaceae   89
Areca catechu   239, 247–252
Arecaceae   36–37
Arenga pinnata   239
Argania spinosa   70, 149, 151, 152
Argyreia nervosa   79
Arisaema triphyllum   189
Aristolochia trilobata   193
Armoracia rusticana   67
Arnica montana   88
Aronia melanocarpa   52
Artemisia   18, 87, 160, 229, 243, 252–258
  A. absinthium   87, 88, 252, 253, 254, 256
  A. annua   18, 21, 87, 160, 243, 254, 255
  A. arborescens   87, 254
  A. dracunculus   254, 256
  A. pontica   252, 254
  A. princeps   254, 256
  A. spicigera   254, 256
  A. vulgaris   252, 254
Artocarpus   57–59
  A. altilis   57–58, 193, 239, 312, 315
  A. camansi   58
  A. heterophyllus   58–59
  A. lacucha   263
Asclepiadaceae   78
Asclepias   79, 186
Ascophyllum nodosum   202
Asimina triloba   29, 186
Aspalathus linearis   155

Asparagaceae   34–36, 38, 98, 203–208
Asparagus   35–36, 98
  A. densiflorus   173
  A. officinalis   282
Asteraceae   87–88, 160–166
Astracantha gummifera   232
Astragalus   50–52
Atropa belladonna   82, 83, 292
Attalea cohune   36
Avena
  A. sativa   39
  A. sterilis   232
Averrhoa carambola   239
Avicennia germinans   94
Azadirachta   64, 272–276
  A. excelsa   272
  A. indica   64, 267, 272–273
  A. siamensis   272

b

Banisteriopsis caapi   196
Barringtonia asiatica   315
Berberidaceae   44
Berberis   44
Bertholletia excelsa   195
Beta   94, 223, 226, 232
Betula utilis   267–272
Betulaceae   268
Bixa orellana   193, 194, 195
Boehmeria nivea   58, 240, 242
Bombacaceae   65
Borassus   38, 239
  B. flabellifer   38
Boswellia   149–150, 233
Brassica   279
  B. juncea   230
  B. napus   232, 233, 279, 282
  B. nigra   232, 282
  B. oleracea   232, 282
  B. rapa   242, 279, 282, 285
Brassicaceae   66–68
Bromeliaceae   38
Brugmansia   83, 196
Brunfelsia   83, 196
Brya ebenus   193
Bryonia alba   171
Bunium persicum   230

Bursera   190, 192
Burseraceae   192, 233

C

Cactaceae   68, 98, 100
Caesalpinia spinosa   51
Caesalpiniaceae   49
Cajanus cajan   152
Calendula   87
Calocedrus   25
Calophyllum inophyllum   313, 316
Camellia   71, 91, 138, 235, 242
Campomanesia   60
Camptotheca acuminata   27, 46, 70
Cananga odorata   29, 239
Canella winterana   161, 165
Cannabaceae   55, 231
Cannabis   55, 68, 230–231, 235
Capparis spinosa   280
Capsicum   33, 80–82, 189–191
   C. annuum   65, 80, 189–191
   C. baccatum   191
   C. chinense   191
   C. fructescens   191
   C. pubescens   191
Carapichea ipecacuanha   17, 73–74,
      196–197
Carica papaya   191–192, 260, 265,
      266, 312
Carissa macrocarpa   156
Carthamus tinctorius   88, 232, 281–282
Carum carvi   88
Carya   106, 186
Casimiroa edulis   63
Cassia   92, 161, 229
   C. fistula   161, 165
Cassytha filiformis   315
Castanea
   C. dentata   19, 186
   C. sativa   282
Castanospermum australe   50
Catha edulis   46, 147, 148, 149, 156
Catharanthus roseus   79, 154, 155
Ceanothus   55
Cedrus   23
   C. deodara   267
   C. libani   149

Ceiba   65, 79
Celastraceae   46
Celtis   55
Centella asiatica   155, 262, 263
Ceratonia siliqua   51
Cerbera manghas   315
Chaenomeles japonica   52
Chamaecyparis
   C. nootkatensis   200
   C. obtusa   24
Chamaemelum nobile   283
Chenopodium quinoa   195, 209–215,
      222–226
Chondrodendron tomentosum   196, 285
Chrysanthemum   87, 242
Chrysophyllum cainito   70, 190, 193
Chrysopogon zizanioides   236
Chukrasia tabularis   235
Cicer   51, 232
Cichorium   88, 282
Cinchona   14–16, 73, 77, 160, 195,
      197, 283
Cinnamomum   30–32, 304
Cirsium   87
Citrullus lanatus   60, 152, 156
Citrus   63–64
   C. medica   63, 232, 241–242, 244
   C. sinensis   63, 235, 304
Claviceps   40, 90
Clematis   45
Clusiaceae   47
Cocos nucifera   37, 104, 239, 310, 312, 316
Coffea   73–77, 90, 104, 147, 156
Cola   65, 147, 153, 156
   C. nitida   153, 156
Colchicaceae   34
Colchicum   34, 283
Colocasia   33–34, 98, 152, 312
Commiphora   149, 150, 233, 282
Conium maculatum   88, 282, 284
Convolvulaceae   79, 90, 92, 193
Copernicia prunifera   38
Corchorus   156, 157
Coriandrum sativum   88, 149, 191, 280
Coriaria arborea   292–293
Cornaceae   68, 70
Cornus amomum   186

Cornus
   C. florida 161
   C. mas 232
Corylus avellana 232
Corypha umbraculifera 239
Cota tinctoria 281, 282
Crataegus 52–53, 99
Crinum asiaticum 239
Crocus 34, 232, 279, 282
Croton lechleri 139
Cucumis
   C. aculeatus 60
   C. heptadactylus 60
   C. melo 232
   C. sativus 102, 235, 239, 260
Cucurbita 185, 190, 195
   C. foeti 90
   C. foetidissima 94
   C. maxima 263
Cucurbitaceae 94, 105, 147
Cuminum cyminum 88, 149, 232
Cupressaceae 23–25, 96
Cupressus 24
Curcuma
   C. australasica 316
   C. longa 40–41, 234–235, 263
Cuscuta reflexa 260–262
Cyamopsis tetragonoloba 51
Cyclopia genistoides 155
Cydonia oblonga 52, 230
Cymbopogon 304
Cynara 88, 232
Cyperus esculentus 149
Cyrtosperma merkusii 34, 312

d

Daemonorops 38
Dahlia 87
Datisca glomerata 161
Datura 7, 82–83, 282, 292–293
   D. metel 235
   D. stramonium 82
Daucus carota 88, 94, 232
Delphinium 45, 104
Dendrobium 35
Dermatophyllum 52, 185
Derris 50–51, 315
Digitalis 85, 93–94, 283–284

Dimocarpus longan 61
Dioscorea 147, 192, 239, 312
   D. deltoidea 239
   D. mexicana 192
   D. polystachya 242
Diplopterys cabrerana 196
Dipteryx odorata 195
Dracaena 38, 167
Drimia maritima 283
Duboisia hopwoodii 83
Durio 65–67, 239
Dysphania ambrosioides 193

e

Echinacea purpurea 186
Echinochloa 242
Echinopsis 68, 196
Elaeis guineensis 147, 154, 178, 237
Eleocharis dulcis 242
Elettaria 40–41, 234–235
Eleusine coracana 39, 147, 156
Eleutherococcus senticosus 89
Elliottia paniculata 291
Ensete 40, 156, 239
Entada gigas 50
Ephedra 23, 28, 243
Ephedraceae 27–28
Epipremnum pinnatum 247
Eragrostis tef 147, 156
Ericaceae 72–73, 291
Eriobotrya japonica 52
Eruca vesicaria 232
Eryngium foetidum 88, 190–191
Erythroxylaceae 46–47
Erythroxylum coca 46, 195–196, 209
Eucalyptus 60–61, 83, 304, 306, 310, 313
Eugenia 60, 307
Euphorbia 47, 97
Euphorbiaceae 47–48, 97–98
Eurya acuminata 248
Eutrema japonicum 240, 242
Excoecaria agallocha 47

f

Fabaceae 49–52, 83, 185, 295–303
Fagaceae 267–272
Fagopyrum esculentum 230, 242
Ferula marmarica 89, 149, 151

Ficus   57–58
  F. altissima   239
  F. aurea   95
  F. bengalensis   94
  F. carica   58, 105, 232, 281
  F. elastica   58, 239
  F. religiosa   235, 239
Foeniculum vulgare   88
Fortunella   63
Fragaria   53, 290
Fritillaria   35
Fucus   199–203
Funtumia elastica   149, 156

g

Galagania fragrantissima   230
Galium odoratum   73, 283
Garcinia × *mangostana*   47, 239
Gaylussacia   72, 186
Gelsemium sempervirens   291
Gevuina   45
Ginkgo biloba   23, 101, 229
Gloriosa   34
Glycine max   51, 242
Glycyrrhiza glabra   51, 149, 232,
    236, 243
Gmelina arborea   235, 244–247
Gnetum   23, 27–28
Gossypium   65–66
  G. arboreum   234–235
  G. barbadense   66, 195
  G. herbaceum   66, 147, 156
  G. hirsutum   66, 190
Grevillea   83
Grossularia   45
Grossulariaceae   45
Guarea   64
Guizotia abyssinica   147

h

Haematoxylum campechianum   51
Harpagophytum procumbens   155
Hedychium coronarium   239
Helianthus   88, 103, 186
Heliotropium   14
Helleborus   45
Hevea brasiliensis   48–49, 58, 92, 97,
    193, 197

Hibiscus   65–66, 147
  H. cannabinus   156
  H. elatus   193
  *H. esculentus, see Abelmoschus
    esculentus*
  H. sabdariffa   156
  H. tilliaceus   312, 315
Hierochloe odorata   186
Hippocastanaceae   63
Hopea shingkeng   19
Hordeum vulgare   38, 103, 151, 232, 279
Humulus   55–56, 279, 281, 282, 285
Hydrastis canadensis   45
Hylocereus   68–69, 98, 190
Hyoscyamus niger   82, 282, 284
Hypericaceae   47
Hypericum perforatum   47, 283
Hypoxis   173–175
Hyssopus   229, 283

i

Ilex   86–87
  I. aquifolium   86
  I. opaca   86
  I. paraguariensis   86–87, 195, 197
Indigofera   68
  I. arrecta   295
  I. tinctoria   51, 235, 281, 282, 295–303
Inocarpus fagifer   312
Inula helenium   283
Iochroma fuchsioides   83
Ipomoea   79–80
  I. alba   79
  I. aquatica   79
  I. batatas   94, 158, 195, 310, 312
  I. violacea   79
Iris   98, 103, 261
Isatis tinctoria   68, 230, 281–282,
    295–303

j

Jatropha curcas   48, 247
Juglandaceae   106
Juglans   106
  J. jamaicensis   193
  J. nigra   186
  J. regia   230
Juniperus   24, 97, 101, 186, 267

k

Kaempferia galanga   40
Kalmia latifolia   291
Kunzea ericoides   61

l

Lactuca   88, 149, 232
Lagenaria siceraria   147–148, 156,
    190, 204
Lamiaceae   85–86, 285–290
Laminaria   200
Larix   23
Latua pubiflora   196
Lauraceae   30–32
Laurus   30, 32, 282
Lavandula   85–86, 285–290
    L. angustifolia   86, 283, 286, 288
    L. dentata   286, 288
    L. latifolia   286, 288
    L. stoechas   285–290
Lawsonia inermis   229, 235, 280
Lecythidaceae   195
Ledum   72
Lens culinaris   51, 230, 232, 279
Leontopodium nivale   87
Leptospermum scoparium   61
Lessertia frutescens   173
Leucanthemum   87
Ligusticum porteri   186
Liliaceae   34–35
Lilium   279
Linum usitatissimum   232
Liparis   35
Liriodendron tulipifera   161
Litchi chinensis   61, 242
Lonicera involucrate   200
Lophophora williamsii   68, 190, 193
Ludwigia octovalvis   316
Luffa cylindrica   60
Lupinus   50–51
*Lycopersicon esculentum, see Solanum
    lycopersicum*

m

Macadamia   45, 310, 312
Maclura   105, 190
Macrocystis   199

Magnolia baillonii   235
Malus   52
    M. domestica   52, 105
    M. pumila   230
Malvaceae   65–66
Mammea americana   193
Mandragora   6, 7, 82
Mangifera   61–62, 102, 104, 193, 312
Manihot esculenta   49, 97, 195, 312
Manilkara   70, 97, 190, 192
Marrubium   85–86, 283
Matricaria chamomilla   283
Medicago sativa   51, 230
Melaleuca   61
Melastoma malabathricum   315
Melia azedarach   274
Meliaceae   64
Melissa officinalis   86, 229, 283
Mentha   85–86
    M. longifolia   229
    M. spicata   236
    M. × piperita   86, 304
Merwilla plumbea   173
Michelia champaca   239
Mimosa pudica   261, 262
Mimosaceae, *see* Fabaceae
Momordica   60, 155, 262
Monarda   85, 186
Monodora myristica   30
Moraceae   57–58
Morinda citrifolia   73, 312, 314
Morus   57, 101, 105
Mukia maderaspatana   263
Murraya koenigii   63
Musa   40, 152–153, 244, 317–323
    M. acuminata   318, 320
    M. balbisiana   318
    M. × paradisiaca   261, 321
    M. sapientum   320
Musaceae   40
Myristica fragrans   237, 239
Myrothamnus flabellifolia   172–174
Myrtaceae   33, 60–61, 304

n

Nardostachys jatamansi   282
Nelumbo nucifera   239

Nemopanthus   86
Nepeta   85–86, 283
  N. cataria   86, 283
Nephelium lappaceum   61, 239
Nereocystis luetkeana   200
Nerium oleander   79
Nesiota elliptica   19
Nicotiana   80, 82, 186–187, 200
Nigella sativa   149
Nux vomica   17
Nypa fruticans   248

o

Ochroma   65, 66
Ocimum   85–86
  O. basilicum   86, 229
  O. gratissimum   262
  O. tenuiflorum   239, 262, 265, 267
Olea   83–85, 104, 232, 280
Oleaceae   83–85
Oncosiphon piluliferum   163
Opuntia   68, 98, 186, 223
Orchidaceae   35, 101
Oreomunnea   106
Oreosyce africana   60
Origanum   85, 86
Ormosia   52
Oroxylum indicum   262
Oryza
  O. glaberrima   147, 156–157
  O. sativa   38, 151, 237, 239
Osmanthus fragrans   267
Oxalis tuberosa   98, 99
Oxytropis   50

p

Pachyrhizus erosus   190
Paederia foetida   247
Panax   89–90, 229, 243
Pandanus   312–314, 317
Panicum miliaceum   242
Papaver   41–44
Papaveraceae   41–44
Parkia biglobosa   147, 156
Parkinsonia   97
Paspalum   40
Passiflora edulis   47–48, 190–191, 312

Passifloraceae   47
Pastinaca sativa   88, 89, 232
Paullinia
  P. cupana   63, 195, 197, 198
  P. yoco   197
Pausinystalia johimbe   73, 155
Pavetta indica   261, 262
Pelargonium sidoides   155
Pennisetum glaucum   39, 147, 156
Peperomia   32
Persea   30, 32, 190
  P. americana   32, 104, 190
Persoonia   45
Petroselinum crispum   88, 232
Phalaenopsis   35, 102
Phaseolus   51, 189, 190, 195
Philenoptera cyanescens   295
Phlogacanthus thyrsiformis   262
Phoenix dactylifera   37, 232
Phrynium   244
Phyllanthus   98, 236, 260–262, 265
Physalis   82, 190, 191
Picea   23
Pilocarpus   63, 195, 196, 284
Pimenta dioica   33, 60, 193, 194
Pimpinella anisum   232
Pinaceae   23–24, 96
Pinus   23–24, 96
  P. aristata   96
  P. balfouriana   96
  P. longaeva   96
  P. pseudostrobus   192
  P. roxburghii   267
Piper   32–33
  P. betle   33, 36, 247–248, 251
  P. longum   32, 265
  P. methysticum   20, 33, 312, 316,
    323–328
  P. nigrum   32, 234–235, 262–263
Piperaceae   32–33
Piscidia piscipula   50
Pistacia   61, 62, 230, 281
Pisum sativum   51, 230, 232, 279
Pittosporum ferrugineum   316
Plantaginaceae   85
Plantago   85, 103
Pleurothallis   35

Plumeria rubra   239
Poaceae   38–40, 103
Podophyllum peltatum   44, 186
Pogostemon   85, 86
Porphyra   199–200
Pouteria lucuma   70, 195
Premna serratifolia   315
Proteaceae   45, 83
Protium copal   192
Prunus   52
  P. armeniaca   52, 242
  P. avium   52
  P. cerasus   52, 232
  P. domestica   52, 282
  P. dulcis   52, 104, 230
  P. persica   52, 242, 268
  P. serotina   52
  P. virginiana   52, 104
Pseudotsuga   23
Psidium
  P. dumetorum   19
  P. guajava   60, 190–192, 312
Psophocarpus tetragonolobus   312
Psychotria viridis   196
Pueraria   51
Punica granatum   104
Pyrus   52–53, 97, 230, 242, 263

q

Quercus   101, 103, 186, 267–272

r

Ranunculaceae   44–45
Ranunculus   45
Raphanus
  R. raphanistrum   67, 282
  R. sativus   94, 232
Rauvolfia serpentina   79, 239
Reseda luteola   296
Rhamnaceae   55
Rhamnus   55
Rhaponticum carthamoides   210
Rheum officinale   282
Rhizophora mucronata   315
Rhododendron   16, 73, 291–295
  R. arboreum   267
  R. flavum   292

R. luteum   294
R. ponticum   73
Rhodymenia   200
Rhus chinensis   248
Ribes   45, 188, 282
Ricinus communis   48, 97, 152
Rivea   79
Rorippa sarmentosa   315
Rosa   52, 99, 104, 283
Rosaceae   52–55
Rosmarinus officinalis   86, 283, 304–305
Rubia tinctorum   73, 281, 282
Rubiaceae   73–78
Rubus   53–55, 105, 188, 282
Ruscus aculeatus   98
Rutaceae   63–64

s

Saccharum   39, 151, 235
  S. officinarum   39, 235, 239, 260, 312
  S. robustum   312
  S. spontaneum   235
Salicaceae   49
Salix   49, 101, 283, 284
Salvia   21, 85–86
  S. apiana   186
  S. divinorum   193
  S. officinalis   21, 86, 280, 283
  S. sclarea   229
Sambucus   104
Sandersonia   34
Santalum album   150, 235, 236
Sapindaceae   61–63
Sapotaceae   70
Saraca asoca   235
Sassafras   30, 186, 188–189
Scaevola taccada   316
Schlumbergera truncata   98
Scrophulariaceae   85
Scutellaria   85, 86
Secale
  S. anatolicum   232
  S. cereale   38
Senecio   87
Senegalia senegal   51
Senna   51, 83, 149
Sequoia   23–25

Sequoiadendron 24
Sesamum indicum 149
Setaria italica 39
Shorea robusta 235
Silphium laciniatum 186
Sinapis alba 67, 232
Smilacaceae 35
Smilax 35
Solanaceae 33, 80–83
Solandra 83
Solanum
  S. americanum 82
  S. lycopersicum 80, 82, 102, 104, 191, 195
  S. melongena 82, 102
  S. tuberosum 80, 82, 98, 195
  S. virginianum 236
Sorghum 39, 147, 156
Spilanthes acmella 161
Spinacia oleracea 230
Spondias purpurea 193
Sterculiaceae 65
Strophanthus gratus 79
Strychnos 196
Styrax officinalis 282
Sutherlandia frutescens 173, 175
Swainsona 50
Swietenia 64, 193
Symphytum officinale 283
Symplocarpus foetidus 33
Syzygium
  S. aromaticum 60, 236, 237, 239, 304
  S. jambos 312

t

Tacca leontopetaloides 312
Tagetes 87
Tamarindus indica 51, 147, 156, 263
Tanacetum 229, 283
Taraxacum 87
Taxaceae 26–27
Taxodium 23, 24, 94
Taxus 26–27, 186, 239
Tectona grandis 235
Terminalia 236, 262, 265
Theaceae 70–72
Theobroma cacao 65, 195, 197–198

Thuja 24–25, 171
  T. occidentalis 171
  T. plicata 186
Thymus 85, 86, 283, 304
Thysanolaena maxima 244
Tinospora sinensis 236
Tournefortia argentea 316
Toxicodendron 61, 240, 249
Trachyspermum ammi 149
Trifolium 51
Triticum 38, 103, 151, 230, 232, 279
  T. boeoticum 232, 279
  T. dicoccum 232
  T. timopheevii 279
Triumfetta procumbens 317
Tsuga 23
Turbina corymbosa 79, 291

u

Uapaca 156
Ulmus 103
Ulva 200
Urtica dioica 58
Urticaceae 58

v

Vaccinium 72, 104, 186, 188–189
Valeriana officinalis 282
Vanda 35
Vandopsis 35
Vanilla 35, 156, 166–171
  V. odorata 170
  V. planifolia 35, 65, 166–171, 190, 191
  V. tahitiensis 170
Ventilago viminalis 83
Verbascum thapsus 283
Vernicia fordii 48
Veronicaceae, *see* Plantaginaceae
Vicia
  V. ervilia 232, 279
  V. faba 51, 232
Vigna
  V. marina 315
  V. mungo 263
  V. radiata 235, 263
  V. subterranea 147, 156
  V. unguiculata 51, 147, 156, 159

Virola   196
Viscum album   94
Vitaceae   45–46
Vitellaria paradoxa   70, 147, 156
Vitis vinifera   45–46, 149, 279

w

Wasabia japonica   67, 241
Welwitschia   23, 27–28
Withania somnifera   236

x

Xanthorrhoeaceae, *see* Liliaceae
Xylopia aethiopica   29–30

y

Yucca   35, 36, 103

z

Zea mays   38, 94, 101, 185
Zehneria scabra   60
Zingiber officinale   40, 98, 235, 236
Zingiberaceae   40
Zizania   186
Ziziphora clinopodioides   230
Ziziphus   55, 193

Subject Index

a

abaca fiber   319
absinthe   87, 253–255
absorbance   131
acacia, gum   50, 51
acai   219, 249
acetate pathway   114, 120, 123–125
Acetobacter xylinum   37
acetogenins   29
acetylandromedol, *see* grayanotoxin
acetyloleanolic acid   268
acetylsalicylic acid   49
Achaemenid empire   231
achene   55, 87, 103–105
achillifolin   162
aconitine   45, 120, 243
acorns   103, 186
adaptogen   89, 209–213
adhesives   24, 61, 192, 313
aescin   63
aesculetin   196
agave   22, 35–36, 191, 203–208
ajmalicine   79, 154
alfalfa   51, 230, 231
algae   199–203
alizarin   73
alkaloids   120–123
  acridine   112, 120, 121
  bis-benzyltetrahydroisoquinoline
      112, 121, 285
  carbazole   63
  ergoline   79–80, 193
  ergometrine   40
  furoquinoline   63

  glyco   80–81
  imidazole   63, 112, 122, 123, 284
  indole   46, 68, 73, 77, 79, 112, 120, 121,
      154, 196, 217, 284, 293, 300–301
  indolizidine   35, 50, 122
  isoquinoline   44–45, 68, 74, 112, 120,
      121, 196
  isoquinolone   44
  phenethylisoquinoline   34, 112, 121
  phenylethylamine   28, 68
  piperidine   80–81, 89, 120, 122
  purine   72, 76, 112, 122, 123
  pyridine   80–81, 112, 122, 250, 251
  pyridylpiperidine   81
  pyridylpyrrolidines   81
  pyrrolidine   80–81, 112–113, 122
  pyrrolidone   32
  pyrrolizidine   35, 50, 112, 122
  pyrroloindole   112, 120, 121
  pyrroloquinoline   112, 120, 121
  quinazoline   112, 120, 121
  quinoline   70, 77–78, 112, 120, 121, 217
  quinolizidine   35, 50, 52, 122, 185
  steroidal   35, 80–81, 120
  terpenoid   35, 74, 77, 115, 120
  tropane   80, 83, 112–113, 122, 293
N-alkylamides   161
S-allylcysteine   34
allicin   34
alliin   34
allspice   33, 60, 193, 194
almond   52, 103, 104, 230
aloe   6, 35–36, 155, 171
amaranth   103, 195, 209–215

Ethnobotany: A Phytochemical Perspective, First Edition. Edited by B. M. Schmidt and D. M. Klaser Cheng.
© 2017 John Wiley & Sons Ltd. Published 2017 by John Wiley & Sons Ltd.

amaryllis   34
amino acids   66, 111, 114, 120, 122, 224, 292
ammoniac gum   89, 149, 151
amphetamines   46, 150
amygdalin   52
amylopectin   179–180
amylose   33, 179–180
anabasine   81, 82
anacardic acid   61
analgesic   14, 16, 44, 49, 82, 83, 186, 233, 282, 284
anatabine   81, 82
Andes   183, 189, 195, 209–215
andrographiside   264, 265
andrographolide   264, 265
andromedotoxin, *see* grayanotoxin
anesthetic   6, 7, 82, 197, 288
angel's trumpet   83
angelicin   89
angiosperm   27–28
Anglicus, Bartholomaeus   9
aniline   301
anise   232, 253
annatto   193–195
annonacin   29
anthocyanins   115, 116, 217–219
anthranilic acid   112, 120, 121, 301
anthraquinone   47, 73, 123
antibacterial, *see* antimicrobial
antibiotic   34, 82, 272, 320
anticancer drugs   26–27, 46, 79, 239
anticholinergic   16, 80
anticonvulsant   86, 254
antidepressant   47, 86, 197, 286, 289
antidiabetic   215–220, 312
antidote   45, 282
antifeedant, *see* insecticide
antifungal, (*see also* antimicrobial)   245, 251, 286, 289
antihelminthic   186, 249, 254, 270
antiinflammatory   34, 40, 88, 200–201, 210–211, 218–219, 284
antimalarial, *see* artemisinin; quinine
antimicrobial   25, 34, 56, 61, 86, 245–247, 270, 288–289, 320
antioxidant   47, 52, 55, 200, 211–212, 218–219, 270

antipsychotic   79, 239, 268, 270, 284
antiretroviral   139, 171–176
antiseptic   61, 155, 247, 254, 268
antispasmodic   162, 254, 288
antitussive   35, 44, 189, 191, 243, 254
antiviral   50, 171–176, 218, 270
anxiety   286, 325, 327
aphrodisiac   82, 155
apigenin   289
apothecary   12, 14, 127
appetite
   stimulant   55, 229, 253
   supressant   197
apple   52–53, 230
apricot   52, 104, 242
arecaidine   250
arecatannins   139, 249, 250
arecoline   250
argan oil   70, 149, 151, 152
arginine   113
aril   26, 63, 66, 104, 198, 237
aristone   217, 218
aristoteline   217, 218
aristotelinine   217, 218
aristotelone   217, 218
Aristotle   6
aromatherapy   129, 286–289
arrow poison   26, 45, 60, 79, 196, 285
arrowroot   312
arteannuin, *see* artemisinin
artemisinin   17–18, 87, 160, 240, 243, 255–256, 256
arthritis   243
artichoke   88, 186, 232
artocarpin   58
arylethylene-α-pyrones   33
asclepin   79
asebotoxin, *see* grayanotoxin
asparagus   35–36, 98, 173, 282
asparagusic acid   36
aspartic acid   112
aspirin, (*see also* willow)   49, 283
asthma   28, 82, 191, 243, 254
astringent   16, 52, 54, 280
Atharvaveda   252
atherosclerosis   286
atropine   16, 80, 82, 196

aubergine 82, 102, 157, 236
autoimmune 256, 259
Avicenna 9, 231, 303
avocado 30, 32, 104, 190
Ayahuasca 196
Ayurveda 8, 138, 229, 234–236,
    258–266, 272
azadirachtin 273–275
azeotropic mixture 129
Aztec, *see* Native Americans

b

baccatin 26
Badianus Manuscript 14–15, 83
Badianus, Juannes 14–15
Baeyer, Alfred Von 299, 301
baicalein 86
baicalin 86
balchanolide 162, 163
balsa 6, 65, 66, 150
bamboo 242, 244
banana 40, 53, 103, 104, 152–153, 207,
    237, 239, 244, 260, 312, 317–323
Bangladesh 233, 245, 319–320
banj 267–272
Banks, Sir Joseph 15–16
Bantu 153, 157, 189
banyan tree 94
baobab tree 65, 66
barberry 44
barcoding, DNA 137
barley 6, 19, 38–39, 58, 103, 151, 232,
    233, 279
basil 85, 86, 229, 239, 267
batik 73
Baytar, al- 9, 282–283
bean, (*see also* legumes) 49–52, 185,
    188–190
bee 291–294
beebalm 85
beer 56, 73, 205, 279
beet 94, 223, 225, 232
benzo(a)pyrene 5
benzopyrone 31
benzoxazinoids 39
berberine 44–45, 112, 120, 121, 240
betacyanin 223–244

betalain 120, 139, 222–226
betalamic acid 224
betanidin 224
betanin 223–225
betaxanthin 223–224
betulin 268, 270
betulinic acid 268, 270
bhojpatra 267–271
biennial 93–94
bilberry 72
bilirubin 259, 263
bioassay 133–139, 161, 164
    guided fractionation 70, 136, 161
bioavailability 175, 205, 212
biodiversity 4, 18–20, 134, 196, 237
biofuel 21, 22, 48, 49, 51, 178–181, 206
bioprospecting 19
birch 8, 267–272
blackberry 53–54, 105, 188
Blackwell, Elizabeth 15
Bligh, William 15–16, 57–58
blight 19, 48
blind-your-eye 47–48
blistering 48, 89, 312
blue mahoe 193
blueberries 72, 188–189
Bock, Hieronymus 11
bolting 204
borneol 86, 150, 255
Bower Manuscript 8
bract 105
bramble 54
brassinosteroids 115
bread 188, 189, 197
breadfruit 57–58, 312–313, 315–316
breadnut 58
breadroot 189
broccoli 67
bromelain 38
bromeliad 38
bronchitis 28, 186, 254
Bronze Age 279–281
broomcorn millet 242
browning, (*see also* oxidation) 71, 249
bruises 88, 316
Brunfels, Otto 11
Brussels sprouts 67

Büchner, Johann   16, 285
buckthorn   55
buckwheat   230, 242
buds   60, 71, 96, 99, 288, 304
Buddha's hand   241, 242
Buddhism   34, 35, 234–235, 237, 239, 240,
    242–243
bulb   35, 98, 100, 173, 205
butcher's broom   98
buttercup   44–45
buttress roots   58, 94

c

cabbage   67, 232, 240, 242, 282
cactus   68, 69, 98, 100, 186, 190, 193,
    196, 223
caffeic acid   77, 114
caffeine   16, 63, 65, 72, 75–77, 86–87, 122,
    123, 153, 197
calcium oxalate   28, 33, 189
CAM plants   38
cambium   96
camphene   17
camphor   17, 32, 86, 150, 255, 288
camptothecin   27, 46, 70
canadine   45
canavanine   175
candlenut, see nut
candling   260
candy   235
cannabinoids
    cannabichromene   55
    cannabigerol   55
    cannabinol   55
    receptors   55
    Δ9-tetrahydrocannabinol   55
capers   280
capitulum   87, 102
capsaicin   81, 82, 121
capsule   41–42, 103, 166–167, 219
caraway   6, 88
β-carboline   120, 196
carcinogen   5, 30, 83, 250, 301
cardamom   40–41, 234, 235
cardiotonic   78, 85
cardiotoxic   26, 243
cardol   61

carmalols   200
carminic acid   223
carob   51
β-carotene   119, 223
carotenoids   17, 34, 82, 115, 119, 126, 194
carpel   28, 49, 101, 103, 241
carrot   17, 88, 94, 103, 232
caryophyllene   86, 115
caryopsis   103
cashew, see nut
cassava   49, 97, 195, 237, 312
castor bean   48, 97, 152
catechins   72, 115, 116, 138–139, 218,
    249, 270
catechols   61
catharanthine   120
cathinone   46, 150
catnip   85, 86, 283
cauliflower   67
Caventou, Joseph Bienaimé   16
cedar   23–25, 149–150, 186
celeriac   88
celery   88, 236, 282
cellulose   37, 66, 206–207
cephaeline   74
cereal   11, 38–40, 103, 147, 151, 209, 231,
    279–280
α-chaconine   81, 82
chalcones   33, 114, 117, 325
chamomile   281–283
chemotype   86, 255, 288, 325
cherimoya   28, 195
cherry   52, 188, 232, 236
chestnut, see nut
chickpea   51, 232
chicle/chicklets   70, 97, 190, 192
chicory   88, 282
childbirth   40, 200, 216, 315
chinaberry   274
chitinases   48
chives   34, 282
chlorogenic acid   77, 87
chocolate, (see also cocoa)   65, 167
chokeberry   52
chokecherry   188
Christianity   236, 247, 259
chromatography

affinity 130
column 129–130
counter current 131
gas (GC) 131
high performance liquid (HPLC) 131, 132
ion-exchange 130
tandem 131, 161
thin layer (TLC) 130, 131, 162
chromophore 38, 224
chrysanthemum 87, 242
cigarette, (*see also* tobacco) 83, 288
cilantro 88, 191
cinchona bark, (*see also* quinine) 14, 15,
77–78, 195, 197, 283
cinchonidine 77
cinchonine 77
cineole 61, 86, 255, 288, 304, 306, 310
cinnamaldehyde 31
cinnamic acid 114, 117
cinnamon 30–31, 216, 235, 304–305
citral 17
citric acid cycle 111, 113, 122, 123
citrinellal 61
citron 63, 232, 241, 242
citronella 304–305
citronellal 306
citrus 53, 63–64, 66, 104, 151, 241, 249
cladodes 98
clary sage 229
Claviceps 40
clove 60–61, 237, 304, 305
clover 51
coca 14, 46, 195–197, 209
Coca-Cola 47, 153
cocaine 46–47, 122, 196–197
cocoa 46, 65–66, 195, 197
coconut 37, 310, 312–316, 324
codeine 43–44
coffee 16, 73–77, 147, 237
coir 37, 207
colchicine 34, 283
colic 155, 286, 313
columbine 45
Columbus, Christopher 38, 185
comfrey 283
coneflower 186
cones, *see* strobilus

Confucianism 237, 239, 240, 242
γ-coniceine 89
conifer 23–27
coniferin 169
coniferyl 115
coniine 89
Cook, Captain James 3, 15, 310, 323–324
copal 190, 192
copra 37, 313
cordage 36, 97, 313, 314
coriander, *see* cilantro
corm 33, 34, 40, 98, 189, 317
corn, *see* maize
cornelian cherry 232
corolla 66, 101
Cortés, Hernán 167
costunolide 118, 163
cotton 41, 42, 65, 66, 147, 155, 156, 190,
195, 234–235, 298
cough, *see* antitussive
coumarin 31, 73, 90, 112, 114, 117, 170,
196, 288
p-coumaryl alcohol 114, 117
coumestans 50
cowpea 152, 153, 157, 159
crabapple 52
cranberry 72, 186
crazyweed 50
creosote 5
crepidamine 35
crepidine 35
crocus 34, 279, 283
Crohn's disease 256
Cronquist, Arthur John 49
croton 139
cucumber 102, 235, 239
cucurbitacin 60, 161
cucurbits 60
culantro 88, 190, 191
Culpeper, Nicholas 12–13, 252
cumin 88, 149, 230–232
curare 18, 196
curcumin 40
currant 45, 188, 282
cuscohygrine 82
cyanide 49, 122
cyanidin 137, 218

cyanogenic glycosides 39, 49, 52, 111, 114, 120, 122
p-cymene 86
cypress 23, 94

d

daidzein 50
dammarane 89
Darwin, Charles 16, 28
10-deacetylbaccatin 26–27
10-deacetyltaxol 26
decoction 128, 254, 260–263, 287
DEET (*N,N*-diethyl-m-toluamide) 304
dehiscence 103, 167
de Jussieu family 14
de la Cruz, Martinus 14–15
delphinidin 217–219
delphinine 45
demethyloleuropein 83
dendrobine 35
dendroprimine 35
dendrowardine 35
dendroxine 35
dengue fever 50
denim 299, 302
dental care, *see* tooth
8-deoxygartanin 47
depression, *see* antidepressant
dereplication 20
dermatitis 61, 89
desmethoxyyangonin 326
dessert 29, 37, 51, 53, 66, 318
desulfosinigrin 67
dextran 205
dhurrin 39
diabetes, *see* antidiabetic
diallylthiosulfinate 34
dianthrones 123
diarrhea 139, 246, 317–323
dibenzodioxin 200
digitoxin 85
digoxin 85
dihydrokavain 325, 326
dihydromethysticin 325, 326
3,6-dihydronicotinic acid 81
dihydroparthenolide 162
dihydroxyphenylalanine 224

Dinawari, Abu Hanifa al- 282
diode array detector (DAD) 131, 132
dioecious plant 26, 101, 216
Dioscorides, Pedanios 6, 7, 9, 11–12, 34, 282
diosgenin 192, 239
dissolution 135–136
distillation 129, 204, 303, 304
diterpenoids, *see* terpenes
diuretic 36, 63, 66, 197
divination 83, 193, 291
DMT (*N,N*-dimethyltryptamine) 196
Doctor Universalis, *see* Magnus, Albertus
dogbane 78
dogwood 68
Dover's Powder 74, 197
dracorhodin 38
dragon's blood 38, 139
dragon's eye 63
dragonfruit 68, 69, 190
dropsy 82, 85
drupe 87, 104, 105
DSHEA (Dietary Supplement Health and Education Act) 138
duckweed 178–181
durian 65–67, 239, 248
dye
  annatto 194
  barberry 44
  betalains 222–225
  dragon's blood 38
  fustic 190
  indigo 68, 281, 285, 288–302
  madder 73, 281
  noni 73, 313
  Oceanian 316–317
  raspberry 54
  Rhamnus 55
  saffron 34–35, 281
  woad 68, 281, 295–300
  Tyrian purple 150
dysentery, (*see also* diarrhea) 74, 191, 246, 253, 320

e

ebony 190
α-ecdysone 210, 211, 274
ecdysteroids 115, 210–212

edema   34, 40, 63
eggplant, *see* aubergine
einkorn, *see* wheat
elucidation   131, 162
elute   130
emetic   40, 44, 74, 155, 186, 197, 236
emetine   17, 74, 120, 121, 163, 196
emodin   123, 125
empyreumatic oil of coconut   248–249
endive   88
endocarp   32, 37, 61, 103, 104, 317
endocrocin   123
ent-trachylobane   30
ephedrine   28, 121, 243
epicatechin   72, 218, 249
epicatechin gallate (ECG)   72
epigallocatechin (EGC)   72
epigallocatechin gallate (EGCG)   72
epilepsy   86, 254
ergine   79, 193
ergotamine   40, 120
essential oils
   Apiaceae   88
   artemisia   87, 252, 255
   citrus   63
   cypress   25
   eucaluptus   60, 310
   extraction   129
   historical   17, 30
   insecticides   303–306
   Lamiaceae   85–86, 286, 288–289, 305
   Lauraceae   30–32
   Myrtaceae   60–61
eucalyptol   32, 288, 306
eucalyptus   60–61, 304, 306, 310, 313
eugenol   86, 169, 170, 263, 264, 288,
      304, 306
evaporators   128
expectorant   35, 186, 197
eye health, *see* ophthalmology

f

fabric, *see* fiber, plant
Farnsworth, Norman   18
farro, *see* wheat
fenchone   288
fennel   88, 253

fenugreek   236
fermentation, (*see also* oxidation)
   beverages   38, 71–72, 75, 153, 204–205
   biofuel   179
   dye   249, 297, 300–301
   food   32, 33, 37, 65, 313
ferruginol   25
ferulate   81, 114
ferulic acid   114, 117, 170, 211
festivals   159, 244–247
fever   14, 16, 34, 77, 243, 252–255
feverfew   13, 229, 283
fiber, plant   97
   banana   40, 319
   coir   37, 324
   cotton   66, 300
   hemp   55
   Hibiscus tilliaceus   315
   kapok   65
   milkweed   79
   pandan   313
   ramie   58, 240
   sisal   204–208
fiberglass   208
fingerprinting   137
fish poison   47, 50, 51, 82, 315
flavokavin   325–327
flavonoids   17, 72, 112, 114–115, 126, 211
   flavonols   115, 116, 137
   flavones   289
   isoflavonoids   50
flax   13, 232, 233
flour   49, 51, 195, 209, 244
flower anatomy   101–103
flu   74, 310
fluorescence   135
fossil   37, 279–281
foxglove   85, 94, 283, 284
fraction   27, 131, 134, 136, 161–162,
      170, 172
fragrance, *see* perfume
frankincense   149, 150, 233
freeze-drying, *see* lyophilization
fructose   113, 205, 292
frugoside   79
fruit anatomy   101–105
Fuchs, Leonard   11

fucophlorethol   200, 201
fungi   40, 79–80, 114, 156
fungicides   306
fustic (dye)   190

g

galangal   40, 41
galactosamine   263
Galen, Claudius   6, 9, 281
gallic acid   72, 115, 116, 270
gangrene   40
garbanzo bean, *see* chickpea
garcinone   47
garlic   6, 34, 149, 230, 236
gastrointestinal, (*see also* diarrhea; dysentery)
       55, 139, 188, 254, 270, 317–323, 324
GC-MS   131, 246
gelsemine   293
genistein   50
Gerard, John   12, 14
Gerrard, A. W.   284
Gessner, Conrad   11–12
Ghafiqi, al-   9
gibberellic acid   115
gin   324
ginger   40, 41, 98, 235, 236
gingerols   40
ginkgo   23, 101, 229
ginseng   89–90, 229, 243
ginsenosides   89–90
girinimbine   63
gitoxin   85
glaucoma   63, 197, 284
glochids   68, 98
glucoamylase   197
glucose   113, 122, 123, 179–180, 205–206
glucosidase   50, 122, 169, 180, 201, 218, 300
glucosinolates   67, 111, 112, 114, 120, 122
glue, *see* adhesives
glycitein   50
glycolipids   161
glycoproteins   36, 50
glycyrrhizin   51
goatgrass   232
Goethe, Johann Wolfgang   16
gofruside   79
goldenseal   45

gonorrhea   268, 312
gooseberry   236
goosefoot   45
gossypol   66
gotu kola   155
gourd   87, 147, 148, 156, 190, 204
gout   34, 252, 267, 270
grafting   46, 52
grain, (*see also* cereal)   39, 103, 222
grape   6, 45–46, 104, 149, 232, 236, 252,
       279, 281, 282
grapefruit   63
grass   38–40, 103–104
   goat   232
   lemon   304
   tiger   244
   sweet   186
   vetiver   236
grayanotoxin   72–73, 118, 291–293
groundnut *see* nut
guaiacol   169–170
guanábana   29
guarana   63, 195, 197, 198
guava   19, 60, 191–193, 248, 312
guavasteen   60
guinea pepper   29
gums
   ammoniac   89, 151
   arabic   147, 151, 155
   acacia   50
   birch   267, 268, 270
   chicle   70, 97, 192
   guar   50
   mastic (tears of Chios)   61
   spruce   24
   tara   50
   tragacanth   50, 232
gumbo   30, 157, 188–189
guvacine   250
guvacoline   250
gymnosperms   23–27
gynecology   45, 188

h

hackberries   55
hallucinogen   18, 40, 52, 79, 82, 83, 185,
       193, 196

Harshberger, John W. 3
Hatshepsut (pharaoh) 5, 6
hazelnut 232
headache 11, 40, 49, 86, 192, 216, 243,
    249, 286, 324
healer 172, 176, 236, 259–265
heartburn, *see* dyspepsia
helenalin 88
hemicellulose 66, 206
hemlock 23, 88, 282, 284
hemorrhoids 63, 216
hemp 7, 55–57, 208, 230
henbane 82, 83, 282, 284
henequen 204, 206, 207
henna 6, 229, 235, 280
hepatitis 86, 254, 263
hepatoprotective 55, 162, 212, 246, 260,
    261, 263–265, 270
hepatotoxic 31, 170, 263, 326–327
herbalists & herbals 5–16
herbarium 107–109
herbicides 170, 306
herbivory 23, 50, 80, 97
Hernández de Toledo, Francisco 15
heroin 44
H&E stain 51
hibiscus 65–66, 312, 315
hickory, *see* nuts
high-performance liquid chromatography
    (HPLC) 131–132
Hildegard of Bingen 9
Hindu 32, 34, 234–237, 239, 259
Hinoki wood 24
hinokitiol 24
Hippocrates 6, 34, 49, 236, 252, 281
histamine 58, 63, 122, 123, 201, 321
histidine 112, 122, 123
HIV 48, 139, 171–176, 270
HMG-CoA 113, 120
holly 86–87, 101
hollyhock 283
homeopathy 138
honey 73, 261, 262, 282, 289, 291–295
honeybush 155
honeydew 291, 293
Hooker, Joseph Dalton 16
Hooker, Sir William Jackson 16

hops 55–56, 279, 281, 282
horehound 85, 86, 283
horseradish 67
Hua Tuo 7
huckleberry 72, 186, 188
humors 236
humulone 56
hydrastine 45
hydrocodone 44
hydrolysis 169, 180
hydroponics 37
4-hydroxybenzaldehyde 168, 169
hydroxyecdysone 210, 211
hyenanchin 293
hygrine 82
hyoscine 81
hyoscyamine 80, 82, 196
hyperbilirubinemia, *see* jaundice
hyperforin 47
hyperglycemia, *see* antidiabetic
hypericin 47, 123, 125
hyperpigmentation 58, 89, 201
hypertension 35, 79, 284, 286, 294, 312
hypnotic 16, 192
hypophyllanthin 263, 264
hypotension 73, 293
hyssop 229, 283

i

Ibn Juljul 9, 282
Ibn Sina *see* Avicenna
Idrisi, al- 9
imidacloprid 83
immunomodulatory 210
imperialine 35
Inca, *see* Native Americans
incense 61, 150, 192
indehiscent 49, 103, 104
indican 68, 300
indicaxanthin 224
indigo 68, 281, 285–302
indigotin 68
indigotine 300
indospicine 301
Indus Valley civilization 233–234
inflorescence 101–102
ingwe tonic 171

insecticide 272–275, 288–289, 303–306
insomnia 11, 47, 315, 327
intoxicant 16, 82, 83, 235, 255, 291–294, 323
inulin 88
ipecac 17, 73–74, 112, 121, 195–197
iridoids 113, 115, 118, 139, 249, 251
iris 103, 261
isabgol 85
isatan 68, 300
Islamic/Muslim
    cultural influence 149, 151
    medicine 235–236, 260
    scholars 6, 8–10, 282
isobetanin 224
isobologram 136
isobutyric acid 36
isoprene units 17, 48, 115
isothiocyanate 67

j

jaborandi 63, 195–197, 284
Jack-in-the-pulpit 189
jackfruit 57–58, 59
Jamu medicine 138
jasmine 291
jatropha 48, 247
jaundice 254, 258–266, 268
Jerusalem artichoke 88
Jesuit 14, 77, 86, 197
jewelry 52, 60, 151, 314
jicama 190
jimadores 203–204
jimsonweed 82, 83
jujube 55, 193
juniper 24, 97, 101, 267
jute 156, 157, 237

k

kaempferol 55, 218, 270
kaffir lime 63
kale 67
Kampo medicine 138, 229, 243
kapok fiber 65, 79
Kaposi's sarcoma 27
kaurene 292

kaurenoic acid 30
kava 312, 316, 323–328
kavain 33, 325, 326
kavalactones 33, 323–328
kenaf 156
Khan, Genghis 235
khat 46, 147–150, 156
kimchi 240
kiwifruit 312
knockout extract 130
koenimbine 63
kohlrabi 67
kola (nut) 147, 151, 153, 155–157
kratom 73, 239
Kräuterbuch 11
Krebs cycle, *see* citric acid cycle
kudzu 51
kumquats 63

l

Labrador tea 72
lacquer 24, 240, 241, 249
lactation 155
lactone 86–88, 114, 139, 162–163, 252, 255–256, 326
Lagos silk 149, 156
lanatoside 85
larch 23
larkspur 45
latex, (*see also* opium; rubber) 96–97
    Apocynaceae 79
    Euphorbiaceae 47, 139, 247, 249
    Moraceae 313
    Sapotaceae 70
laticifers 57, 96
laurel 30, 32, 216, 282, 291
lavender 86, 283, 285–290
laxative 36, 48, 51, 85, 155
LC-MS 131–132
leaf anatomy 99–100
lectin 48
legumes 49–52, 103–104, 157
lemon 63, 219
lemongrass 304
lenticels 99
lentil 51, 230, 232, 233, 279
leprosy 324

lettuce   88, 149, 232
leucine   112, 122
Levant   279
Li Shizhen   11, 240
liana   28, 45, 46, 50
licorice   51, 149, 232, 236, 243
ligand   130, 325
lily   34–35, 239
lily-of-the-valley   283
limes   63, 205, 260, 261
limonene   63, 115, 118, 304, 306
limonin   63
linalool   86, 118, 288
linalyl acetate   86, 288
linoleic acid   55, 125
linolenic acid   55
locoweed   50, 52
locust bean   51, 147, 156
logwood   51
longan   61, 63
loquat   52
lotus   6, 239
lucuma   70, 195
lupenone   268
lupeol   268
lupine   51
lupulone   56
lutein   17, 115
luteolin   263, 289
lychee   61
lycopene   17, 119
lye   83–84
lyophilization   129
lysergic acid   40, 79, 193
lysine   81, 112, 122

m

macadamia nut, *see* nut
maceration   128, 324
madder   73, 281, 282
Magellan, Ferdinand   310
magnolia   235
Magnus, Albertus   9
maguey   204
mahogany   64, 193
maize   21, 38, 39, 65, 94, 101, 181, 185,
   187, 189, 190, 192, 195, 205

malaria (*see also* antimalarial)   16,
   160–166, 315
Malay bushbeech   235
Mandarin orange   63
mandrake   6, 7, 82
mango   61, 62, 102, 104, 193, 235, 312
mangosteen   47, 239, 248
mangrove   47, 94, 229, 237, 310, 315
manioc   157, 195
Maori   310
maple   63, 103, 186–188
maqui   215–222
maracujá   47
marigolds   87
marijuana   55–57, 230, 235
marjoram   85
mass spectrometry   27, 131, 161–162
materia medica   6–7, 9, 11, 231, 240,
   252, 282
Mattioli, Pierandrea   11–12
Mayans, *see* Native Americans
MDMA (ecstasy)   30, 68
melanin   201
melon   147, 151, 152, 155–157, 232
menstrual   54, 324
p-menthane-3,8-diol   306
menthol   86, 115, 118, 288, 306
menthone   86, 288
mercerization   206–207
mescal
   bean   52
   beverage   203
mescaline   68, 185, 193, 196
metabolic syndrome   219, 256
metabolism
   human   21, 34, 136, 175, 256, 259, 327
   plant   38, 111–127
β-methoxypsoralen   150
methyl betulonate   268
methyllicaconitine   45
methysticin   33, 325, 326
mevalonic acid pathway   115, 120, 292
migraines, *see* headache
milkweed   78–79
millet   39, 147, 151–152, 156, 157, 233,
   242, 253
mimosa   261, 262

Ming Dynasty 11, 240
Minoan 279–280
mint 85–87, 229–230, 236, 283, 304–305
mistletoe 94
mitragynine 73
Monardes, Nicolás 11, 12
monasteries 9, 74, 239
Mongol Empire 235, 240
monk's hood 7, 11–12, 44–45, 171, 243,
    282, 284
monoecious plant 23, 101
monoterpenes, *see* terpenes
Montezuma 167
moonseed 236
moonvine 79
mordant 4, 55, 281, 300
Mormon tea 28
mortuary 5, 222
mosquito, (*see also* malaria) 78, 315
mucilage 65, 66, 319
mugwort 252–258
mulberries 57, 101, 105
mummification 58, 149, 240
mung bean 235
murrayamine-D 63
muscarinic acetylcholine receptor 63
Muslim, *see* Islamic
mustard 66–68, 122, 230, 232, 233, 282
mycelium 80, 94
mycorrhizae 94
myrobalan 236
myrosinase 67, 122
myrrh 5, 149, 150, 233, 282
myrtle 60

n

Nagoya Protocol 19–20
naphthodianthrone 47
narcotic 40, 44, 55–57, 68, 83, 120, 197
natal plum 156
Native Americans
    Algonquin 188
    Apache 185
    Aztec 14–15, 36, 38, 65, 83, 167,
        189, 204
    Caribs 193
    Choctaw 189

Guarani 38, 63, 197
Inuit 185, 199–200
Inca 38, 53, 194–195, 209, 215
Iroquois 188, 189
Mapuche 53, 196, 209, 215–219
Mayans 36, 65, 192, 204, 291
Mazatec 193
Nahuatl 15, 65, 189, 190, 192, 198
Navajo 185
Olmec 38, 65, 189
Oaxaca 193
Seminole 205
Teotihuacan 189, 204
Tupi 63, 197
nausea 55, 81, 171, 292, 293, 325
naval stores 24
nectar 292–294
nectarine 52
neem 64, 267, 272–276
nematicidal 218
neoandrographolide 264, 265
neohesperidin 63
nepetalactones 86
nettle 58
neuroprotective 212, 219, 270
neurotoxic 154, 243, 255, 301
Newberry crater 188
nicotine 81–83, 186
nicotinic acid 81, 112, 122
nightshade 80, 82, 83
N-nitrosamine carcinogens 250
NMR 27, 132, 133, 162, 170
nomadic 185, 229
noni 73, 312, 313
non-protein amino acids 34, 120, 122
noog 147
noscapine 43
nuts 103, 186
    betel 36, 247
    buckeye 63
    candlenut 239, 316
    cashew 61, 62, 193–195
    chestnut 19, 50, 63, 186, 242, 282, 312
    gevuina 45
    groundnut 147, 156
    hickory 186
    macadamia 45, 310, 312

peanut   50, 51, 147, 195, 244
pecans   186
pistachio   61, 62, 230
walnut   186, 193, 230, 240
nutmeg   5, 30, 61, 237, 239

o

oak   101, 103, 267–272
oats   39, 232
obesity   55, 219, 286, 289
β-ocimene   288
oils, (*see also* essential oils)   38, 55, 66, 67, 88, 178, 279
ointment   47, 138, 216
okra   65, 66, 147, 152, 153, 156–158, 189
oleander   79
oleandrigenin   79
oleandrin   79
oleic acid   55, 125
oleuropein   83
oligomers   132, 133, 201, 249
oligosaccharides   179, 321
olive   19, 83–85, 104, 232, 267, 280, 281
omics   133–135
onion   34, 98, 149, 230
oolong tea   71
ophthalmology   16, 47–48, 63, 80, 197, 259, 315
opiate/opioid   44, 193, 270, 321
opium   6, 7, 41–44, 73, 74, 97, 197, 233
opium war   42
oracle, Delphi   83, 282, 291
orange   63, 64, 235, 304, 306
orchid   35, 101–102, 156, 166–171
oregano   85, 86
osage orange   105
Ottoman Empire   41, 285
ovary/ovule   28, 101, 103–105
oxidation   71–72, 83, 200, 211, 218, 249, 297, 300
oxycodone   44

p

paan   247
paclitaxel   26–27, 119, 239
pain relif, *see* analgesic
paleobotany   279

Palheta, Francisco de Mello   74
palm   36–38
   areca   36, 247, 251
   carnauba   38
   coconut   37, 239, 313, 316
   cohune   36
   date   37, 205, 232
   nipah   248
   oil   5, 36–37, 147, 153, 154, 156, 178, 181, 237
   Palmyra   38, 239
   sugar   38, 239
   talipot   239
   wine   38, 314
Panama disease   318
pandan/screwpine   312–314, 317
papaverine   43, 120
papaya   191–193, 260, 263, 265
Papyrus Ebers   5
parenchyma   94, 96, 99
Parkinson's disease   40, 80
parsley   88, 232
parsnip   88, 89, 232
parthenocarpic fruit   38, 103, 104, 318
paspalicine   40
paspaline   40
paspalinine   40
passionflower   47, 48
passionfruit   190, 191, 312
patchouli   85, 86
Pavón Jiménez, José Antonio   15
pawpaw   29, 186
peach   52, 104, 242
peanut *see* nut
pear   52, 53, 97, 230, 242
peas   157, 159, 233, 279
pectin   179, 180, 206, 321
pediculicidal   251
peimine   35
peiminine   35
peimisine   35
Pele   324
pentose phosphate pathway   112, 114
pepper, chili   80–82, 102, 189–191, 195
peppercorns   32–33, 234, 235
percolation   128
perennial   93

performance-enhancers   213, 294
perfume   30, 53, 61, 233, 279–281, 316
perianth   101, 106
pericarp   103, 104
periwinkle   79, 154, 155
peroxidases   71–72
persin   32
pesticide, *see* insecticide
petals   101
petiole   99, 100, 317
petroleum   22, 169, 178, 251, 301
peyote   68, 190, 193
pharmacopeia   9, 20, 40, 200, 229, 243,
   249, 315, 324
α-phellandrene   30
phloem   23, 55, 96
phloroglucinol   47, 123, 200, 201
phlorotannin   123, 199–203
Phoenicians   150–151
photosensitive   47
photosynthesis   96, 111
phototherapy   259
phototoxicity   237
phygrine   82
phyllanthin   263, 264
phylloclades   98
phylogeny   28
physoperuvine   82
phytase   205
phytobezoars   321
phytolith   188, 279
pigeon pea   152, 153
pigment   47, 55, 68, 120, 223, 249, 281, 296
pilocarpine   63, 196, 197, 283, 284
pine   23–24
pineapple   38, 105, 195, 207
pinene   17, 25, 86
piñon   186
Δ^1-piperidinium   81
piperine   32
pipermethystine   325, 326
piscicide, *see* fish poison
pistachio *see* nut
pistil   101, 103–105
plantain   40, 317–323
Plasmodium falciparum   78, 160–166, 255
Platearius, Matthaeus   9

pleurisy root   186
Pliny the Elder   6, 34, 252, 291
plum   52, 156, 193, 242, 281
plumeria   239
pneumatophores   94
podophyllotoxin   44, 186
Pohnpei   315
poi   33
pointsettia   47, 97
poison ivy/oak   61
poisons   78–79, 83, 88–89, 189, 282
   arrow   26, 45, 60, 79, 196, 285
   honey   291–293
   livestock   45, 47, 66
polarity   130
pollen   23, 28, 107, 168, 188, 189, 279, 294
pollination   28, 168
polyketide   29, 40, 113, 114, 120,
   123–125, 200
polymerization   27, 78, 123, 132, 133
polyphenol oxidase   71–72, 249
pomegranate   104, 231, 248
pomelo   63
poppy, (*see also* opium)   41–44, 97,
   232–234
popweed   199
potato   80–82, 98, 195
poultice   85, 128, 173, 189, 197, 200, 216,
   254, 324
prebiotics   321
precipitate   43, 135
pregnane glycosides   186
preservative   56, 86, 192, 280
prickles   99–100
proanthocyanidins   72, 126, 132, 133, 137
progesterone   192
prostaglandins   113, 123, 124, 201, 321
protease inhibitor   38
protein   48, 66, 111, 122, 135, 153
protozoa   46, 78, 255–256
pseudostem   317
psychotropics/psychoactives
   Cactaceae   68, 193, 196
   Cannabis   55, 230
   Claviceps   40, 79
   Convolvulaceae   79
   honey   291–294

Piper 33
Solanaceae 80–83, 196, 282
South American 196–197
psyllium 85
pteridophytes 93, 94
pterocarpans 50
pulegone 288
pulque 36, 204, 205, 207
puqiedine 35
puqietinone 35
purgative 60, 197
purpurin 73
purpuroxanthin 73

q

Quakers 188
quercetin 34, 55, 137, 211, 218–219, 270
quid 33, 83, 248–250
quince 52
quinic acid 77, 172
quinidine 77
quinine 77–78
  discovery 14, 196–197
  isolation 16, 160
  phytochemistry 120
  drug resistance 160, 256
quinizarin 73
quinoa 195, 209–215, 222–226

r

radical 200, 255
radicle 94
radish 67, 94, 232, 282
ragweed 87
raisin 45–46
Raj (British) 237
rambutan 61, 239
ramie 58, 240, 242
rapeseed 232, 233, 279, 282
raphides 33–34
raspberries 54–55
redwood 23–24, 25, 96
Renaissance 12, 253
repellents 303–307
resin 23–24, 61, 96, 149–151, 192, 196, 233, 267–268, 280–281
resurrection plant 173, 174

resveratrol 46, 114, 117
reticuline 43
Réunion Island 167, 168, 170
rhamnetin 55
rhamnose 66
rhizome 40, 68, 74, 89, 98, 196, 261, 317, 323–326
rhododendron 16, 73, 267, 291–295
rhodotoxin 292
rhubarb 103, 282
rice 38, 157, 237–238
ricin 48
rooibos tea 155
root anatomy 93–96
rose 52, 99, 104, 105, 280, 283
rosemary 85, 86, 283, 304, 305
rosette 35, 99, 100, 296
rosin 24, 96
rosmarinic acid 288
rotenoids 50, 51
rotenone 50, 51, 315
Royal Botanic Gardens Kew 16, 20
rubber
  chemistry 97
  Ficus elastica 58, 239
  Hevea brasiliensis 48, 97, 193, 197
  substitutes 79, 192
Ruiz López, Hipólito 15
Runge, Friedlieb Ferdinand 16
rutin 17, 219

s

safflower 88, 232, 281, 282
saffron 34, 35, 232, 279, 281, 282
safrole 30
sage 21, 85, 86, 186, 229, 280, 283, 305
sagebrush 252
sailcloth 55
Saint John's wort 47, 229, 283
salannin 274
salicin 16, 283, 284
salicylic acid 49, 112, 114, 116
salmonberry 188
salve, (*see also* poultice) 5, 32
salvinorin 193
samara 103
samhitas 8, 234

Sanamahism   259
sandalwood   150, 235, 236
Sanskrit texts   8, 234, 267, 272
sap   24, 36–38, 63, 186–188, 232, 240, 313, 316, 317
sapodilla   70
sapogenin   239
saponins
    biosynthesis   115
    poison   48, 315
    quinoa   211–212
    soap   36
    triterpene   63, 89
sapote   63
sapwood   96, 97
sarsaparilla   35
sassafras   30, 186, 188–189
schaftoside   47
schistosomiasis   256
schizocarp   103
schizophrenia, (see also antipsychotic) 284
Schultes, Richard Evans   18
sclereid   97
sclerenchyma   97
scopolamine   80–81, 82, 122, 196, 293
scopoletin   196
screwpine, see pandan
scurvy   17, 24
seaweed, see algae
secologanin   74, 120
sedative   33, 44, 50, 53, 63, 82, 282, 312
seizures, (see also epilepsy)   255, 292
senna   51, 83, 149
sepals   101
separation methods   129–131
sequoia   24
Sertürner, Friedrich   16, 44
sesame   149, 233, 244
sesquiterpenes, see terpenes
shallots   34
shaman   83, 192–193, 196, 243
shampoo   6, 36, 251, 289
shea butter   70, 147, 155, 156
Shen Nong (Emperor)   6, 240
Shigella/shigellosis   245, 320
shikimic acid pathway   35, 114, 121
shingles   173, 315

shipbuilding   48, 55, 64, 150
shogaols   40
Siddha   234–235, 261
silica gel   130, 131, 162
silicle   67, 104, 105
silique   67, 104, 105
silk road   229, 233
simples   9
sinalbin   67
sinigrin   67
sisal   204–208
β-sitosterol   119, 211, 268
skullcap   85–86
skunk cabbage   33
skunk vine   247
slaked lime   43, 248–249
slavery   57, 151, 153, 156–159, 188, 298–301
sleep, (see also sedative)   16, 44, 74, 81, 324
slobber-mouth plant   63
smallpox   155
smoke/smoking   30, 43, 68, 82, 83, 186, 192, 254
smudging   185–186
snail   150
snake bite   79, 186, 189, 229, 268
snakeroot   79, 239
snapdragon   101, 283
soap   32, 36, 225, 272, 286, 289
soapberry   61
soaptree   36
Socrates   89, 282
solanidine   81, 82, 120
solanine   81, 82
solubility   39, 136, 256
solvents   24, 44, 128–131, 162
sonication   128
Sonnenorden   315
sorcerers' tree   196
sorghum   22, 39, 147, 151, 152, 156, 157
soursop   29, 193, 312
soybean   50, 51, 242
spermicide   267, 272
Spice Islands   60, 237
spikenard   282
spinach   230
spleen   254, 259–261

spruce 23–24
squash, (see also gourd) 185, 189, 190, 195
squill 283
staggers 40
stamen 28, 101
standardization 137
Staphylococcus 246
starch 33–34, 49, 58, 153, 157, 179–181, 321
starfruit 239
steatosis 263
stem anatomy 96–99
steroids 35, 80, 81, 115, 120, 210–213, 239
stigma 34, 101, 168
stimulants 28, 46, 63, 76, 83, 86, 149–150, 153, 197, 243, 250
stipule 100
Stockholm Papyrus 296
stoleniferous 98
stomach tonic 186, 192, 252, 270, 312
Stone, Reverend Edward 16
Strauss, Levi 302
strawberry 53–54, 102, 105, 282
strobili/strobilus 23, 56
strychnine 17
succulent 68, 203, 317
sugar 36–39, 111, 188, 235
sugarcane 39, 157, 178, 235, 260, 263, 312
suicide 26, 45, 78–79, 282
Sun Simiao 9
sunflower 87–88, 103, 186
supplement, dietary 45, 127, 137–138, 202, 219, 260, 289, 326–327
surgery 11, 82
Suri, al- 9
sweetener, (see also sugar) 51, 186, 239
sweetsop 29
symmetry, floral 101
synergism 136, 172, 212, 219, 304, 321
synthesis, chemical 27, 44, 81, 169, 274, 301
syphilis 35
syringic acid 249
syrup 37, 58, 63, 74, 186, 188, 197
Szent-Györgyi, Albert 17

t
tamarind 51, 152, 157, 263
tangerine 63
tannins
  condensed 115, 116, 249, 250, 270
  hydrolyzable 115, 116
tansy 283
Taoism 237, 240, 242
tapioca 49, 195
taproot 88, 94
taro 33–34, 98, 152, 235, 239, 312
tarragon 87
tattoos 316
taxine 26
taxol 26–27, 186
taxonomy 3, 14, 23
TCM 35, 51, 86, 138, 229, 240, 243, 258–259
tea
  Camellia sinensis 70–72, 235
  tree oil 61
teak 235
teff 147, 156
tenderizers 38
tendrils 45, 56, 96
teosinte 189
tequila 36, 203, 207
terpenes/terpenoids 115, 118–119
  diterpenes 25, 27, 30, 40, 44–45, 48, 72, 88, 115, 118, 186, 193, 239, 265, 292
  monoterpenes 24, 32, 61, 115, 118, 255, 270, 288
  sesquiterpenes 66, 72, 87, 88, 115, 118, 139, 161–163, 252, 255–256, 288, 293, 304
  triterpenes 35, 60, 63, 89, 90, 115, 119, 161, 175, 268, 270, 273
  tetraterpenes 115
γ-terpinene 270
terpineol 288
tetanus 73
thatch 36, 313
theanine 122
thebaine 43, 120
theobromine 65, 72, 76–77, 87, 122, 123, 197
Theophrastus 6, 280, 281

thermoplastic   207
thickeners   66
thimbleberry   188
thin layer chromatography (TLC)
    130, 162
thistle   87
thorns   97, 99, 100
thujaplicin   24–25
thujone   86, 255
thyloses   96
thyme   85, 86, 283, 304, 305
thymol   86, 304, 306
timber   23, 61, 63, 64, 193, 245
tinctures   128
tobacco   11, 12, 68, 80–82, 83, 186, 187,
    200, 250
toddy   38, 314
tomatillos   82, 190, 191
tomatine   81, 82
tomatoes   17, 80, 102, 191
tonics   33, 45, 52, 58, 74, 78, 155, 171, 176
tonka bean   170, 195
tooth
    ache   11, 186, 191, 249
    blackening   247–252
    brushes   317
    decay   249
    enamel   250
    filling   192
    paste   251, 272, 317
Tournefort, Joseph Pitton de   14
tragacanth, gum   50, 51, 232
trance   192, 216, 282
transcriptomics   134–135
trichomes   55, 58, 86, 255
triterpenes, *see* terpenes
tropine   82
tryptophan   112, 114, 116, 120–122
Tu Youyou   17–18, 243, 255
tuber   33, 49, 81, 98–99, 155
tuberculosis   34, 45, 171, 301
tubocurarine   196, 283, 285
tubulin   27, 154
tung oil   48, 240
turmeric   40, 41, 234–236
Turner, William   12, 14
turnip   67, 189, 279, 282

turpentine   6, 24, 96
Tutankhamun   295
tutin   293
twinberry   200
twine   204, 205
Tyrian purple   150
tyrosinase   201, 202
tyrosine   112, 114, 116, 120–122

u
ulcer
    oral   216
    peptic   40, 61, 254, 256, 286
    skin   197
umbelliferone   114
Unani medicine   229, 236
unisexual plants   101
urinary health   72, 155
urine   36, 267, 296, 300, 301
utricle   103

v
vacuole   120, 122, 169
valine   81, 112, 122
vanilla   35, 65, 139, 156, 166–171,
    190, 191
vanillic acid   168
vanillin   35, 114, 117, 131, 139
varnish   24, 48, 61
vegetative tissue   50, 93, 200
velvetleaf   316
venom   11, 171
vermouth   253–255
Vespucci, Amerigo   185
vetch   232, 279
vinblastine   79, 120, 154
vincristine   79, 154
Vishnu   239
vitexin   47
vitiligo   150
vomiting   48, 53, 74, 159, 204, 291–293, 320
vulcanization   197

w
waldmeister   73
Wallach, Otto   17
walnut, *see* nut

warts   44, 139, 186
wasabi   67, 240–242
wastewater   84, 180–181
watermelon   152, 156
wax   22, 24, 36, 38, 61, 68, 111
weaving   227, 234
wheat   19, 21, 38, 39, 58, 103, 151,
    230–232, 233, 279
whortleberries   188
wildcrafters   197
willow   6, 16, 49, 101, 186, 283, 284, 316
Willstätter, Richard Martin   17
wine
    chokeberry   52
    fortifiers   219, 252–253
    grape   7, 45–46, 279, 280
    maqui (colorant)   219
    palm   38, 153, 314
    wormwood   252
wingbean   312
woad   68, 281, 295–300
wolfsbane   44
wood   23–25, 48, 66, 96–97, 150, 193,
    216, 315
woodcuts   11
woodruff   73, 283
wool   208, 223
World War
    WWI   58, 299
    WWII   160, 205, 236
worms, intestinal   11, 253, 254, 256
wormwood   252–258

wrappers/wrappings
    food   60, 244, 313, 319
    mummy   58, 240
    quid   36
Wrigley, William   192

x
x-ray crystallography   27
xanthine   65, 76, 86, 197
Xenophon   291
xylem   23, 28, 96

y
yagé vine   196
yam   147, 157–159, 192, 239, 312
yangonin   33, 326
Yangtze River   237
yarrow   161, 283
yeast   46, 135, 180, 205
Yellow Emperor's Cannon   7, 243
yerba mate   86–87, 195, 197
yew   26–27, 239
ylang-ylang   29, 239
yoco   197

z
zapote   70
Zhang Zhongjing   7
zinc   205, 263, 301
zingerone   40
zingiberene   115
zygomorphic   101, 102

Printed in the USA
CPSIA information can be obtained
at www.ICGtesting.com
LVHW051557051023
759904LV00049B/577